T0134956

Studies in Computational Intelligence

Volume 1081

The series "Studies in Computational Intelligence" (SCI) publishes new developments and advances in the various areas of computational intelligence—quickly and with a high quality. The intent is to cover the theory, applications, and design methods of computational intelligence, as embedded in the fields of engineering, computer science, physics and life sciences, as well as the methodologies behind them. The series contains monographs, lecture notes and edited volumes in computational intelligence spanning the areas of neural networks, connectionist systems, genetic algorithms, evolutionary computation, artificial intelligence, cellular automata, self-organizing systems, soft computing, fuzzy systems, and hybrid intelligent systems. Of particular value to both the contributors and the readership are the short publication timeframe and the world-wide distribution, which enable both wide and rapid dissemination of research output.

Indexed by SCOPUS, DBLP, WTI Frankfurt eG, zbMATH, SCImago.

All books published in the series are submitted for consideration in Web of Science.

Roussanka Loukanova · Peter LeFanu Lumsdaine ·
Reinhard Muskens
Editors

Logic and Algorithms in Computational Linguistics 2021 (LACompLing2021)

Editors
Roussanka Loukanova
Institute of Mathematics and Informatics
Bulgarian Academy of Sciences
Sofia, Bulgaria

Peter LeFanu Lumsdaine
Department of Mathematics
Faculty of Science
Stockholm University
Stockholm, Sweden

Reinhard Muskens
Faculty of Science
ILLC—Institute for Logic, Language
and Computation
University of Amsterdam
Amsterdam, The Netherlands

ISSN 1860-949X ISSN 1860-9503 (electronic)
Studies in Computational Intelligence
ISBN 978-3-031-21782-1 ISBN 978-3-031-21780-7 (eBook)
https://doi.org/10.1007/978-3-031-21780-7

This Springer imprint is published by the registered company Springer Nature Switzerland AG
The registered company address is: Gewerbestrasse 11, 6330 Cham, Switzerland

Preface

The chapters in this book evolved from the symposium *Logic and Algorithms in Computational Linguistics 2021 (LACompLing2021)* and its predecessors, the workshop on Logic and Algorithms in Computational Linguistics 2017 (LACompLing2017) and the symposium Logic and Algorithms in Computational Linguistics 2018 (LACompLing2018).

The workshop LACompLing2017, August 16–19, 2017 and the symposium LACompLing2018, 28–31 August 2018, took place at the Department of Mathematics, Stockholm University, Stockholm, Sweden:

http://staff.math.su.se/rloukanova/LACompLing17.html

http://staff.math.su.se/rloukanova/LACompLing2018-web/

The symposium LACompLing2021, 15–17 December 2021, was part of the week Mathematical Linguistics (MALIN), 13–17 December 2021, Université de Montpellier, LIRMM, CNRS, France, via online streaming by CNRS, Montpellier, France:

https://staff.math.su.se/rloukanova/LACompLing2021-web/

Similarly to the previous LACompLing events, the symposium LACompLing2021 was a distinguished forum, at which researchers from different subfields of Computational Linguistics presented and discussed their work.

Computational linguistics studies natural language in its various manifestations from a computational point of view, both on the theoretical level (modeling grammar modules dealing with natural language form and meaning, and the relation between these two) and on the practical level (developing applications for language and speech technology). Right from the start of the 1950s, there have been strong links between natural and formal languages, computer science, logic, and many areas of mathematics—one can think of Chomsky's contributions to the theory of formal languages and automata, or Lambek's logical modeling of natural language syntax.

The book *Logic and Algorithms in Computational Linguistics 2021 (LACompLing2021)* takes forward the initiation of its previous editions. The initial idea of assessment of the place of computer science, mathematical logic, and other subjects of mathematics, in the versatile areas of Computational Linguistics, has been a fruitful resource of research. The research presented in the chapters of this book varies from classic to newly emerging theories and applications.

The topics of the book are in the intersection and interconnections between Logic, Language, Syntax, and Semantics, from computational perspectives.

The book focuses mainly on logical approaches to Computational Linguistics and computational processing of natural language, and on the applicability of methods and techniques from the study of formal languages in computational linguistics, e.g., mathematical logic, theory of computation and algorithms, formal languages, programming, and other specification languages. It presents work from other approaches to language and computational linguistics, as well, especially because they inspire new work and approaches.

While Computational Linguistics is a relatively young discipline, decades of research on theoretical work and practical applications have demonstrated that it is a distinctively interdisciplinary area. There is also a sense that computational approaches to linguistics can benefit from research on the nature of human language, from the perspective of its usage and evolution.

The chapters of the book cover topics on computational theories of human language, across grammar, syntax, and semantics, in theories and applications. The common threads of the research in the chapters are the roles of computer science, mathematical logic, and other subjects of mathematics, in Computational Linguistics and Natural Language Processing.

Goal and Intended Readers of this Collection of Works

The goal of this collection of works is to promote intelligent approaches to Computational Linguistics and Natural Language Processing (NLP).

The intended readers of the book are researchers in Computational Linguistics, theory and applications of Artificial Intelligence (AI) and Natural Language Processing (NLP), and related subjects.

Sofia, Bulgaria — Roussanka Loukanova
Stockholm, Sweden — Peter LeFanu Lumsdaine
Amsterdam, The Netherlands — Reinhard Muskens
October 2022

Contents

Complexity of the Lambek Calculus and Its Extensions 1
Stepan L. Kuznetsov

Categorial Dependency Grammars: Analysis and Learning 31
Denis Béchet and Annie Foret

Diamonds Are Forever 57
Michael Moortgat, Konstantinos Kogkalidis, and Gijs Wijnholds

A Hybrid Approach of Distributional Semantics and Event Semantics for Telicity 89
Hitomi Yanaka

Generalized Computable Models and Montague Semantics 107
Artem Burnistov and Alexey Stukachev

Multilingual Text Generation for Abstract Wikipedia in Grammatical Framework: Prospects and Challenges 125
Aarne Ranta

Decomposing Events into GOLOG 151
Symon Jory Stevens-Guille

Generating Pragmatically Appropriate Sentences from Logic: The Case of the Conditional and Biconditional 171
Renhao Pei and Kees van Deemter

White Roses, Red Backgrounds: Bringing Structured Representations to Search 191
Tracy Holloway King

Rules Are Rules: Rhetorical Figures and Algorithms 217
Randy Allen Harris

Integrating Deep Neural Networks with Dependent Type Semantics 261
Daisuke Bekki, Ribeka Tanaka, and Yuta Takahashi

Meaning-Driven Selectional Restrictions in *Remember* Versus *Imagine Whether* 285
Kristina Liefke

A Unified Cluster of Valence Resources 311
Lars Hellan

Complexity of the Lambek Calculus and Its Extensions

Stepan L. Kuznetsov

Abstract We survey results on algorithmic complexity of extensions of the Lambek calculus. We classify these extensions as "harmless," which do not increase complexity, and "dangerous," which make decision problems undecidable. For the latter, we focus on extensions with subexponentials and with Kleene star.

Keywords Lambek calculus · Complexity · Undecidability · Categorial grammars

1 Introduction

In this survey, we shall discuss *mathematical* results concerning categorial grammar formalisms based on the Lambek calculus. We shall focus on *algorithmic complexity* results. Linguistic motivations will be provided in a very limited way, just in order to make the formalisms more understandable. For deeper discussions of the usage of Lambek-style categorial grammars in linguistics, we refer the reader to monographs by Carpenter [1], Moot and Retoré [2], and Morrill [3]. In general, the landscape here is as follows. Once a formal grammar framework gets introduced, and duly linguistically motivated, there arise mathematical questions about this framework. Most importantly, these are algorithmic questions. Indeed, possible usage of such a formalism in automated analysis of texts in natural language, in systems like CatLog [4] or Grail [5], requires the corresponding algorithmic problems to be solvable and, moreover, solvable in reasonable time. This is the point where mathematicians enter the area, and this is the focus of this survey paper.

The Lambek calculus was introduced in 1958 in a seminal paper by Joachim Lambek [6]. Following ideas of Ajdukiewicz [7] and Bar-Hillel [8], Lambek proposed

S. L. Kuznetsov (✉)
Steklov Mathematical Institute of RAS, 8 Gubkina St., Moscow, Russia
e-mail: sk@mi-ras.ru

R. Loukanova et al. (eds.), *Logic and Algorithms in Computational Linguistics 2021 (LACompLing2021)*, Studies in Computational Intelligence 1081,
https://doi.org/10.1007/978-3-031-21780-7_1

an algebraic-logical approach for defining grammatical validity. The central algebraic operations used in Lambek's approach are *divisions*, or residuals. We shall postpone formal definitions and formulation of the calculus to the next section, and start with some simple examples. These examples actually go back to Lambek's original work.

Each word (lexeme) in a Lambek grammar gets associated with one or several algebraic expressions, which are also called *syntactic types* or *Lambek formulae*. These expressions are constructed using two division operations, \ and /, and multiplication $\cdot$ (also called product). Two divisions are necessary due to the non-commutative nature of natural language: one may not freely shuffle words in a sentence. In extensions of the Lambek calculus, which we shall discuss further, the set of operations will be extended. Having, as a basic algebraic law, that $A \cdot (A \setminus B)$ reduces to B, and, symmetrically, $(B \mathbin{/} A) \cdot A$ also reduces to B, one can parse simple sentences. In the standard example, *"John likes Mary,"* the grammar uses two basic syntactic types, N for noun phrase (*John* and *Mary*) and S for sentence. The transitive verb *likes* gets the following syntactic type: $(N \setminus S) \mathbin{/} N$. Intuitively this means that *likes* would become a sentence, if it receives two noun phrases, one on the left and one on the right. Formally this is supported by the fact that $N \cdot ((N \setminus S) \mathbin{/} N) \cdot N$ can be reduced to S or, in the logical terms, the following *sequent* is derivable in the Lambek calculus:

$$N, (N \setminus S) \mathbin{/} N, N \to S.$$

The reduction schemata $A, A \setminus B \to B$ and $B \mathbin{/} A, A \to B$ are available even in weaker formalisms of *basic categorial grammars,* the idea of which goes back to aforementioned works of Ajdukiewicz and Bar-Hillel. These reductions could be treated as variants of the *modus ponens* logical rule, if \ and / are considered as directed forms of implication. The Lambek calculus adds the possibility of *hypothetical reasoning.* Hypothetical reasoning corresponds to so-called *extraction,* which is used in analysis of sentences with dependent clauses. The standard example here is as follows: *"Pete met the girl whom John likes."* The dependent clause here is *"John likes,"* and it lacks a noun phrase to become a complete sentence like *"John likes Mary."* Thus, a hypothetical reasoning may be performed: if x has syntactic type N, then *"John likes x"* has syntactic type S. The Lambek calculus then allows to proceed in the style of deduction theorem and assign the syntactic type $S \mathbin{/} N$ to *"John likes"*:

$$N, (N \setminus S) \mathbin{/} N \to S \mathbin{/} N.$$

Further analysis is based on the following type assignment. The word *girl* receives syntactic type CN, which stands for common noun, i.e., noun without article; *the* receives type $N \mathbin{/} CN$. Most importantly, the syntactic type for *whom* is $(CN \setminus CN) \mathbin{/} (S \mathbin{/} N)$. This allows assigning the syntactic type CN to *"girl whom John likes."* Finally, the whole sentence *"Pete met the girl whom John likes"* gets assigned the syntactic type S (grammatically correct sentence), which is justified by the following sequent:

$$N, (N \setminus S) \mathbin{/} N, N \mathbin{/} CN, CN, (CN \setminus CN) \mathbin{/} (S \mathbin{/} N), N, (N \setminus S) \mathbin{/} N \to S.$$

As one can see from this example, Lambek grammars are capable of handling quite involved phenomena of natural language syntax. Moreover, as shown by Pentus [9], the Lambek calculus is the atomic theory of language algebra. This means that the Lambek calculus derives exactly the sequents which are generally true under interpretations of syntactic types as formal languages, $\to$ meaning language (set-theoretic) inclusion and Lambek operations being interpreted in a natural way. (An accurate formulation is given below in Sect. 2.) This result is called language completeness, or L-completeness.

Language completeness is an argument to prefer Lambek grammars to other closely related categorial grammar frameworks, like combinatory categorial grammars [10], categorial dependency grammars [11], and others. However, for extensions of the Lambek calculus which we consider below L-completeness usually either is unknown, or even fails.

However, the usage of Lambek grammars faces severe theoretical limitations, both in *expressive power* and in *algorithmic complexity.* We shall discuss these limitations in detail below in Sect. 2. These limitations motivate the study of various *extensions* of the Lambek formalism. Practical usefulness of such extensions, however, depends on the following trade-off: when proposing such an extension, one should *simultaneously* try to increase expressivity and to harness complexity.

In what follows, we shall survey extensions of the Lambek calculus, focusing on their algorithmic properties: decidability and complexity. As a starting point, we take the survey [12] of complexity results for the Lambek calculus and its fragments given by Mati Pentus at AiML 2010. This paper could be considered as a sequel of Pentus' survey; at the same time, it is self-contained. Unlike Pentus [12], we do not give mathematical proofs of results cited, but rather refer to corresponding original articles.

The rest of the paper is organised as follows. In Sect. 2 we define the Lambek calculus and Lambek grammars. We formulate Pentus' theorem that Lambek grammars can generate only context-free languages. This is a serious limitation of Lambek's original formalism which motivates considering its extensions. In Sect. 3 we briefly survey complexity results for the original Lambek calculus and its fragments, reminding Pentus' survey [12]. In Sect. 4, we consider the first extension, by means of additive connectives. These connectives correspond to lattice-theoretic meet and join, and they increase both expressive power and complexity of the Lambek calculus. In Sect. 5, we go further and try to classify extensions of the Lambek calculus as "harmless" (which allow the same proof-search procedures and therefore do not increase complexity) and "dangerous" (which increase complexity). "Harmless" extensions include such important mechanisms as brackets and discontinuous operations. Sections 6, 7 are devoted to two important "dangerous" extensions, subexponentials and Kleene star. The final Sect. 8 is for concluding remarks.

2 The Lambek Calculus and Its Limitations

In this section we shall discuss the original Lambek calculus, as a starting point for further extensions.

2.1 *The Lambek Calculus*

The Lambek calculus **L** is formulated as a sequent system in the following way. Formulae (syntactic types) of **L** are built from a countable set of variables (primitive types) $\mathrm{Var} = \{p_1, p_2, \ldots\}$ using three operations: $\backslash$ (left division), $/$ (right division), and $\cdot$ (product). The set of formulae is denoted by Fm. Formulae are denoted by capital Latin letters; capital Greek letters stand for sequences of formulae (including the empty sequence Λ). Sequents are expressions of the form $\Pi \to A$.

Axioms and inference rules of the Lambek calculus are as follows:

$$\frac{}{A \to A}\ Id \qquad \frac{\Gamma, A, B, \Delta \to C}{\Gamma, A \cdot B, \Delta \to C}\ {\cdot}L \qquad \frac{\Pi \to A \quad \Delta \to B}{\Pi, \Delta \to A \cdot B}\ {\cdot}R$$

$$\frac{\Pi \to A \quad \Gamma, B, \Delta \to C}{\Gamma, \Pi, A \backslash B, \Delta \to C}\ \backslash L \qquad \frac{A, \Pi \to B}{\Pi \to A \backslash B}\ \backslash R, \text{ where } \Pi \text{ is non-empty}$$

$$\frac{\Pi \to A \quad \Gamma, B, \Delta \to C}{\Gamma, B / A, \Pi, \Delta \to C}\ / L \qquad \frac{\Pi, A \to B}{\Pi \to B / A}\ / R, \text{ where } \Pi \text{ is non-empty}$$

There is also a special rule called Cut:

$$\frac{\Pi \to A \quad \Gamma, A, \Delta \to C}{\Gamma, \Pi, \Delta \to C}\ Cut$$

This rule is handy for constructing derivations, but is not in fact necessary, since any derivable sequent has a derivation without cut [6]. Therefore, by default we do not regard Cut as an official rule, but rather consider it an *admissible* rule. For each calculus we consider further, we shall make a notice on admissibility of Cut.

From a modern point of view, the Lambek calculus can be considered as a multiplicative-only, intuitionistic, and non-commutative version of Girard's [13] linear logic; this connection was noticed by Abrusci [14]. A subtle difference is the non-emptiness restriction imposed on $\backslash R$ and $/ R$, which we shall discuss below. However, the Lambek calculus was introduced much earlier than linear logic. From the algebraic point of view, Lambek's division operations are called *residuals*. Residuals obey the algebraic rules for division, but w.r.t. a *partial order* $\preceq$ rather than equality:

$$b \preceq a \setminus c \iff a \cdot b \preceq c \iff a \preceq c \,/\, b.$$

The partial order corresponds to the sequential arrow $\to$ of the Lambek calculus, and these equivalences for residuals are derivable in the Lambek calculus. Algebraic structures with residuals were studied before Lambek's paper, in the works of Krull [15] and Ward and Dilworth [16]. Finally, algebraic structures with residuals provide natural algebraic semantics for the family of *substructural* logics, the Lambek calculus being one of the most basic ones of them; see the monograph of Galatos et al. [17] for details.

The "Π is non-empty" constraint on the $\setminus R$ and $/\, R$ rules is the so-called *Lambek's non-emptiness restriction.* It ensures that for any derivable sequent its left-hand side is non-empty. Lambek's non-emptiness restriction is motivated by linguistic applications of the Lambek calculus: otherwise it *overgenerates,* i.e., declares grammatically incorrect phrases as valid ones. An example is given in [2], and we shall discuss it below. A variant of **L** without Lambek's non-emptiness restriction is also considered [18]. We shall denote it by $\mathbf{L}^{\Lambda}$, following the notation from [19]. (The more common notation $\mathbf{L}^{*}$ yields a collision with Kleene star, which will be introduced below.)

2.2 Language Models

Natural semantics for the Lambek calculus is given by *language models* (L-models for short).[1] In a language model, each formula is interpreted by a formal language, i.e., a set of words over a given alphabet Σ. Lambek's non-emptiness restriction, which distinguishes **L** from $\mathbf{L}^{\Lambda}$, is reflected by the status of the empty word: in L-models for **L** it is not allowed to be used in the languages which interpret formulae. Thus, an L-model is a pair (Σ, w), where $w \colon \mathrm{Fm} \to \mathcal{P}(\Sigma^*)$ for $\mathbf{L}^{\Lambda}$ and $w \colon \mathrm{Fm} \to \mathcal{P}(\Sigma^+)$ for **L**. (Here $\mathcal{P}$ stands for "powerset" and Σ^* and Σ^+ are, respectively, the set of all words and the set of all non-empty words over alphabet Σ.) The interpreting function w is defined arbitrarily on variables, and for complex formulae it should obey the rules stated below.

Product corresponds to pairwise concatenation:

$$w(A \cdot B) = \{uv \mid u \in w(A) \text{ and } v \in w(B)\}.$$

The interpretation of division operations corresponds to the idea that $w(A \setminus B)$ (resp., $w(B \,/\, A)$) should be the set of all words which lack a word from $w(A)$ on the left (resp., on the right) to become a word from $w(B)$. However, the definitions are

[1] Another well-known class of models for the Lambek calculus is formed by R-models, i.e., models on algebras of binary relations [20]. However, L-models are more relevant to linguistic applications.

essentially different depending on whether we allow the empty word (i.e., whether we impose Lambek's non-emptiness restriction). For example, the empty word, if allowed, would always belong to $w(A \setminus A)$ and $w(A \mathbin{/} A)$; otherwise, such a language could be empty. Thus, at this point the definitions of L-models for **L** (without the empty word) and for $\mathbf{L}^\Lambda$ (with the empty word) are different. For **L**, we have

$$w(A \setminus B) = \{u \in \Sigma^+ \mid (\forall v \in w(A))\ vu \in w(B)\},$$
$$w(B \mathbin{/} A) = \{u \in \Sigma^+ \mid (\forall v \in w(A))\ uv \in w(B)\},$$

and for $\mathbf{L}^\Lambda$ in this definition Σ^+ is replaced by Σ^*.

A sequent of the form $A_1, \ldots, A_n \to B$, where $n > 0$, is true in an L-model (Σ, w), if $w(A_1 \cdot \ldots \cdot A_n) \subseteq w(B)$. For sequents with empty antecedents (which are allowed only in the case of $\mathbf{L}^\Lambda$), of the form $\Lambda \to B$, the definition is different. Such a sequent is declared true if the empty word belongs to $w(B)$.

Pentus [9, 21] proved completeness w.r.t. L-models, both for **L** and $\mathbf{L}^\Lambda$:

Theorem 1 (M. Pentus 1995, 1996) *A sequent is derivable in* **L** *(resp., in* $\mathbf{L}^\Lambda$*) if and only if it is true in all L-models without the empty word (resp., in all L-models with the empty word).*

In other words, **L** and $\mathbf{L}^\Lambda$ are atomic theories for the corresponding variants of L-models.

2.3 *The Unit Constant*

Lambek's non-emptiness restriction being lifted, one can extend the Lambek calculus (i.e., $\mathbf{L}^\Lambda$) with the unit constant **1** with the following axiom and rule

$$\frac{\Gamma, \Delta \to C}{\Gamma, \mathbf{1}, \Delta \to C}\ \mathbf{1}L \qquad \frac{}{\to \mathbf{1}}\ \mathbf{1}R$$

This gives $\mathbf{L_1}$, the Lambek calculus with the unit. Admissibility of *Cut* for $\mathbf{L_1}$ is proved in the same way as for **L** and $\mathbf{L}^\Lambda$.

Adding the unit constant makes completeness w.r.t. L-models fail. Indeed, the natural interpretation of **1** would be $\{\varepsilon\}$, the singleton of the empty word. But then interpretations of $\mathbf{1} \mathbin{/} A$ trivialise. Namely, $w(\mathbf{1} \mathbin{/} A)$ becomes $\varnothing$ if $w(A)$ contains a non-empty word. For $w(A) = \varnothing$ and $w(A) = \{\varepsilon\}$ we have $w(\mathbf{1} \mathbin{/} A) = \Sigma^*$ and $w(\mathbf{1} \mathbin{/} A) = \{\varepsilon\}$ respectively. This means that for formulae of the form $\mathbf{1} \mathbin{/} A$ we actually have a three-valued logic (possible values are only $\varnothing$, $\{\varepsilon\}$, and Σ^*; see [22] for details), which validates more sequents than $\mathbf{L_1}$. For example, the sequent $\mathbf{1} \mathbin{/} p \to (\mathbf{1} \mathbin{/} p) \cdot (\mathbf{1} \mathbin{/} p)$ is true in all L-models, but it is not derivable in $\mathbf{L_1}$. This means incompleteness.

2.4 Lambek Grammars

Let us now see how the Lambek calculus is used for modelling language syntax. A Lambek grammar is a triple $G = (\Sigma, \triangleright, H)$, where Σ is the alphabet (in linguistic applications, elements of Σ are lexemes rather than letters), H is a designated formula (called the target type), and $\triangleright \subset \Sigma \times \mathrm{Fm}$ is a finite binary correspondence between elements of Σ and formulae. A word $w = a_1 \dots a_n$ over alphabet Σ (in linguistic applications, w corresponds to a phrase formed of lexemes $a_1, \dots, a_n$) belongs to the language defined by G if there exist such formulae $A_1, \dots, A_n$ that:

1. $a_1 \triangleright A_1, \dots, a_n \triangleright A_n$;
2. the sequent $A_1, \dots, A_n \to H$ is derivable.

It is important to notice that the language defined by G depends on the calculus used: **L** or $\mathbf{L}^{\Lambda}$ (i.e., whether Lambek's non-emptiness restriction is imposed). Furthermore, **L** or $\mathbf{L}^{\Lambda}$ can be replaced by a more elaborate calculus, where the set of formulae is extended with new operations, and the calculus itself is extended with new axioms and inference rules.

For example, in the sentence *"Pete met the girl whom John likes"* each word (i.e., *Pete*, *met*, ...) will be an element of Σ; $\triangleright$ is defined as follows:

$$\begin{aligned} \textit{Pete, John} &\triangleright N \\ \textit{girl} &\triangleright CN \\ \textit{met, likes} &\triangleright (N \setminus S) / N \\ \textit{the} &\triangleright N / CN \\ \textit{whom} &\triangleright (CN \setminus CN) / (S / N) \end{aligned}$$

Finally, $H = S$. The sequent

$$N, (N \setminus S) / N, N / CN, CN, (CN \setminus CN) / (S / N), N, (N \setminus S) / N \to S$$

is derived in **L** as follows:

$$\dfrac{\dfrac{\dfrac{\dfrac{N \to N \quad \dfrac{N \to N \quad S \to S}{N, N \setminus S \to S}\,\setminus L}{N, (N \setminus S) / N, N \to S}\,/L}{N, (N \setminus S) / N \to S / N}\,/R \quad \dfrac{CN \to CN \quad CN \to CN}{CN, CN \setminus CN \to CN}\,\setminus L}{CN, (CN \setminus CN) / (S / N), N, (N \setminus S) / N \to CN}\,/L \quad \dfrac{N \to N \quad \dfrac{N \to N \quad S \to S}{N, N \setminus S \to S}\,\setminus L}{N, (N \setminus S) / N, N \to S}\,/L}{N, (N \setminus S) / N, N / CN, CN, (CN \setminus CN) / (S / N), N, (N \setminus S) / N \to S}\,/L$$

This shows that the sentence belongs to the language defined by the grammar.

Notice that the "left" rules $\setminus L$ and $/ L$ correspond to reduction schemata of basic categorial grammar, while the "right" rules $\setminus R$ and $/ R$ formalise hypothetical reasoning. Thus, the language generated by a categorial grammar depends not only on the type assignment, but also on the calculus being used. Namely, our derivation essentially uses the $\setminus R$ rule, thus in the formalism of basic categorial grammars, which does not include "right" rules, the same type assignment would not lead to validation of "*Pete met the girl whom John likes*" as being of type S.

Another example showing the dependence on the calculus is connected to Lambek's non-emptiness restriction and is taken from the book by Moot and Retoré [2]. Consider the following type assignment:

$$\textit{book} \rhd CN \qquad \textit{interesting} \rhd CN \,/\, CN$$
$$\textit{very} \rhd (CN \,/\, CN) \,/ (CN \,/\, CN)$$

The phrase *"very interesting book"* receives the syntactic type CN, which is validated by the following derivation, which works both in $\mathbf{L}^{\Lambda}$ and in $\mathbf{L}$:

$$\dfrac{CN \,/\, CN \to CN \,/\, CN \qquad \dfrac{CN \to CN \quad CN \to CN}{CN \,/\, CN, CN \to CN} \,/\, L}{(CN \,/\, CN) \,/ (CN \,/\, CN), CN \,/\, CN, CN \to CN} \,/\, L$$

Let us take a look, however, at the following derivation:

$$\dfrac{\dfrac{CN \to CN}{\to CN \,/\, CN} \,/\, R \qquad \dfrac{CN \to CN \quad CN \to CN}{CN \,/\, CN, CN \to CN} \,/\, L}{(CN \,/\, CN) \,/ (CN \,/\, CN), CN \to CN} \,/\, L$$

Though the left-hand side of the goal sequent is non-empty, this derivation is valid only in $\mathbf{L}^{\Lambda}$, not in $\mathbf{L}$, since the usage of $/\, R$ does not obey the non-emptiness condition.

Derivability of this sequent corresponds to recognising *"very book"* as a valid phrase of type CN. Intuitively, when one uses $\mathbf{L}^{\Lambda}$, empty syntactic structures are allowed. In particular, the empty word can serve as an "adjective," being of type $CN \,/\, CN$, and allows attaching adverbs like *very*. This is against the rules of grammar, therefore Lambek's non-emptiness restriction should be applied whenever possible; unfortunately, it is incompatible with some extensions of the Lambek calculus, e.g., the unit constant or (see Sect. 6 below) subexponential modalities.

The linguistic examples presented above are very natural and go back to the work of Lambek [6] or even earlier. In what follows, we shall discuss more sophisticated linguistic examples, which are mostly taken from works of Morrill [3, 4, 23].

2.5 Lambek Grammars and Context-Free Grammars

What is the expressive power of Lambek grammars? In 1960 Gaifman [24] showed that already basic categorial grammars are capable of defining all context-free languages, except those which include the empty word. (An empty sequence of syntactic types just cannot be reduced to a goal type.) Buszkowski [25] noticed that Gaifman's result can be used for Lambek grammars also; a more modern and straightforward proof is based on Greibach normal form [26] for context-free grammars (see [27]).

Theorem 2 (C. Gaifman 1960, W. Buszkowski 1985) *For any context-free language M without the empty word there exists a Lambek grammar G such that M is exactly the language defined by G. This grammar G defines the same language M* *both for* **L** *and* $\mathbf{L}^{\Lambda}$.[2] *Moreover, formulae used in G are constructed using only one division* $\backslash$, *without* $\cdot$ *and* $/$.

A more fine-grained result was obtained by Safiullin [29]:

Theorem 3 **(A. Safiullin 2007)** *For any context-free language M without the empty word there exists a Lambek grammar G which defines language M, where for any* $a \in \Sigma$ *there exists a unique* $A \in \mathrm{Fm}$ *such that* $a \rhd A$. *Such grammars are called grammars with unique type assignment.*[3]

Safiullin's result shows that in Lambek grammars one can always get rid of *syntactic homonymy,* where a word has several syntactic roles. Notice that for basic categorial grammars this is not possible.

The inverse question of whether any language defined by a Lambek grammar is context-free, forms the so-called *Chomsky conjecture,* and it remained open for decades. Finally, Pentus [30] gave a positive answer:

Theorem 4 **(M. Pentus 1993)** *Any language defined by a Lambek grammar is context-free. This holds both for the case of* **L** *and for the case of* $\mathbf{L}^{\Lambda}$.

Pentus' theorem imposes an essential limitation on the expressive capacity of Lambek grammars. Indeed, context-freeness of natural language had been an important linguistic question [31] discussed in the research community until Shieber [32] presented an evidence against it by pointing out a construction in Swiss German which is not context-free. Examples like Shieber's one may be regarded as exotic. However, even in cases where a natural language phenomenon is formally context-free, its context-free description might be unnatural from the linguistic point of view and/or too complicated. These considerations motivate the study of *extensions* of Lambek-style grammar formalisms, which are usually accomplished by adding new operations to the Lambek calculus. The practical usefulness of such extensions is, however, limited by algorithmic complexity issues: with extra operations, algorithmic problems could become too hard or even undecidable. Thus, one has to balance between extending expressive power and keeping efficient parsing procedures (as we know for context-free grammars). In the rest of the paper, we shall discuss algorithmic complexity questions for the Lambek calculus and its extensions.

[2] In the case of $\mathbf{L}^{\Lambda}$, it becomes possible to define languages which include the empty word. In fact, any context-free language (including those with the empty word) can be defined by a Lambek grammar based on $\mathbf{L}^{\Lambda}$. This is performed by a technical trick which adds the empty word to a Lambek grammar [28].

[3] Safiullin's original construction works for **L** only; its modification for $\mathbf{L}^{\Lambda}$ is given in [19].

3 Complexity of the Lambek Calculus and Its Fragments

3.1 NP-Completeness of the Lambek Calculus

Derivations of sequents in the Lambek calculus (we consider only cut-free ones) have the following property: each rule application in the derivation introduces a *designated* occurrence of the corresponding operation in the goal sequent. Indeed, if we take such an occurrence and trace it upwards in the derivation, this trace could never branch. Thus, it ends either at a unique rule application, where this operation is introduced, or at an instance of the Id axiom.

This yields the following consequence: a cut-free derivation of any derivable sequent has polynomial size w.r.t. the size of the goal sequent. This property of the Lambek calculus is unusual among other logical systems: in more classical systems, cut elimination could lead to dramatic increase of derivation size, see [33]. Using this property, we obtain an upper bound on its complexity.

In what follows, we use the term "*derivability problem*" for the algorithmic question of checking whether a given sequent is derivable. Categorial grammars give rise to more complicated *parsing problems:* given a word and a categorial grammar, determine whether the word belongs to the language generated by the grammar. The parsing problem requires, additionally, to guess the correct type assignment.

Proposition 1 *The derivability problems (i.e., whether a given sequent is derivable) in* $\mathbf{L}$ *and* $\mathbf{L}^{\Lambda}$ *belong to the* NP *complexity class. Moreover, the same holds for the parsing problems.*

Indeed, the derivation itself serves as a polynomial size witness. For the parsing problem, the witness would also include the needed type assignment.

This upper complexity bound is tight:

Theorem 5 (M. Pentus 2006) *The derivability problems (and, therefore, the parsing problems) in* $\mathbf{L}$ *and* $\mathbf{L}^{\Lambda}$ *are* NP*-complete.* [34].

What are the practical consequences of this NP-completeness result? Let us start with something relatively positive. Unless $\mathsf{P} = \mathsf{NP}$, there is no polynomial-time translation of Lambek grammars into context-free ones. The existence of such a translation, without complexity conditions, is guaranteed by Pentus' theorem, but this theorem cannot be made efficient. On the other hand, the converse translation (Gaifman's theorem) is polynomial. This means that, while Lambek grammars define exactly the same class of languages as context-free ones do, for concrete languages Lambek's approach could offer better, more succint ways of defining these languages than the context-free approach.[4] To this extent, Lambek grammars restore their status of being in some sense more powerful than context-free ones.

[4] This could be compared with the discussion by Loukanova [35] on another formalism, which also provides essential reducing of the grammar size, if compared to context-free grammars.

On the negative side, of course, NP-completeness means that (again, unless P = NP) the best possible algorithms for derivability and parsing problems require exponential time, in fact as brute-force search. This might sound as a killer argument against using Lambek grammars in natural language analysis. In other words, NP-complete problems are commonly considered too hard. However, this is not exactly the case: the input size of the parsing algorithm is proportional to the length of an individual *sentence* rather than the whole *text*. An English sentence could typically include up to 10–20 words. On such small values of the input length exponential growth can be coped with by optimising the searching algorithms. Such ideas really work in the CatLog parser [4], which is based on an extension of the Lambek calculus which has even a higher level of complexity; see below for details.

3.2 Polynomial and Pseudo-polynomial Algorithms

Despite the argument that exponential-time algorithms could be practically useful for parsing in Lambek grammars, NP-completeness of the Lambek calculus still motivates seeking for fragments of the Lambek calculus where polynomial-time algorithms are possible. First of all, this is the case for basic categorial grammars, which can be considered as Lambek grammars where $\backslash R$ and $/ R$ are not used. A more interesting result was obtained in 2010 by Savateev [36]:

Theorem 6 (Yu. Savateev 2010) *The derivability problem and, moreover, the parsing problem for the fragment of the Lambek calculus with only one division (i.e., with* $\backslash$*, but without* $\cdot$ *and* $/$*) are decidable in polynomial time.*

Savateev's result can be extended to an efficient translation of Lambek grammars with one division into context-free grammars. In order to formulate this extension, let us define the *size* of a grammar in a natural way. For a context-free grammar its size is the sum of lengths of all its production rules (the length of a rule is counted as the number of symbols except $\to$, e.g., the length of $A \to aBC$ is 4). For a Lambek grammar, we take the sum of lengths of pairs in $\triangleright$ (counted as the number of symbols except brackets in formulae and the $\triangleright$ symbol itself, e.g., the length of $a \triangleright p / (q \backslash r)$ is 6) plus the size of the target type H.

Theorem 7 (S. K. 2016) *For any Lambek grammar G which uses only one division there exists a context-free grammar G' which defines the same language, whose size is polynomial w.r.t. the size of G. Moreover, G' can be constructed from G by a polynomial-time algorithm* [37].

The one-division fragment of the Lambek calculus is already capable of defining all context-free languages, so Savateev's result looks quite interesting. However, as one can see from examples in the Introduction, even simple examples require two divisions, $\backslash$ and $/$, for *linguistically natural* modelling. And, as shown in another work of Savateev [38], the fragment of the Lambek calculus with two divisions (but without product) is already NP-complete.

Another result in this direction, which might be more significant, was obtained in 2010 by Pentus [39]. This result features a *pseudo-polynomial* algorithm for the Lambek calculus and Lambek grammars. The idea which makes such an algorithm possible is as follows. The proofs of NP-completeness [34, 38] proceed by encoding the well-known 3-SAT problem. Lambek sequents used in these encodings involve formulae with *deeply nested* division operations. In contrast, sequents used when parsing with Lambek grammars are constructed from syntactic types taken from a fixed lexicon $\triangleright$. For such sequents with bounded depth one could hope to construct a polynomial-time algorithm.

This intuition can be made formal in the following way. First, Pentus [39] defines the notion of *order* $\mathrm{ord}(A)$ of a formula. For product-free formulae (built using only $\backslash$ and $/$), its order is the nesting depth of divisions, where "nesting" means putting into the denominator. For example, $\mathrm{ord}((q \,/ (r \,/\, s)) \,\backslash\, p) = 3$, while $\mathrm{ord}(s \,\backslash ((p \,/\, q) \,/\, r)) = 1$. In the presence of product, the definition of order is more complicated and includes an auxiliary function prod:

$$\mathrm{prod}(A) = \begin{cases} 1, & \text{if } A \text{ is of the form } B \cdot C, \\ 0 & \text{otherwise;} \end{cases}$$

$$\mathrm{ord}(p) = 0 \qquad \text{for } p \in \mathrm{Var};$$

$$\mathrm{ord}(A \cdot B) = \max\{\mathrm{ord}(A), \mathrm{ord}(B)\};$$

$$\mathrm{ord}(A \,\backslash\, B) = \mathrm{ord}(B \,/\, A) = \max\{\mathrm{ord}(A) + 1, \mathrm{ord}(B) + 2 \cdot \mathrm{prod}(B)\}.$$

For example, $\mathrm{ord}(p \,\backslash ((q \,\backslash\, q) \cdot p)) = 3$. The order of a sequent is defined as follows:

$$\mathrm{ord}(C_1, \ldots, C_{k-1}, C_k \to D) = \mathrm{ord}(C_k \,\backslash (C_{k-1} \,\backslash \ldots \backslash (C_1 \,\backslash\, D) \ldots)).$$

For a Lambek grammar, its order is defined as the maximal order of a type used in this grammar (counting both types in $\triangleright$ and the target type H). By $\mathrm{poly}(n, m)$ we denote a value bounded by a polynomial of n and m. Now we are ready to formulate Pentus' result.

Theorem 8 (M. Pentus 2010) *For both* **L** *and* $\mathbf{L}^{\Lambda}$*, there exist algorithms for checking derivability of a sequent whose running time is* $\mathrm{poly}(n, 2^d)$*, where n is the size and d is the order of the sequent in question. Moreover, the parsing problem for Lambek grammars is solvable by an algorithm whose running time is* $\mathrm{poly}(n, m, 2^d)$*, where n is the length of the input word, m is the size of the grammar and d is the order of the grammar.*

Extending this result to a pseudo-polynomial translation of Lambek grammars to context-free ones is still an open problem.

We finish this section by mentioning that the derivability problem from finite sets of hypotheses (extra, so-called non-logical axioms), also called the consequence relation, is much harder. The use of *Cut* in such situations is unavoidable, and as shown by Buszkowski [40], it is undecidable even for the one-division fragment.

However, if these axioms have a simple form, $p_1, \ldots, p_n \to q$, then this problem is decidable and belongs to NP. (These axioms can be transformed into Gentzen-style rules, and the resulting system enjoys cut elimination.) For a generalisation of this result to subexponential extensions of the Lambek calculus, see Sect. 6 below.

4 Additive Operations

As noticed above, the Lambek calculus in its original formulation has essential limitations in expressive power. Therefore, real life applications require *extensions* of the Lambek calculus with extra operations and syntactic mechanisms. Let us first consider the extension of the Lambek calculus with *additive operations,* which are *additive conjunction* $\wedge$ and *additive disjunction* $\vee$. Both are binary operations, and the rules of inference for them are as follows:

$$\frac{\Gamma, A, \Delta \to C}{\Gamma, A \wedge B, \Delta \to C} \quad \frac{\Gamma, B, \Delta \to C}{\Gamma, A \wedge B, \Delta \to C}\ \wedge L \qquad \frac{\Pi \to A \quad \Pi \to B}{\Pi \to A \wedge B}\ \wedge R$$

$$\frac{\Gamma, A, \Delta \to C \quad \Gamma, B, \Delta \to C}{\Gamma, A \vee B, \Delta \to C}\ \vee L \qquad \frac{\Pi \to A}{\Pi \to A \vee B} \quad \frac{\Pi \to B}{\Pi \to A \vee B}\ \vee R$$

The Lambek calculus extended with $\wedge$ and $\vee$ is called the *multiplicative-additive Lambek calculus,* **MALC**. A standard argument shows that **MALC** also enjoys admissibility of Cut.

Lambek's non-emptiness restriction is compatible with additive operations, and one can consider a version of **MALC** with this restriction imposed.

The idea of using additive operations in the Lambek calculus is extremely natural; a non-exhaustive list of references includes Lambek [18], van Benthem [41], Kanazawa [42]. Additive conjunction and disjunction exactly correspond to additive operations in linear logic.[5] In algebraic semantics, adding $\vee$ (join) and $\wedge$ (meet) corresponds to transforming the partial order $\preceq$ into a lattice, join and meet being pairwise supremum and infimum.

Extending the Lambek calculus with additive operations increases the expressive power of Lambek grammars. Namely, as shown by Kanazawa [42], Lambek grammars with additive conjunction are capable of defining finite intersections of context-free languages (without the empty word). This class includes, e.g., the well-known non-context-free language $\{a^n b^n c^n \mid n \geq 1\}$. A strengthening of this result was proved in [43]: Lambek grammars with additive conjunction can define any language defined by a conjunctive context-free grammar, as defined by Okhotin [44].

[5] The notation in linear logic is different: Girard uses $\otimes$, &, and $\oplus$ instead of $\cdot$, $\wedge$, and $\vee$ respectively. As linear logic is commutative, two divisions (linear implications) there coincide, and they are denoted by $\multimap$.

This is a class broader than finite intersections of context-free language: e.g., it includes $\{a^{4^n} \mid n \geq 0\}$ [45].

In the $\wedge R$ and $\vee L$ rules much of the sequent (e.g., the whole Π in case of $\wedge R$) gets *copied* to both premises of the rule. This means that in the presence of additive operations we do not have an injective mapping of rule applications inside the derivation to occurrences of operations in the goal sequent. Such an injective mapping was the key to the NP upper bound.

However, it is still possible to obtain a polynomial (actually linear) upper bound on the *height* of a cut-free derivation tree. Indeed, if we consider a path from the goal sequent to an axiom instance, then on this path each rule application maps to a distinguished occurrence of an operation in the goal sequent. This gives a PSPACE upper bound, as sketched by Lincoln et al. [46].[6]

The corresponding lower bound, PSPACE-hardness, was shown by Kanovich [50] and independently by Kanazawa [51], using a method of encoding QBFs (quantified Boolean formulae) closely related to the proof of PSPACE-completeness for the multiplicative-additive fragment of linear logic by Lincoln et al. [46].

Theorem 9 (M. I. Kanovich 1994, M. Kanazawa 1999) *The derivability problem for* **MALC** *is* PSPACE*-complete.*

A more fine-grained version of this result is presented in [52]: for PSPACE-hardness it is sufficient to have only two operations: $\backslash$ and $\wedge$.

The natural interpretation of $\wedge$ and $\vee$ on formal languages is intersection and union, $\cap$ and $\cup$. This interpretation is sound, but completeness fails due to non-derivability of the distributivity law, $(p \vee q) \wedge (p \vee r) \to p \vee (q \wedge r)$, in **MALC**. It is possible to extend **MALC** with this law as an additional axiom. For this extension, Kozak [53] developed a hypersequential calculus with eliminable cut. For this calculus, Kozak proved decidability; the exact complexity bound, however, remains unknown.

5 "Harmless" and "Dangerous" Extensions

Let us go further and consider other extensions of the Lambek calculus by extra operations and other syntactic mechanisms. The number and diversity of such extensions is huge. For example, Morrill's CatLog uses about 45 operations. We are not going to discuss all known extensions of the Lambek calculus in detail, we shall rather give a more general perspective.

[6] This upper bound can be obtained in two ways. One way is to construct an *alternating* proof search algorithm, and alternating polynomial algorithms form a computation model equivalent to PSPACE [47]. Another way is to construct a non-deterministic polynomial space algorithm based on depth-first search, and then make it deterministic using Savitch's theorem [48]. The second strategy is sketched in [46] and accurately presented in [49] (for an extension of **MALC**).

5.1 Classifying Extensions of the Lambek Calculus

We shall try to classify the operations added to the Lambek calculus from the algorithmic point of view. Fortunately, most of these operations are "*harmless*," in the sense that adding these operations does not increase algorithmic complexity, provided it is already NP or higher. Extending polynomial or pseudo-polynomial algorithms to such "harmless" extensions is much harder, even if possible. Other operations, however, can be considered "*dangerous*," because adding these operations leads to growth of complexity or even makes the system algorithmically undecidable. These "dangerous" extensions are, of course, more interesting for us, and we shall focus on them in the next two sections. In this section, we shortly discuss "harmless" extensions.

A toy example of such an extension is the extension obtained by adding the reversal operation, denoted by A^R. In the language interpretation, this operation reverses the order of letters in each word:

$$w(A^R) = \{a_n \dots a_1 \mid a_1 \dots a_n \in w(A)\}.$$

For $\Gamma = A_1, \dots, A_n$ let $\Gamma^R = A_n^R, \dots, A_1^R$. Then the rules for ^{R} are formulated as follows [54, 55]:

$$\frac{\Gamma \to C}{\Gamma^R \to C^R}\ {}^R M \qquad \frac{\Gamma, A^{RR}, \Delta \to C}{\Gamma, A, \Delta \to C}\ {}^{RR}LE \qquad \frac{\Pi \to A^{RR}}{\Pi \to A}\ {}^{RR}RE$$

("M" stands for monotonicity, "LE" and "RE" stand for left and right elimination).

Adding the reversal operation with such rules to the Lambek calculus keeps completeness w.r.t. L-models, both with and without Lambek's non-emptiness restriction; Cut is admissible [54, 55]. From the point of view of complexity, adding reversal is a "harmless" extension: it keeps complexity in the NP class for the Lambek calculus and in PSPACE for **MALC**.

Polynomiality of the one-division fragment [36], however, is not robust: in the presence of ^{R} one can express the other division: $B \,/\, A$ is equivalent to $(A^R \setminus B^R)^R$, and the fragment of the Lambek calculus with two divisions is already NP-complete [38]. This is going to be a typical situation with "harmless" extensions: starting from the NP level, such extensions do not increase complexity. In contrast, preserving polynomial and pseudo-polynomial decidability is a separate issue.

5.2 The Displacement Calculus

A more interesting example of a "harmless" extension is the *displacement calculus* by Morrill, Valentín, and Fadda [56]. Displacement calculus extends the Lambek calculus with *discontinuous operations*. In this paper, we give only informal definitions, and refer the reader to [56] for an accurate presentation of the syntax for the displacement calculus. Following the language interpretation, now we consider words with designated points of discontinuity; natural examples are expressions which form non-continuous parts of a phrase, like the idiomatic *"give ... the cold shoulder"* in English or the *"ne ... pas"* negation construction in French. Notice that this language interpretation is only informal; the author is not aware of a completeness result for displacement calculus w.r.t. a corresponding extension of L-models.

In order to handle *infixation,* i.e., putting a phrase inside a discontinuous one, Morrill et al. introduce discontinuous multiplication operations $\odot_i$, indexed by natural numbers. Infixation $A \odot_i B$ means putting B to the i-th point of discontinuity in A. For example, $\odot_1$ applied to *"gave ... the cold shoulder"* and *"Mary"* yields *"gave Mary the cold shoulder."* (Correctness of the number of points of discontinuity is controlled by a syntactic mechanism.) Each discontinuous multiplication is equipped with residuals $\downarrow_i$ and $\uparrow_i$, just like normal multiplication's residuals are $\backslash$ and $/$:

$$B \to A \downarrow_i C \iff A \odot_i B \to C \iff A \to C \uparrow_i B.$$

The calculus is also equipped with a discontinuous unit denoted by J.

As mentioned above, the sequent calculus syntax for displacement calculus is quite involved (for this reason we do not present it here). However, this extension of the Lambek calculus is also "harmless," i.e., does not increase complexity (starting from the NP level).

5.3 *The Lambek Calculus with Brackets*

Yet another "harmless" extension is the Lambek calculus with *brackets,* proposed by Morrill [57] and Moortgat [58]. Brackets may be put into antecedents of sequents and they introduce *controlled non-associativity.* Brackets are operated by two bracket modalities, $[]^{-1}$ and $\langle\rangle$.

For simplicity, let us consider the bracketed extension of $\mathbf{L}^\Lambda$ (i.e., the system without additives). This calculus will be denoted by $\mathbf{L}^\Lambda\mathbf{b}$. (One may also consider a variant with Lambek's non-emptiness restriction imposed.) Formulae are now built using three unary connectives: $\backslash, /, \cdot$, and two unary ones: $[]^{-1}$ and $\langle\rangle$. The interesting syntactic feature is the language of antecedents of sequents. Each antecedent is a *meta-formula,* which is built using sequential comma (as in the original Lambek calculus) and brackets. Formally, the set of meta-formulae is defined recursively:

- Each formula A is a meta-formula;
- The empty meta-formula Λ is a meta-formula;
- If Γ and Δ are meta-formulae then Γ, Δ is a meta-formula;[7]
- If Γ is a meta-formula then $[\Gamma]$ is a meta-formula.

[7] As there are no parentheses here, comma, as a meta-connective, is associative.

A sequent is an expression of the form $\Pi \to C$, where C is a formula and Π is a meta-formula. By $\Delta(\Gamma)$ we denote a meta-formula Δ with a designated sub-meta-formula occurrence Γ in it.

Axioms and inference rules of $\mathbf{L}^{\Lambda}\mathbf{b}$ are listed below. Notice that, due to the modified syntax of sequents, we need to modify the rules for "old" connectives ($\backslash$, /, $\cdot$) also, not only add rules for bracket modalities.

$$\frac{}{A \to A}\; Id \qquad \frac{\Delta(A, B) \to C}{\Delta(A \cdot B) \to C}\; \cdot L \qquad \frac{\Pi \to A \quad \Delta \to B}{\Pi, \Delta \to A \cdot B}\; \cdot R$$

$$\frac{\Pi \to A \quad \Delta(B) \to C}{\Delta(\Pi, A \setminus B) \to C}\; \setminus L \qquad \frac{A, \Pi \to B}{\Pi \to A \setminus B}\; \setminus R$$

$$\frac{\Pi \to A \quad \Delta(B) \to C}{\Delta(B \,/\, A, \Pi) \to C}\; / L \qquad \frac{\Pi, A \to B}{\Pi \to B \,/\, A}\; / R$$

$$\frac{\Delta(A) \to C}{\Delta([[]^{-1}A]) \to C}\; []^{-1}L \qquad \frac{[\Pi] \to A}{\Pi \to []^{-1}A}\; []^{-1}R$$

$$\frac{\Delta([A]) \to C}{\Delta(\langle\rangle A) \to C}\; \langle\rangle L \qquad \frac{\Pi \to A}{[\Pi] \to \langle\rangle A}\; \langle\rangle R$$

The Cut rule here has the following form:

$$\frac{\Pi \to A \quad \Delta(A) \to C}{\Delta(\Pi) \to C}\; Cut$$

Admissibility of Cut in $\mathbf{L}^{\Lambda}\mathbf{b}$ was proved by Moortgat [58].

Controlled non-associativity is useful when it is needed to avoid ungrammatical attempts of extraction from dependent clauses which are *islands*. A standard example includes *and*-coordinated sentences like *"John likes Mary and Pete likes Ann."* In the original Lambek calculus, this phrase receives type S, provided that

$$and \rhd (S \setminus S) \,/\, S.$$

Furthermore, *"John likes Mary and Pete likes"* receives type $S \,/\, N$. This is an unwanted result, because it would justify phrases like *"the girl whom John likes Mary and Pete likes,"* which are ungrammatical.

Making the *and*-coordinated group a bracketed one resolves this issue. Let

$$and \rhd (S \setminus []^{-1}S) \,/\, S,$$

and then *"John likes Mary and Pete likes Ann"* will be required to be put in brackets:

$$[N, (N \setminus S) / N, N, (S \setminus []^{-1} S) / S, N, (N \setminus S) / N, N] \to S,$$

which cannot be penetrated by extraction.[8] Namely, extraction would give the sequent

$$[N, (N \setminus S) / N, N, (S \setminus []^{-1} S) / S, N, (N \setminus S) / N], N \to S,$$

which is not derivable.

The extension of the Lambek calculus with brackets and bracket modalities is a "harmless" one: the derivability problem for $\mathbf{L}^{\wedge}\mathbf{b}$ is in NP, and in the presence of additive operations (with a natural modification of the rules for $\vee$ and $\wedge$ for bracketed syntax) it is in PSPACE.

Moreover, there exists a pseudo-polynomial algorithm for $\mathbf{L}^{\wedge}\mathbf{b}$ [59]. This algorithm is an extension of Pentus' one. However, its complexity estimation is worse: it is $\mathrm{poly}(n, 2^d, n^b)$, where n is the size of the goal sequent, d is the order, and b is bracket nesting depth. The exponential dependence on b is unfortunate, because sentences with deeply nested dependent structures (and, therefore, deeply nested brackets) do exist. The question whether there exists an algorithm with complexity $\mathrm{poly}(n, 2^d)$ is still open.

Finally, for systems with brackets there is a specific question of *bracket guessing,* or *bracket induction* [60]. This issue arises, since the bracketing (island) structure is not given as an input. Therefore, besides guessing the correct type assignment and finding the derivation (these two constitute the parsing problem), in the bracketed situation a parsing algorithm should also guess (induce) the correct bracketing. In the non-deterministic case this is not a problem, since the maximal possible number of bracket pairs is bounded by the number of bracket modality occurrences, which is less than the size of the goal sequent [60]. For attempts to construct a pseudo-polynomial algorithm, this is a separate issue, which is by now an open problem.

5.4 "Harmless" Extensions as "Linear" Ones

"Harmless" extensions, being different from a first glance, actually share an important common property. The new inference rules maintain *linearity,* in the sense that each formula occurrence in the conclusion of a rule maps to *at most one* occurrence in the premise(s). This maintains polynomiality for cut-free proofs. Without additives, the size of the whole cut-free proof is polynomial w.r.t. the size of the goal sequent, which yields an NP upper bound. With additives, the polynomial bound is available

[8] In real-life grammars, the type for *and* is $(S \setminus []^{-1} []^{-1} S) / S$, making the phrase a double-bracketed, or *strong* island. The distinction of strong and weak (single-bracketed) islands is needed for correct handling of so-called parasitic extraction, which we briefly discuss in the next section.

for the height of a cut-free proof (while the proof itself could be of exponential size); this gives PSPACE.

Thus, from a bird-eye view, the rules for "harmless" operations just perform some reorganisation of the same material (formulae); the differences are only in the structure of meta-formulae which form a sequent. Recently Pshenitsyn [61] proposed a general framework for such situtations, called the *hypergraph Lambek calculus* (**HL**). The syntax of **HL** is quite involved: formulae in this system are hypergraphs. However, it is still in the NP class, and an extension of **HL** with additive operations belongs to PSPACE.

The hypergraph Lambek calculus absorbs many of the known "harmless" extensions of the Lambek calculus. These include the displacement calculus and the Lambek calculus with reversal, and also variations of the structural rules of the Lambek calculus: adding commutativity, removing associativity etc. As for the Lambek calculus with brackets, the possibility to embed it into **HL** is still an open question. From this point of view, **HL** could probably become an "umbrella logic" for "harmless" extensions of the Lambek calculus.

6 (Sub)exponentials

Now let us discuss "dangerous" extensions which lead to undecidability of various levels. Let us start with exponential and subexponential modalities. Like additive operations, the exponential also comes from linear logic. The idea is as follows. Unlike intuitionistic or classical logic, the Lambek calculus does not have *structural* rules of contraction, weakening, and permutation. This is reflected by the fact that antecedents of sequents are ordered sequences of formulae, not multisets or sets. The exponential, denoted by !, is a modality which allows structural rules in a controlled way.

The exponential modality is incompatible with Lambek's non-emptiness restriction [62], so in this section we consider systems without this restriction.

In the formula language, ! is added as an extra unary operation in the prefix notation: $!A$. Inference rules for ! are as follows:

$$\frac{\Gamma, A, \Delta \to C}{\Gamma, !A, \Delta \to C}\ !L \qquad \frac{!A_1, \dots, !A_n \to B}{!A_1, \dots, !A_n \to !B}\ !R \qquad \frac{\Gamma, \Delta \to C}{\Gamma, !A, \Delta \to C}\ !W$$

$$\frac{\Gamma, !A, !A, \Delta \to C}{\Gamma, !A, \Delta \to C}\ !C \qquad \frac{\Gamma, B, !A, \Delta \to C}{\Gamma, !A, B, \Delta \to C} \quad \frac{\Gamma, !A, B, \Delta \to C}{\Gamma, B, !A, \Delta \to C}\ !P$$

As shown by Lincoln et al. [46], adding the exponential modality makes propositional linear logic undecidable, both in its commutative and non-commutative variants. Moreover, in the non-commutative situation undecidability holds even without additive operations [46], see also [63] for the exponential extension of the Lambek calculus.

Besides the "full-power" exponential ! which allows all structural rules (weakening, contraction, and permutation), it makes sense to consider weaker modalities which allow only some of these rules. For example, a modality ! which allows only permutation (i.e., the set of rules for ! is $!L$, $!R$, $!P$) can be used for handling *medial extraction* from dependent clauses, in examples like *"the girl whom John met yesterday."* Here *"John met yesterday"* is neither of type $S \,/\, N$, nor of type $N \setminus S$, since the gap which has to be filled by a noun phrase (like in *"John met Mary yesterday"*) is in the middle of the phrase. Using the $!P$ rule, however, allows the following derivation:

$$
\dfrac{N \to N \qquad \dfrac{N \setminus S \to N \setminus S \qquad \dfrac{N \to N \quad S \to S}{N, N \setminus S \to S}\,\setminus L}{N, N \setminus S, (N \setminus S) / (N \setminus S) \to S}\,\setminus L}{\dfrac{N, (N \setminus S) / N, N, (N \setminus S) / (N \setminus S) \to S}{\dfrac{N, (N \setminus S) / N, !N, (N \setminus S) / (N \setminus S) \to S}{\dfrac{N, (N \setminus S) / N, (N \setminus S) \setminus (N \setminus S), !N \to S}{N, (N \setminus S) / N, (N \setminus S) \setminus (N \setminus S) \to S / !N}\, / R}\, !P}\, !L}\, / L
$$

This derivation justifies *"John met yesterday"* as a phrase of type $S \,/\, !N$, provided that *yesterday* receives the type $(N \setminus S) \setminus (N \setminus S)$ and types for *John* and *met* have been assigned above.[9]

Now we modify the type for *whom:*

$$whom \rhd (CN \setminus CN) / (S / !N)$$

and derive that *"the girl whom John met yesterday"* is of type N.

The subexponential modality which allows only permutation gives a "harmless" extension; however, polynomiality with one division gets ruined, and the corresponding fragment becomes NP-complete.

Another such modality, which is useful for linguistic applications, is the *relevant* modality [64], for which the set of rules is as follows: $!L, !R, !C, !P$. Thus, the relevant modality allows contraction and permutation, but not weakening ($!W$). The term "relevant" here reflects the fact that formulae under this modality behave like those in relevant logic [65, 66]. The relevant modality can be used for modelling a more subtle phenomenon of *parasitic extraction,* where multiple gaps in the dependent clause are to be filled with the same N. An example is *"the paper that the author of signed without reading,"* where there are three gaps: *"the author of signed without reading"* to be filled with the same *the paper.* This is handled by copying $!N$ using contraction ($!C$) and then moving it to the needed places by permutation ($!P$). In contrast, weakening ($!W$) would justify ungrammatical phrases, and therefore is not allowed. Similar to the case of full-power exponential, adding the relevant modality makes the Lambek calculus undecidable [64].

[9] As noticed by one of the reviewers, *yesterday* could also receive the type $S \setminus S$, i.e., be regarded as a modifier for the whole sentence, not just the verb phrase. This does not change anything in our analysis.

A more general approach to structural modalities involves considering *polymodal* extensions of the Lambek calculus with a family of such modalities, which are called *subexponentials.* Subexponentials in commutative linear logic are due to Danos et al. [67] and also Nigam and Miller [68]. The theory of subexponentials in non-commutative linear logic (in particular, in the Lambek calculus) is presented in [69]. Each subexponential is marked by a label s from a finite set of labels; the subexponential is denoted by $!^s$. For each $!^s$, the calculus includes introduction rules $!L$ and $!R$, and a subset of structural rules. Thus, the power of a subexponential may range from a modality without any structural features to the full-power exponential. Moreover, subexponentials are in general not canonical: even if two subexponentials share the same subset of structural rules, they are in general not provably equivalent. Finally, there is a preorder on subexponentials which allows postulating entailments among them.

Formally, the *subexponential signature* Σ is a tuple $(\mathcal{I}, \preceq, \mathcal{W}, \mathcal{C}, \mathcal{E})$, where $\mathcal{I}$ is the finite set of labels, $\preceq$ is a preorder on $\mathcal{I}$ (which could be trivial, i.e., the equality relation), and $\mathcal{W}$, $\mathcal{C}$, and $\mathcal{E}$ are subsets of $\mathcal{I}$, upwardly closed w.r.t. the preorder. The extension of **MALC** with a family of modalities $\{!^s \mid s \in \mathcal{I}\}$ of a subexponential signature Σ is denoted by $\mathbf{SMALC}_\Sigma$ ("**S**" stands for "subexponential"). Each subexponential $!^s$ is equipped with $!L$ and $!R$ rules. The $!L$ rule is the same as for the exponential:

$$\frac{\Gamma, A, \Delta \to C}{\Gamma, !^s A, \Delta \to C}\ !L$$

As for $!R$, the polymodal version of this rule takes into account the preorder on subexponential labels:

$$\frac{!^{s_1} A_1, \ldots, !^{s_n} A_n \to B}{!^{s_1} A_1, \ldots, !^{s_n} A_n \to !^s B}\ !R, \text{ where } s_j \succeq s \text{ for each } j$$

Next, if $s \in \mathcal{W}$, then $!^s$ also allows weakening, and if $s \in \mathcal{E}$, then it allows permutation (also called exchange):

$$\frac{\Gamma, \Delta \to C}{\Gamma, !^s A, \Delta \to C}\ !W, \text{ where } s \in \mathcal{W}$$

$$\frac{\Gamma, B, !^s A, \Delta \to C}{\Gamma, !^s A, B, \Delta \to C} \quad \frac{\Gamma, !^s A, B, \Delta \to C}{\Gamma, B, !^s A, \Delta \to C}\ !P, \text{ where } s \in \mathcal{E}$$

The situation with contraction is a bit more subtle. In the absence of permutation, adding the $!C$ rule as presented above leads to failure of cut elimination [69]. In order to mitigate this issue, $!C$ is replaced by a more powerful rule (actually, two rules) of *non-local contraction,* which are as follows:

$$\frac{\Gamma, !^s A, \Phi, !^s A, \Delta \to C}{\Gamma, !^s A, \Phi, \Delta \to C} \quad \frac{\Gamma, !^s A, \Phi, !^s A, \Delta \to C}{\Gamma, \Phi, !^s A, \Delta \to C} \; !NC, \text{ where } s \in \mathcal{C}$$

Since $!P$ is derivable using $!W$ and $!NC$, we further suppose that $\mathcal{W} \cap \mathcal{C} \subseteq \mathcal{E}$. The *Cut* rule in its standard form is admissible in $\mathbf{SMALC}_\Sigma$ [69].

As shown in [69], in $\mathbf{SMALC}_\Sigma$ non-local contraction is exactly the rule which makes the subexponential "dangerous":

Theorem 10 (M. Kanovich et al. 2019) *If $\mathcal{C} \neq \varnothing$ (i.e., at least one subexponential allows $!NC$), then the derivability problem in $\mathbf{SMALC}_\Sigma$ is undecidable; more precisely, it is Σ_1^0-complete. Moreover, this holds even without additive operations. If $\mathcal{C} = \varnothing$, then the complexity of the derivability problem is the same as for the basic system without subexponentials:* PSPACE*-complete with additives and* NP*-complete without additives.*

This undecidability result is unfortunate, since the relevant modality (see above) is a particular case of a subexponential from $\mathcal{C}$. This motivates the search for restrictions on the usage of $!^s$, where $s \in \mathcal{C}$, in order to make it harmless. Such results were indeed obtained for systems with one modality, which is either the full-power exponential ($s \in \mathcal{C} \cup \mathcal{E} \cup \mathcal{W}$) or the relevant modality ($s \in \mathcal{C} \cup \mathcal{E}$).[10] First, if $!^s$ may be applied only to variables (primitive types), then it is "harmless," i.e., keeps NP without additives and PSPACE with them [64]. This covers the example presented above, where ! is applied to the primitive type N. This result can be strengthened in the following way:

Theorem 11 (S.M. Dudakov et al. 2021) *Let ! be the full-power exponential or the relevant modality. Let us allow to apply it only to formulae of order 0 or 1, built using two divisions ($\backslash$ and $/$). Then this modality is "harmless" in the sense that the corresponding extension of $\mathbf{L}^\Lambda$ is still in* NP *and the extension of* **MALC** *is in* PSPACE. *However, for the relevant modality this fragment is* NP*-complete even with one division (while the original system in polynomially decidable)* [70].

This result enables the usage of ! for extraction of, e.g., noun phrases (N, order 0) or verbs ($N \backslash S$, order 1). This covers most of the demands. However, there are examples of extraction for more complex types. One of such examples is due to Michael Moortgat[11]: *"tell me how the prisoners escaped."* Here in the type for *how* we shall have ! over $(N \backslash S) \backslash (N \backslash S)$ (verb phrase modifier). If this could be combined with parasitic extraction, this would be a truly problematic case: here we have order 2, and order 2 is sufficient for undecidability.

In the end of this section, let us briefly mention interaction of subexponentials and brackets. In the bracketed situation, Morrill [4, 23] suggests a rather subtle version of contraction which interacts with the bracketing structure. The idea is that parasitic extraction should be performed from weak islands, and no island can be used for this

[10] Extending the results mentioned below to polymodal systems with many subexponentials should probably be not so hard, but it has not been done yet.

[11] Private communication with the author.

purpose more than once. Moreover, the two articles by Morrill feature significantly different versions of bracket-aware contraction rules. A complete analysis of Morrill's systems, including issues with Cut, is given in [49, 71]. The first of these articles features undecidability results. The second one is more optimistic, and it shows that the boundaries under which contraction for a subexponential is "harmless," in the bracketed case, are much broader than in the non-bracketed one. For details, we refer to [4, 23, 49, 71].

7 Kleene Star

We conclude our survey by considering iteration, or *Kleene star.* This is one of the most intriguing operations, due to its inductive (fixpoint) nature. Substructural logics involving Kleene star, being propositional at the first glance, behave closer to more powerful systems like arithmetic. We shall see some details below.

Kleene star is a unary operation, and traditionally (unlike subexponentials) it is written in the postfix form: A^*. In language models, A^* includes words which are concatenations of *arbitrary* numbers of copies of words from A. Thus, Kleene star can be represented as an infinite disjunction (union):

$$A^* = \bigvee_{n=0}^{\infty} A^n.$$

(Here $A^0 = \mathbf{1}$.) Completeness w.r.t. L-models, in general, fails due to issues with distributivity; partial completeness results we obtained in [72].

This suggests axiomatising Kleene star with *infinitary rules:*

$$\frac{\left(\Gamma, A^n, \Delta \to C\right)_{n=0}^{\infty}}{\Gamma, A^*, \Delta \to C}\ {*L_\omega} \qquad \frac{\Pi_1 \to A \quad \ldots \quad \Pi_n \to A}{\Pi_1, \ldots, \Pi_n \to A^*}\ {*R_n},\ n \geq 0$$

One of the rules, $*L_\omega$, is an infinitary one, which makes proofs possibly infinite (but they should still be well-founded). The extension of **MALC** with Kleene star governed by the aforementioned rules is called *infinitary action logic* and denoted by $\mathbf{ACT}_\omega$. Admissibility of Cut, in its standard form, was proved by Palka [73] by a transfinite modification of the standard cut-elimination argument.

Theorem 12 (W. Buszkowski, E. Palka 2007) *The derivability problem in* $\mathbf{ACT}_\omega$ *is* Π_1^0*-complete.* [73, 74].

In the case of $\mathbf{ACT}_\omega$, we deal with another sort of undecidability, compared to $\mathbf{SMALC}_\Sigma$. Namely, the Π_1^0 complexity class is dual to the Σ_1^0 one, i.e., the class of recursively enumerable (r.e.) sets. For the latter, the fact that an element belongs to an r.e. set can be justified by a finite witness; in the case of $\mathbf{SMALC}_\Sigma$, such a witness is a derivation. For $\mathbf{ACT}_\omega$ this will not work, since derivations involving

$*L_\omega$ are infinite. However, if a sequent is *not* derivable in $\mathbf{ACT}_\omega$, then this has a finite witness. Namely, $\mathbf{ACT}_\omega$ posesses the finite model property (FMP), i.e., for each non-derivable sequent there exists a finite model which falsifies it; this model is the needed witness for non-derivability.[12] (This model is not a language model, but an abstract algebraic one.) Thus, for $\mathbf{ACT}_\omega$ both the upper complexity bound (the fact that it belongs to Π_1^0) [73] and the lower bound (the fact that it is Π_1^0-hard, i.e., is one of the hardest problems in Π_1^0) [74] are non-trivial. Moreover, the fragment of $\mathbf{ACT}_\omega$ without additive operations (i.e., the extension of $\mathbf{L}^\Lambda$ with Kleene star) is also Π_1^0-complete [19].

Using Kleene star, one can define *positive iteration,* or *Kleene plus,* in the following way:

$$A^+ = A \cdot A^*.$$

One advantage of Kleene plus is the fact that it may be added to systems with Lambek's non-emptiness restriction imposed, while Kleene star cannot ($*R_0$ is an axiom with an empty antecedent).[13]

The rules for Kleene plus, derived from rules for * and $\cdot$, are as follows:

$$\frac{\left(\Gamma, A^n, \Delta \to C\right)_{n=1}^{\infty}}{\Gamma, A^+, \Delta \to C}\ {}^+L_\omega \qquad \frac{\Pi_1 \to A \quad \ldots \quad \Pi_n \to A}{\Pi_1, \ldots, \Pi_n \to A^+}\ {}^+R_n,\ n \geq 1$$

Undecidability (Π_1^0-completeness) for the version of $\mathbf{ACT}_\omega$ with Lambek's non-emptiness restriction and Kleene plus instead of Kleene star can be proved by the same arguments of Palka and Buszkowski.

Kleene plus is used in categorial grammars for modelling *iterated coordination,* as shown by Morrill [4].[14] An example is *"John dislikes, Mary likes, and Bill loves London,"* where

$$\textit{and} \rhd ((S \,/\, N)^+ \backslash (S \,/\, N)) \,/\, (S \,/\, N).$$

The coordinator *and* takes an arbitrary non-zero number of groups of type $S \,/\, N$ as its left argument.

The essential property of such usage of Kleene plus is the fact that, due to polarity considerations, we do not need the ${}^+L_\omega$ rule, only ${}^+R_n$. And the latter rule is "harmless"! Therefore such usage of iteration does not lead to undecidability.

The Π_1^0-hardness result means that the set of sequents derivable in $\mathbf{ACT}_\omega$ is not only undecidable, but also not recursively enumerable. Therefore, it is impossible to obtain a finitary axiomatisation of $\mathbf{ACT}_\omega$, without ω-rules or other infinitary mecha-

[12] Usually, the FMP yields decidability, but this requires the logic to be r.e., which is not the case for $\mathbf{ACT}_\omega$.

[13] Interestingly, in his pioneering article [75] Kleene also avoided emptiness, by defining a binary operation $A * B$, meaning $A^* \cdot B$, instead of a unary iteration operation.

[14] Morrill calls this operation "existential exponential" and uses the notation $?A$ instead of A^+. The set of rules for this operation in Morrill's works is formally different, but equivalent to the one presented here.

nisms. Strictly weaker systems, however, could be obtained by replacing $*L_\omega$ by an induction (fixpoint) rule; $*R_n$ is also modified accordingly:

$$\frac{\to B \quad A, B \to B}{A^* \to B} \, *L_{ind} \qquad \frac{}{\to A^*} \, *R_0$$

$$\frac{\Pi \to A \quad \Gamma, A, \Delta \to C}{\Gamma, \Pi, \Delta \to C} \, Cut \qquad \frac{\Pi \to A \quad \Delta \to A^*}{\Pi, \Delta \to A^*} \, *R_{ind}$$

This gives *action logic* **ACT**, which goes back to Pratt [76] and Kozen [77]. Notice that Cut here is included as an official rule of the system, and it is not eliminable.

As noticed above, **ACT** is strictly weaker than $\mathbf{ACT}_\omega$. (This resembles Gödel's incompleteness theorem: ω-rules are stronger than induction.) Being in a sense simpler than $\mathbf{ACT}_\omega$, **ACT** is also undecidable, namely, Σ_1^0-complete [78]. Moreover, the result of [78] shows Σ_1^0-completeness for any recursively enumerable system between **ACT** and $\mathbf{ACT}_\omega$. An interesting example of such an intermediate system is given by the following "induction-in-the-middle" rule:

$$\frac{\to B \quad A \to B \quad A, B, A \to B}{A^* \to B} \, *L_{mid}$$

Finally, if the system includes both Kleene star and a subexponential allowing nonlocal contraction, then the system becomes Π_1^1-hard [79]. This is a very high complexity level, which is even not a hyperarithmetical one. In fact, it is the maximal complexity level possible for a system with infinitary rules with computable sets of premises. There is a series of interesting theoretical questions of finding fragments of this system with intermediate complexity levels. We do not discuss these questions here, since such high complexities clearly have nothing to do with linguistic applications.

8 Conclusion

In this paper, we have presented a survey of algorithmic complexity results for extensions of the Lambek calculus, which could be used in categorial grammars for natural language syntax. There is a vast variety of such extensions, but they can be classified into "harmless" and "dangerous" ones, depending on whether they raise a threat of undecidability or keep the complexity at the same level as for the original systems. In order to make the paper short enough and not to overload it with extra material, we have omitted some important variants of Lambek-style systems. Namely, we kept ourselves in the lines of associative and purely propositional calculi. The non-associative Lambek calculus, on one hand, and extensions of the Lambek calculus with quantifiers, on the other hand, are out of the scope of this survey.

Acknowledgement The author would like to acknowledge Mati Pentus, Max Kanovich, Andre Scedrov, Glyn Morrill, and Stanislav Speranski, joint work and discussions with whom made this survey possible. The author is also grateful to Michael Moortgat and Roussanka Loukanova for helpful comments and discussions.

References

1. Carpenter, B.: Type-Logical Semantics. MIT Press, Cambridge, MA (1997)
2. Moot, R., Retoré, C.: The Logic of Categorial Grammars: A Deductive Account of Natural Language Syntax and Semantics, Lecture Notes in Computer Science, vol. 6850. Springer, Berlin (2012). https://doi.org/10.1007/978-3-642-31555-8
3. Morrill, G.V.: Categorial Grammar: Logical Syntax, Semantics, and Processing. Oxford University Press, Oxford (2011)
4. Morrill, G.: Parsing/theorem-proving for logical grammar CatLog3. J. Logic, Lang., Inf. **28**, 183–216 (2019). https://doi.org/10.1007/s10849-018-09277-w
5. Moot, R.: Grail: an interactive parser for categorial grammars. In: Proceedings of VEXTAL'99, pp. 255–261. University Ca' Foscari, Venice (1999). https://www.labri.fr/perso/moot/vextal.pdf
6. Lambek, J.: The mathematics of sentence structure. Am. Math. Monthly **65**, 154–170 (1958). https://doi.org/10.1080/00029890.1958.11989160
7. Ajdukiewicz, K.: Die syntaktische Konnexität. Stud. Philos. **1**, 1–27 (1935)
8. Bar-Hillel, Y.: A quasi-arithmetical notation for syntactic description. Language **29**(1), 47–58 (1953). https://doi.org/10.2307/410452
9. Pentus, M.: Models for the Lambek calculus. Ann. Pure Appl. Logic **75**(1–2), 179–213 (1995). https://doi.org/10.1016/0168-0072(94)00063-9
10. Steedman, M.: The Syntactic Process. MIT Press, Cambridge, MA (2000)
11. Dekhtyar, M., Dikovsky, A.: Generalized categorial dependency grammars. In: A. Avron, N. Dershowitz, A. Rabinovich (eds.) Pillars of Computer Science. Essays Dedicated to Boris (Boaz) Trakhtenbrot on the Occasion of His 85th Birthday, Lecture Notes in Computer Science, vol. 4800, pp. 230–255. Springer (2008). https://doi.org/10.1007/978-3-540-78127-1_13
12. Pentus, M.: Complexity of the Lambek calculus and its fragments. In: Proceedings of Advances in Modal Logic 2010, Advances in Modal Logic, vol. 8, pp. 310–329. College Publications, London (2010). http://www.aiml.net/volumes/volume8/Pentus.pdf
13. Girard, J.-Y.: Linear logic. Theor. Comput. Sci. **50**(1), 1–101 (1987). https://doi.org/10.1016/0304-3975(87)90045-4
14. Abrusci, V.M.: A comparison between Lambek syntactic calculus and intuitionistic linear propositional logic. Zeitschrift für mathematische Logik und Grundlagen der Mathematik (Math. Logic Q.) **36**(1), 11–15 (1990). https://doi.org/10.1002/malq.19900360103
15. Krull, W.: Axiomatische Begründung der allgemeinen Idealtheorie. Sitzungsberichte der physikalischmedizinischen Societät zu Erlangen **56**, 47–63 (1924)
16. Ward, M., Dilworth, R.P.: Residuated lattices. Trans. Am. Math. Soc. **45**, 335–354 (1939)
17. Galatos, N., Jipsen, P., Kowalski, T., Ono, H.: Residuated Lattices: An Algebraic Glimpse at Substructural Logics, Studies in Logic and the Foundations of Mathematics, vol. 151. Elsevier (2007)
18. Lambek, J.: On the calculus of syntactic types. In: R. Jakobson (ed.) Structure of Language and Its Mathematical Aspects, Proceedings of Symposia in Applied Mathematics, vol. 12, pp. 166–178. AMS (1961). https://doi.org/10.1090/psapm/012
19. Kuznetsov, S.: Complexity of the infinitary Lambek calculus with Kleene star. Rev. Symbolic Logic **14**(4), 946–972 (2021). https://doi.org/10.1017/S1755020320000209
20. Andréka, H., Mikulás, S.: Lambek calculus and its relational semantics: completeness and incompleteness. J. Logic, Lang., Inf. **3**, 1–37 (1994). https://doi.org/10.1007/BF01066355

21. Pentus, M.: Free monoid completeness of the Lambek calculus allowing empty premises. In: J.M. Larrazabal, D. Lascar, G. Mints (eds.) Logic Colloquium 1996, Lecture Notes in Logic, vol. 12, pp. 171–209. Springer (1998)
22. Kuznetsov, S.L.: Trivalent logics arising from L-models for the Lambek calculus with constants. J. Appl. Non-Class. Logics **14**(1–2), 132–137 (2014). https://doi.org/10.1080/11663081.2014.911522
23. Morrill, G.: Grammar logicised: relativisation. Linguist. Philos. **40**(2), 119–163 (2017). https://doi.org/10.1007/s10988-016-9197-0
24. Bar-Hillel, Y., Gaifman, C., Shamir, E.: On the categorial and phrase-structure grammars. Bull. Res. Council of Israel, Section F **9F**, 1–16 (1960)
25. Buszkowski, W.: The equivalence of unidirectional Lambek categorial grammars and context-free grammars. Zeitschrift für mathematische Logik und Grundlagen der Mathematik **31**, 369–384 (1985). https://doi.org/10.1002/malq.19850312402
26. Greibach, S.A.: A new normal-form theorem for context-free phrase structure grammars. J. ACM **12**(1), 42–52 (1965). https://doi.org/10.1145/321250.321254
27. Buszkowski, W.: Lambek calculus and substructural logics. Linguist. Anal. **36**(1–4), 15–48 (2010)
28. Kuznetsov, S.: Lambek grammars with one division and one primitive type. Logic J. IGPL **20**(1), 207–221 (2012). https://doi.org/10.1093/jigpal/jzr031
29. Safiullin, A.N.: Derivability of admissible rules with simple premises in the Lambek calculus. Moscow Univ. Math. Bull. **62**(4), 168–171 (2007). https://doi.org/10.3103/S0027132207040092
30. Pentus, M.: Lambek grammars are context free. In: 1993 Proceedings Eighth Annual IEEE Symposium on Logic in Computer Science, pp. 429–433. IEEE (1993). https://doi.org/10.1109/LICS.1993.287565
31. Pullum, G.K., Gazdar, G.: Natural languages and context-free languages. Linguist. Philos. **4**(4), 471–504 (1982). https://doi.org/10.1007/BF00360802
32. Shieber, S.M.: Evidence against the context-freeness of natural languages. Linguist. Philos. **8**, 333–343 (1985). https://doi.org/10.1007/BF00630917
33. Boolos, G.: Don't eliminate cut. J. Philos. Logic **13**(4), 373–378 (1984). https://doi.org/10.1007/BF00247711
34. Pentus, M.: Lambek calculus is NP-complete. Theor. Comput. Sci. **357**(1–3), 186–201 (2006). https://doi.org/10.1016/j.tcs.2006.03.018
35. Loukanova, R.: An approach to functional formal models of constraint-based lexicalist grammar (CBLG). Fundamenta Informaticae **152**(4), 341–372 (2017). https://doi.org/10.3233/FI-2017-1524
36. Savateev, Y.: Unidirectional Lambek grammars in polynomial time. Theory Comput. Syst. **46**(4), 662–672 (2010). https://doi.org/10.1007/s00224-009-9208-4
37. Kuznetsov, S.L.: On translating Lambek grammars with one division into context-free grammars. Proc. Steklov Inst. Math. **294**, 129–138 (2016). https://doi.org/10.1134/S0081543816060080
38. Savateev, Y.: Product-free Lambek calculus is NP-complete. Ann. Pure Appl. Logic **163**(7), 775–788 (2012). https://doi.org/10.1016/j.apal.2011.09.017
39. Pentus, M.: A polynomial-time algorithm for Lambek grammars of bounded order. Linguist. Anal. **36**(1–4), 441–471 (2010)
40. Buszkowski, W.: Some decision problems in the theory of syntactic categories. Zeitschrift für mathematische Logik und Grundlagen der Mathematik **28**, 539–548 (1982). https://doi.org/10.1002/malq.19820283308
41. van Benthem, J.: Language in Action: Categories, Lambdas and Dynamic Logic. North Holland, Amsterdam (1991)
42. Kanazawa, M.: The Lambek calculus enriched with additional connectives. J. Logic, Lang., Inf. **1**(2), 141–171 (1992). https://doi.org/10.1007/BF00171695
43. Kuznetsov, S., Okhotin, A.: Conjunctive categorial grammars. In: Proceedings of the 15th Meeting on the Mathematics of Language, ACL Anthology, vol. W17-3414, pp. 140–151 (2017). https://doi.org/10.18653/v1/W17-3414

44. Okhotin, A.: Conjunctive grammars. J. Automata, Lang., Combinatorics **6**(4), 519–535 (2001). https://doi.org/10.5555/543313.543323
45. Jeż, A.: Conjunctive grammars generate non-regular unary languages. Int. J. Found. Comput. Sci. **19**(3), 597–615 (2008). https://doi.org/10.1142/S012905410800584X
46. Lincoln, P., Mitchell, J., Scedrov, A., Shankar, N.: Decision problems for propositional linear logic. Ann. Pure Appl. Logic **56**(1–3), 239–311 (1992). https://doi.org/10.1016/0168-0072(92)90075-B
47. Chandra, A.K., Kozen, D.C., Stockmeyer, L.J.: Alternation. J. ACM **28**(1), 114–133 (1981). https://doi.org/10.1145/322234.322243
48. Savitch, W.J.: Relationships between nondeterministic and deterministic tape complexities. J. Comput. Syst. Sci. **4**(2), 177–192 (1970). https://doi.org/10.1016/S0022-0000(70)80006-X
49. Kanovich, M.I., Kuznetsov, S.G., Kuznetsov, S.L., Scedrov, A.: Decidable fragments of calculi used in CatLog. In: R. Loukanova (ed.) Natural Language Processing in Artificial Intelligence—NLPinAI 2021, Studies in Computational Intelligence, vol. 999, pp. 1–24. Springer (2022). https://doi.org/10.1007/978-3-030-90138-7_1
50. Kanovich, M.I.: Horn fragments of non-commutative logics with additives are PSPACE-complete. In: 1994 Annual Conference of the European Association for Computer Science Logic. Kazimierz, Poland (1994)
51. Kanazawa, M.: Lambek calculus: recognizing power and complexity. In: J. Gerbrandy, M. Marx, M. de Rijke, Y. Venema (eds.) JFAK. Essays Dedicated to Johan van Benthem on the Occasion of his 50th Birthday. Vossiuspers, Amsterdam University Press (1999). https://festschriften.illc.uva.nl/j50/contribs/kanazawa/index.html
52. Kanovich, M., Kuznetsov, S., Scedrov, A.: The complexity of multiplicative-additive Lambek calculus: 25 years later. In: WoLLIC 2019: Logic, Language, Information, and Computation, Lecture Notes in Computer Science, vol. 11541, pp. 356–372. Springer (2019). https://doi.org/10.1007/978-3-662-59533-6_22
53. Kozak, M.: Distributive full Lambek calculus has the finite model property. Studia Logica **91**, 201–216 (2009). https://doi.org/10.1007/s11225-009-9172-7
54. Kuznetsov, S.: L-completeness of the Lambek calculus with the reversal operation. In: LACL 2012: Logical Aspects of Computational Linguistics, Lecture Notes in Computer Sciences, vol. 7351, pp. 151–160. Springer (2012). https://doi.org/10.1007/978-3-642-31262-5_10
55. Kuznetsov, S.: L-completeness of the Lambek calculus with the reversal operation allowing empty antecedents. In: Categories and types in logic, language, and physics. Essays dedicated to Jim Lambek on the occasion of his 90th birthday, Lecture Notes in Computer Sciences, vol. 8222, pp. 268–278. Springer (2014). https://doi.org/10.1007/978-3-642-54789-8_15
56. Morrill, G., Valentín, O., Fadda, M.: The displacement calculus. J. Logic, Lang., Inf. **20**(1), 1–48 (2011). https://doi.org/10.1007/s10849-010-9129-2
57. Morrill, G.: Categorial formalisation of relativisation: Pied piping, islands, and extraction sites. Technical Report LSI-92-23-R, Universitat Politècnica de Catalunya (1992)
58. Moortgat, M.: Multimodal linguistic inference. J. Logic, Lang., Inf. **5**(3–4), 349–385 (1996). https://doi.org/10.1007/BF00159344
59. Kanovich, M., Kuznetsov, S., Morrill, G., Scedrov, A.: A polynomial-time algorithm for the Lambek calculus with brackets of bounded order. In: 2nd International Conference on Formal Structures for Computation and Deduction (FSCD 2017), Leibniz International Proceedings in Informatics, vol. 84, pp. 22:1–22:17. Schloss Dagstuhl–Lebniz-Zentrum für Informatik (2017). https://doi.org/10.4230/LIPIcs.FSCD.2017.22
60. Morrill, G., Kuznetsov, S., Kanovich, M., Scedrov, A.: Bracket induction for the Lambek calculus with bracket modalities. In: FG 2018: Formal Grammar, Lecture Notes in Computer Science, vol. 10950, pp. 84–101. Springer (2018). https://doi.org/10.1007/978-3-662-57784-4_5
61. Pshenitsyn, T.: Powerful and NP-complete: hypergraph Lambek grammars. In: ICGT 2021: Graph Transformation, Lecture Notes in Computer Science, vol. 12741, pp. 102–121. Springer (2021). https://doi.org/10.1007/978-3-030-78946-6_6

62. Kanovich, M., Kuznetsov, S., Scedrov, A.: Reconciling Lambek's restriction, cut-elimination, and substitution in the presence of exponential modalities. J. Logic Comput. **30**(1), 239–256 (2020). https://doi.org/10.1093/logcom/exaa010
63. de Groote, P.: On the expressive power of the Lambek calculus extended with a structural modality. In: Language and Grammar, CSLI Lecture Notes, vol. 168, pp. 95–111. Stanford University (2005)
64. Kanovich, M., Kuznetsov, S., Scedrov, A.: Undecidability of the Lambek calculus with a relevant modality. In: FG 2015, FG 2016: Formal Grammar, Lecture Notes in Computer Science, vol. 9804, pp. 240–256. Springer (2016). https://doi.org/10.1007/978-3-662-53042-9_14
65. Anderson, A.R., Belnap, N.: Entailment: The Logic of Relevance and Necessity, vol. 1. Princeton University Press (1975)
66. Maksimova, L.L.: O sisteme aksiom ischisleniya strogoĭ implikatsii [On the system of axioms of the calculus of rigorous implication]. Algebra i logika **3**(5), 59–68 (1964). In Russian
67. Danos, V., Joinet, J.-B., Schellinx, H.: The structure of exponentials: Uncovering the dynamics of linear logic proofs. In: KGC 1993: Computational Logic and Proof Theory, Lecture Notes in Computer Science, vol. 713, pp. 159–171. Springer (1993). https://doi.org/10.1007/BFb0022564
68. Nigam, V., Miller, D.: Algorithmic specifications in linear logic with subexponentials. In: Proceedings of the 11th ACM SIGPLAN Conference on Principles and Practice of Declarative Programming, PPDP '09, p. 129-140. ACM (2009). https://doi.org/10.1145/1599410.1599427. https://doi.org/10.1145/1599410.1599427
69. Kanovich, M., Kuznetsov, S., Nigam, V., Scedrov, A.: Subexponentials in non-commutative linear logic. Math. Struct. Comput. Sci. **29**(8), 1217–1249 (2019). https://doi.org/10.1017/S0960129518000117
70. Dudakov, S.M., Karlov, B.N., Kuznetsov, S.L., Fofanova, E.M.: Complexity of Lambek calculi with modalities and of total derivability in grammars. Algebra and Logic **60**(5), 308–326 (2021). https://doi.org/10.1007/s10469-021-09657-5
71. Kanovich, M., Kuznetsov, S., Scedrov, A.: The multiplicative-additive Lambek calculus with subexponential and bracket modalities. J. Logic, Lang., Inf. **30**, 31–88 (2020). https://doi.org/10.1007/s10849-020-09320-9
72. Kuznetsov, S.L., Ryzhkova, N.S.: A restricted fragment of the Lambek calculus with iteration and intersection operations. Algebra and Logic **59**(2), 129–146 (2020). https://doi.org/10.1007/s10469-020-09586-9
73. Palka, E.: An infinitary sequent system for the equational theory of *-continuous action lattices. Fundamenta Informaticae **78**(2), 295–309 (2007). https://doi.org/10.5555/2366484.2366490
74. Buszkowski, W.: On action logic: equational theories of action algebras. J. Logic Comput. **17**(1), 199–217 (2007). https://doi.org/10.1093/logcom/exl036
75. Kleene, S.C.: Representation of events in nerve nets and finite automata. In: Automata Studies, pp. 3–41. Princeton University Press (1956). https://doi.org/10.1515/9781400882618-002
76. Pratt, V.: Action logic and pure induction. In: JELIA 1990: Logics in AI, Lecture Notes in Artificial Intelligence, vol. 478, pp. 97–120. Springer (1991). https://doi.org/10.1007/BFb0018436
77. Kozen, D.: On action algebras. In: J. van Eijck, A. Visser (eds.) Logic and Information Flow, pp. 78–88. MIT Press (1994)
78. Kuznetsov, S.: Action logic is undecidable. ACM Trans. Comput. Logic **22**(2), article no. 10 (2021). https://doi.org/10.1145/3445810
79. Kuznetsov, S.L., Speranski, S.O.: Infinitary action logic with exponentiation. Ann. Pure Appl. Logic **173**(2), article no. 103057 (2022). https://doi.org/10.1016/j.apal.2021.103057

Categorial Dependency Grammars: Analysis and Learning

Denis Béchet and Annie Foret

Abstract We give an overview on the family of Categorial Dependency Grammars (CDG), as computational grammars for language processing. CDG are a class of categorial grammars defining dependency structures. They can be viewed as a formal system, where types are attached to words, combining the classical categorial grammars' elimination rules with valency pairing rules that are able to define non-projective (discontinuous) dependencies. We discuss both formal aspects of CDG in terms of strength, derivability and complexity and practical issues and algorithms. We also point to some open problems, review results on CDG learnability and discuss CDG as large scale grammars with respect to Mel'čuk principles and to various hypotheses on training corpora.

Keywords Categorial grammar · Dependency grammar · Syntax analysis · Grammatical inference · Incremental learning · Treebanks

1 Introduction to CDG

This paper provides a survey on Categorial Dependency Grammars (CDG). Categorial Dependency Grammars [13] considered in this paper, are a unique class of grammars directly generating unbounded dependency structures (DS), beyond context-freeness, able to define non-projective dependency structures, still well adapted to real NLP applications.

CDG is a formal system combining the classical categorial grammars' elimination rules with valency pairing rules defining non-projective (discontinuous) dependencies, which is a special feature of CDG. Another special feature of CDG is that the

D. Béchet (✉)
Nantes University, Nantes, France
e-mail: Denis.Bechet@univ-nantes.fr

A. Foret
IRISA and University of Rennes 1, Rennes, France
e-mail: Annie.Foret@irisa.fr

R. Loukanova et al. (eds.), *Logic and Algorithms in Computational Linguistics 2021 (LACompLing2021)*, Studies in Computational Intelligence 1081,
https://doi.org/10.1007/978-3-031-21780-7_2

elimination rules are interpreted as local dependency constructors, such that these rules are naturally extended to the so called "*iterated dependencies*". This point needs explanation. A dependency d is *iterated* in a DS D if some word in D governs through dependency d several other words. The iterated dependencies are due to the basic principles of dependency syntax, which concern optional repeatable dependencies [24]: All modifiers of a noun n share n as their *governor* and, similarly, all modifiers of a verb v share v as their *governor*. At the same time, as we explain below, the iterated dependencies are a challenge for grammatical inference.

We discuss both formal aspects of CDG in terms of strength, derivability and complexity and practical issues and algorithms. We also point to some open problems. We also review results on CDG learnability and discuss CDG as large scale grammars with respect to Mel'čuk principles [24] and to various hypotheses on training corpora.

The papers [2, 13] give the main definition and theoretical results on CDG. A statistical parser based on CDG is tested in [22]. Reference [6] describes a graphical environment for the creation of a corpus together with its grammar. References [10, 11] present several results on learning from positive examples in particular for CDG.

The plan of the paper is as follows. Section 1 introduces Categorial Dependency Grammars with its type language, it explains and illustrates languages generated by Categorial Dependency Grammars. Section 2 provides formal definitions including the type calculus. Section 3 discusses the parsing algorithm and its implementation. Sections 4 and 5 review how iterated types are a challenge for learning and present two different approaches. Section 6 discusses open questions.

1.1 Surface Dependency Structures

A CDG is a formal grammar that defines a language of surface dependency structures. With respect to the Meaning-Text Theory,[1] a surface dependency structure [24] is a list of words linked together by dependencies. Each dependency has a name, a starting point called the governor and an ending point called the subordinate. The following example shows a surface dependency structure for the string "*This deal brought more problems than profits.*".

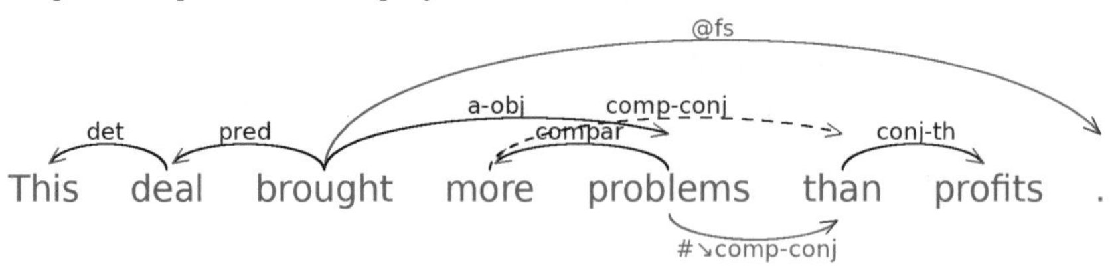

The structure contains eight words (or punctuation symbols) and eight dependencies (or anchors). The arrow between *brought* and *problems* defines a dependency of name $a-obj$ where *brought* is the governor and *problems* is the subordinate.

[1] https://en.wikipedia.org/wiki/Meaning-text_theory.

The root of the structure is the word *brought* (this word isn't the subordinate of any dependency). The structure isn't a tree. In fact, the CDG dependency structures are not necessarily dependency trees because certain dependencies called non-projective dependencies are usually introduced together with an auxiliary dependency called an anchor. In the example, the non-projective dependency is $comp-conj$ and the anchor is $\#\searrow comp-conj$.

In the rest of the paper, we write dependency structure (DS) for surface dependency structure.

1.2 CDG Types Express Dependency Valencies

A CDG has a lexicon that maps each word to a set of CDG types. A CDG type for a word w in a string defines the valencies of the dependencies that start from w or end into w. A positive valency corresponds to the beginning of a dependency, in this case w is the governor. A negative valency corresponds to the end of a dependency, in this case w is the subordinate. A valency defines also the name of the dependency. Thus, a dependency from a word w_1 to a word w_2 can exist only if the type that has been chosen in the lexicon for w_1 contains a positive valency with the name of the dependency and if the type for w_2 contains a negative valency with the same name. Usually, a valency in the type of a word can participate in one and only one dependency (except for the special cases of optional or iterated valencies). There are two kinds of dependencies in CDG, which are the projective dependencies (and anchors) that correspond to the basic dependency types of CDG types and the non-projective dependencies that correspond to the potentials of CDG types.

1.3 Projective Dependencies and Anchors

A projective dependency or an anchor d between a governor *Gov* and a subordinate *Sub* when the governor is on the left and the subordinate is on the right $Gov \xrightarrow{d} Sub$ corresponds to a subtype d or d^* in the right argument list of the type of the governor: $Gov \mapsto [..\backslash../../d/..]^P$ and to the head type d on the type of the subordinate: $Sub \mapsto [..\backslash d/..]^P$.

If the subtype of the governor is d, there is exactly one projective dependency (or anchor) that starts from *Gov*. If the subtype of the governor is d^*, there can be none, one or several projective dependencies that can start from *Gov*. An anchor is very similar to a projective depencency except that d is either an anchor type $\#(\swarrow v)$, $\#(\searrow v)$ (uses to anchor the subordinate of a non-projective dependency) or a name that starts with the character @ (uses to anchor a punctuation symbol). If the governor is on the right and the subordinate is on the left $Sub \xleftarrow{d} Gov$, the subtype d or d^* must be in the left argument list of the type of the governor: $Gov \mapsto [..\backslash d\backslash..\backslash../..]^P$ (the type of the dependent is the same).

The following lexicon defines only projective dependencies and anchors:

$$\begin{array}{ll} in & \mapsto [c\_copul/prepos-l] \\ the & \mapsto [det] \\ beginning & \mapsto [det\backslash prepos-l] \\ was & \mapsto [c\_copul\backslash S/@fs/pred] \\ Word & \mapsto [det\backslash pred] \\ . & \mapsto [@fs] \end{array}$$

There is a dependency structure (that is also a dependency tree) for *In the beginning was the Word .* where the root is the word *was* (it corresponds to the head type *S*):

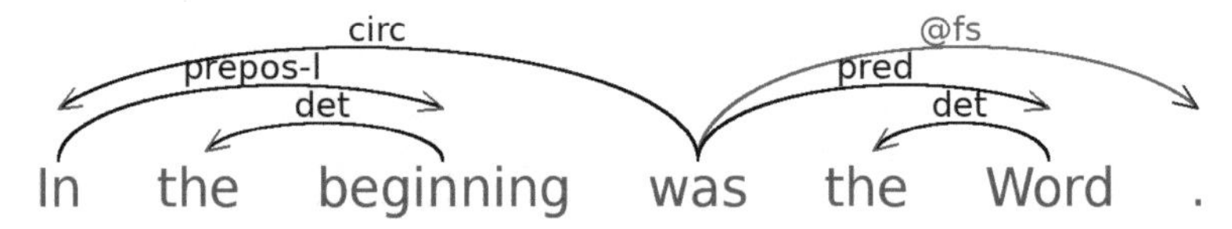

There are five projective dependencies and one anchor (to anchor the full stop). The projective dependencies and anchors define a graph where all the dependencies cannot cross each other.[2]

1.4 Non-projective Dependencies

A non-projective dependency d (also named discontinuous dependency) is associated to a couple of dual polarized valencies $\nearrow d$ and $\searrow d$ if the governor is on the left and the subordinate is on the right or $\swarrow d$ and $\nwarrow d$ if the governor is on the right and the subordinate is on the left. A non-projective dependency $Gov \overset{d}{\dashrightarrow} Sub$ where the governor is on the left and the subordinate is on the right corresponds to the right positive valency $\nearrow d$ in the potential of the type of the governor: $Gov \mapsto [..]^{..\nearrow d..}$ and to the right negative valency $\searrow d$ in the potential of the type of the subordinate: $Sub \mapsto [..]^{..\searrow d..}$.

The following lexicon defines projective dependencies and the non-projective dependency $comp-conj$ but does not use anchors:

[2] The term "projective" comes from the definition of a projective dependency in a dependency structure: A dependency $w_1 \overset{d}{\rightarrow} w_2$ is projective if all the words between w_1 and w_2 are dominated by w_1. Otherwise it is non-projective. In CDG, following the types, a dependency is said "*projective*" or "*non-projective*" but the dependency itself may or may not be really projective in a particular DS. This is the reason why, we can use, in the CDG terminology, the terms "*local*" for "*projective*" and "*discontinuous*" for "*non-projective*".

$$\begin{array}{ll} this & \mapsto [det] \\ deal & \mapsto [det\backslash pred] \\ brought & \mapsto [pred\backslash S/a-obj] \\ problems & \mapsto [compar\backslash a-obj] \\ profits & \mapsto [conj-th] \\ more & \mapsto [compar]^{\nearrow comp-conj} \\ than & \mapsto [\varepsilon/conj-th]^{\searrow comp-conj} \end{array}$$

The following dependency structures for *this deal brought more problems than profits* is based on this lexicon.

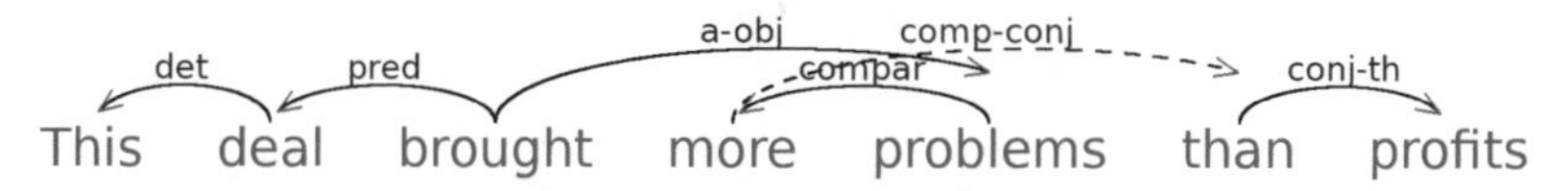

There is no anchor even for punctuation symbols. Thus the type of the subordinate of the non-projective dependency *comp−conj* has an empty head (marked by ε). An empty head is eliminated in the calculus using the left rule $\mathbf{L}^{\mathbf{l}}_{\varepsilon}$. $[\varepsilon]^{P_1}[\beta]^{P_2} \vdash [\beta]^{P_1P_2}$ or the right rule $\mathbf{L}^{\mathbf{r}}_{\varepsilon}$. $[\beta]^{P_1}[\varepsilon]^{P_2} \vdash [\beta]^{P_1P_2}$. The projective dependencies and the non-projective dependency define a dependency tree with *brought* as root. The projective dependencies don't cross each other. However, a non-projective dependency can cross a projective dependency or a non-projective dependency with another name or another orientation. With the pairing rule **FA** (first available), a non-projective dependency cannot cross another non-projective dependency with the same name and the same orientation.

1.5 Non-projective Dependencies with Anchors

In the previous example with projective and non-projective dependencies but without anchor, it isn't possible to anchor punctuation symbols or to anchor the subordinate of a non-projective dependency. In this case, such symbols or words must have a type with an empty head but the types with an empty head create a language whose order is relatively free. Thus, anchors are used to fix such elements in a sentence. Punctuation symbols are fixed using an anchor for punctuation rather than a projective dependency (from a theoretical point of view they are similar). The subordinate of a non-projective dependency is fixed by an anchor to a third word called a host.

In this model, for a non-projective dependency d, there are four polarized valencies $\nearrow d$, $\searrow d$, $\nwarrow d$ and $\swarrow d$ and two anchor valencies $\#\searrow d$, $\#\swarrow d$. For a punctuation symbol, the name of anchor starts with @ like *@fs* for the full stop.

For a non-projective dependency where the governor is on the left, the subordinate in the middle and the host is on the right $Gov \overset{d}{\dashrightarrow} Sub \overset{\#\searrow d}{\longleftarrow} Host$, the type of the governor $Gov \mapsto [..]^{..\nearrow d..}$ has the positive left valency $\nearrow d$ in the potential, the type of the subordinate $Sub \mapsto [..\backslash\#\searrow d/..]^{..\searrow d..}$ has the negative left valency $\searrow d$ in the potential and has the anchor type $\#\searrow d$ as head type. Finally the type of the host $Host \mapsto [..\backslash\#\searrow d\backslash../..]^P$ has the anchor type $\#\searrow d$ in the left argument list.

The following lexicon shows a complete example with all kinds of valencies:

$$\begin{array}{ll}
\textit{this} & \mapsto [det] \\
\textit{deal} & \mapsto [det\backslash pred] \\
\textit{brought} & \mapsto [pred\backslash S/@fs/a-obj] \\
\textit{problems} & \mapsto [compar\backslash a-obj/\#\searrow comp-conj] \\
\textit{profits} & \mapsto [conj-th] \\
\textit{more} & \mapsto [compar]^{\nearrow comp-conj} \\
\textit{than} & \mapsto [\#\searrow comp-conj/conj-th]^{\searrow comp-conj} \\
. & \mapsto [@fs]
\end{array}$$

The following example shows the same sentence as the previous example with a full stop at the end. The full stop is anchored to the word *brought*. The governor of the non-projective dependency $comp-conj$ is the word *more*. The subordinate is the word *than*. It is anchored to the word *problems* using the anchor $\#\searrow comp-conj$.

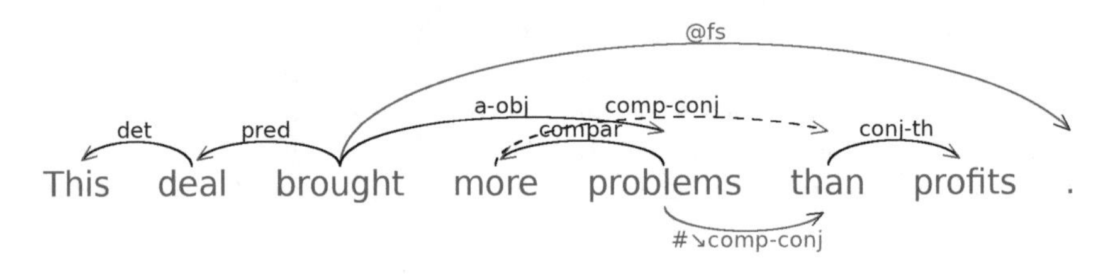

2 CDG Definitions

The section starts with the formal definition of CDG. Then, the relation from lexicon and rules to dependency structures is shown on examples. The final part presents some non-context-free languages that are generated by CDG.

Definition 1 (CDG Types) Let $\mathbf{C}$ be a set of *local dependency names* and $\mathbf{V}$ be a set of *valency names*.

The expressions of the form $\swarrow v, \nwarrow v, \searrow v, \nearrow v$, where $v \in \mathbf{V}$, are called *polarized valencies*. $\nwarrow v$ and $\nearrow v$ are *positive*, $\swarrow v$ and $\searrow v$ are *negative*; $\nwarrow v$ and $\swarrow v$ are *left*, $\nearrow v$ and $\searrow v$ are *right*. Two polarized valencies with the same valency name and orientation, but with opposite arrow directions are *dual*.

An expression of one of the forms $\# \swarrow v$, $\# \searrow v$, $v \in \mathbf{V}$, is called an *anchor type* or just an *anchor*. An expression of the form d^* where $d \in \mathbf{C}$, is called an *iterated dependency type*.

Local dependency names, iterated dependency types and anchor types are *primitive types*.

An expression of the form $t = [l_m\backslash \ldots \backslash l_1\backslash H/r_1/\ldots/r_n]$ in which $m, n \geq 0$, $l_1, \ldots, l_m, r_1, \ldots, r_n$ are primitive types and H is either a local dependency name or an anchor type or is empty (written ε), is called a *basic dependency type*. $l_1, \ldots, l_m$ and $r_1, \ldots, r_n$ are left and right *argument types* of t, H is called the *head type* of t.

A (possibly empty) string P of polarized valencies is called a *potential*.

A *dependency type* is an expression B^P in which B is a basic dependency type and P is a potential. **CAT(C, V)** will denote the set of all dependency types over **C** and **V**.

CDG are defined using the following calculus of dependency types. We give a simplified version of the relativized calculus [11] written with respect to the word positions in the sentence, which allows us to interpret them as construction of Dependency Structures.

Definition 2 (Simplified Calculus of Dependency Types) In this set of rules on lists of types, the symbol C stands for a local dependency name or an anchor type, but cannot be an anchor type in rules $\mathbf{I^l}$, $\mathbf{I^r}$, $\Omega^\mathbf{l}$ and $\Omega^\mathbf{r}$ (anchors are not iterated). The symbol α is a basic dependency type. The symbol β ranges over expressions of the form $l_m\backslash \ldots \backslash l_1\backslash H/r_1/\ldots/r_n$

$\mathbf{L^l}$ $[C]^P[C\backslash\beta]^Q \vdash [\beta]^{PQ}$ $\quad$ $\mathbf{L^r}$ $[\beta/C]^P[C]^Q \vdash [\beta]^{PQ}$
$\mathbf{L^l_\varepsilon}$ $[\varepsilon]^P[\beta]^Q \vdash [\beta]^{PQ}$ $\quad$ $\mathbf{L^r_\varepsilon}$ $[\beta]^P[\varepsilon]^Q \vdash [\beta]^{PQ}$
$\mathbf{I^l}$ $[C]^P[C^*\backslash\beta]^Q \vdash [C^*\backslash\beta]^{PQ}$ $\quad$ $\mathbf{I^r}$ $[\beta/C^*]^P[C]^Q \vdash [\beta/C^*]^{PQ}$
Ω^l $[C^*\backslash\beta]^P \vdash [\beta]^P$ $\quad$ Ω^r $[\beta/C^*]^P \vdash [\beta]^P$
$\mathbf{D^l}$ $\alpha^{P_1(\swarrow V)P(\nwarrow V)P_2} \vdash \alpha^{P_1PP_2}$ $\quad$ $\mathbf{D^r}$ $\alpha^{P_1(\nearrow V)P(\searrow V)P_2} \vdash \alpha^{P_1PP_2}$

In $\mathbf{D^l}$, the potential $P_1(\swarrow V)P(\nwarrow V)P_2$ satisfies the pairing rule **FA**:
FA (*First Available*): P has no occurrence of $\swarrow V$ or $\nwarrow V$.
In $\mathbf{D^r}$, the potential $P_1(\nearrow V)P(\searrow V)P_2$ satisfies the pairing rule **FA**:
FA (*First Available*): P has no occurrence of $\nearrow V$ or $\searrow V$.

Fact. This simplified calculus enjoys a subformula property adapted to CDG types:

- Each type formula without bracket and potential that occurs on the right on a rule (β, α, $C^*\backslash\beta$) also occurs on the left of the same rule;
- Each potential expression on the right of a rule also occurs on the left of the same rule.

Now, when attached to a sentence x and a lexicon, for every proof ρ, represented as a sequence of rule applications, we will define the dependency structure $DS_x(\rho)$ *constructed* in this proof.

Definition 3 (Categorial Dependency Grammar) A Categorial Dependency Grammar (CDG) is a system $G = (W, \mathbf{C}, \mathbf{V}, S, \lambda)$, where W is a finite set of words, **C** is a finite set of local dependency names containing the selected name S (an axiom), **V** is a finite set of non-projective dependency names and λ, called *lexicon*, is a finite substitution on W such that $\lambda(a) \subset \mathbf{CAT(C, V)}$ for each word $a \in W$. λ is extended on sequences of words W^* in the usual way.

For $G = (W, \mathbf{C}, \mathbf{V}, S, \lambda)$, a DS D and a sentence x, let $G[D, x]$ denote the relation:

"$D = DS_x(\rho)$, where ρ is a proof of $\Gamma \vdash S$ for some $\Gamma \in \lambda(x)$".

Then the *language* generated by G is the set $L(G) =_{df} \{w \mid \exists D\ G[D, w]\}$ and the *DS-language* generated by G is the set $\Delta(G) =_{df} \{D \mid \exists w\ G[D, w]\}$. $\mathcal{L}(CDG)$ and $\mathcal{D}(CDG)$ will denote respectively the family of languages and the family of DS-languages generated by these grammars.

An alternative to the **FA** (*first available*) pairing rule has been proposed in [15] as follows:

FC (First Cross) Variant. In rules $\mathbf{D^l}$ and $\mathbf{D^l}$, **FA** is replaced with

In $\mathbf{D^l}$, the potential $P_1(\swarrow V)P(\nwarrow V)P_2$ satisfies the pairing rule **FC** (*first cross*): P_1 has no occurrence of $\swarrow V$ and P has no occurrence of $\nwarrow V$.

In $\mathbf{D^r}$, the potential $P_1(\nearrow V)P(\searrow V)P_2$ satisfies the pairing rule **FC** (*first cross*): P has no occurrence of $\nearrow V$ and P_2 has no occurrence of $\searrow V$.

This variant was shown to handle dutch-crossing dependencies.

2.1 From Rules to Dependency Structures

A CDG defines a lexicon. The rules used in the calculus are fixed. The lexicon maps each word or symbol to one or several CDG types. For instance, the following lexicon gives a unique type to the words *John*, *ran*, *fast* and *yesterday*. The type of *ran* has an iterated dependency type c^* that can introduce several projective dependencies c with the same governor *ran*.

$$\begin{array}{ll} John & \mapsto [pr] \\ ran & \mapsto [pr \backslash S / c^*] \\ fast, yesterday & \mapsto [c] \end{array}$$

The string *John ran fast yesterday* is recognized by the CDG. A proof is given by the following derivation. In this derivation, the words are written just above the type that has been chosen for it in the lexicon. The derivation ends by the axiom S. Each node corresponds to the application of one of the rules of the calculus of dependency types:

$$\dfrac{\overset{John}{[pr]} \qquad \dfrac{\dfrac{\dfrac{\overset{ran}{[pr \backslash S / c^*]}\ \overset{fast}{[c]}}{[pr \backslash S / c^*]}\mathbf{I^r} \qquad \overset{yesterday}{[c]}}{[pr \backslash S / c^*]}\mathbf{I^r}}{[pr \backslash S]}\Omega^r}{[S]}\mathbf{L^l}$$

A derivation determines a dependency structure. Each application of a rule (the nodes of a derivation) can define a dependency. The rules $\mathbf{L}^{\mathbf{l}}$, $\mathbf{L}^{\mathbf{r}}$, $\mathbf{I}^{\mathbf{l}}$ and $\mathbf{I}^{\mathbf{r}}$ define projective dependencies or anchors between the word associated to the types in the premises, the rules $\mathbf{L}^{\mathbf{l}}_{\varepsilon}$, $\mathbf{L}^{\mathbf{r}}_{\varepsilon}$, Ω^{l} and Ω^{r} don't define dependencies and the rules $\mathbf{D}^{\mathbf{l}}$ and $\mathbf{D}^{\mathbf{r}}$ define non-projective dependencies between the words that carry initially the polarized valencies.[3] The dependency structure from the previous derivation is the following:

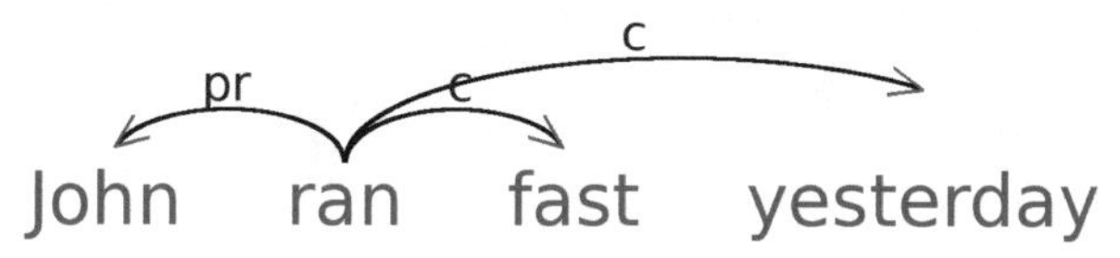

2.2 *Generated Languages*

Every context-free language can be generated by a CDG (with empty potential and no iterated type). In fact CDG with empty potential and no iterated type correspond to classical categorial grammars of order 1 (the argument subtypes are primitive types) and this class of grammars generates all (and only) the context-free langages. CDG with empty potential and with iterative type also generates only context-free languages. But with non-empty potential, it is possible to generate some context-sensitive languages.

2.2.1 CDG Example: MIX [2]

For instance, it is possible to generate the mildly context-sensitive language MIX (with 3 symbols) [26, 27] $\{w, w \in \{a, b, c\}^*, |w|_a = |w|_b = |w|_c > 0\}$. The following lexicon defines a CDG that generates this language. It is also possible to define a CDG avoiding empty heads but this one is simpler. Moreover, the grammar can also be adapted to generate any MIX language with k symbols [13].

$$
\begin{array}{ll}
\mathrm{a} \mapsto [S]^{\nwarrow B \nwarrow C} & \mathrm{b} \mapsto [\varepsilon]^{\swarrow B} \\
\mathrm{a} \mapsto [S \setminus S]^{\nwarrow B \nwarrow C} & \mathrm{b} \mapsto [\varepsilon]^{\searrow B} \\
\mathrm{a} \mapsto [S]^{\nearrow B \nearrow C} & \\
\mathrm{a} \mapsto [S \setminus S]^{\nearrow B \nearrow C} & \mathrm{c} \mapsto [\varepsilon]^{\swarrow C} \\
\mathrm{a} \mapsto [S]^{\nwarrow B \nearrow C} & \mathrm{c} \mapsto [\varepsilon]^{\searrow C} \\
\mathrm{a} \mapsto [S \setminus S]^{\nwarrow B \nearrow C} & \\
\mathrm{a} \mapsto [S]^{\nwarrow C \nearrow B} & \\
\mathrm{a} \mapsto [S \setminus S]^{\nwarrow C \nearrow B} &
\end{array}
$$

[3] A non-projective dependency is introduced only if the polarized valencies don't correspond to the same word. In fact, this situation is possible if, in the lexicon, a word is associated to a potential with a couple of dual polarized valencies like in the type $[...]^{\nearrow d \searrow d}$. This isn't the case if the polarized valencies are in the reverse order $[...]^{\searrow d \nearrow d}$ because $\searrow d$ must interact with a dual polarized valency ($\nearrow d$) on the left and $\nearrow d$ must interact with a dual polarized valency ($\searrow d$) on the right.

The following dependency structure corresponds to the analysis of *caabcb*. The root is the second *a*.

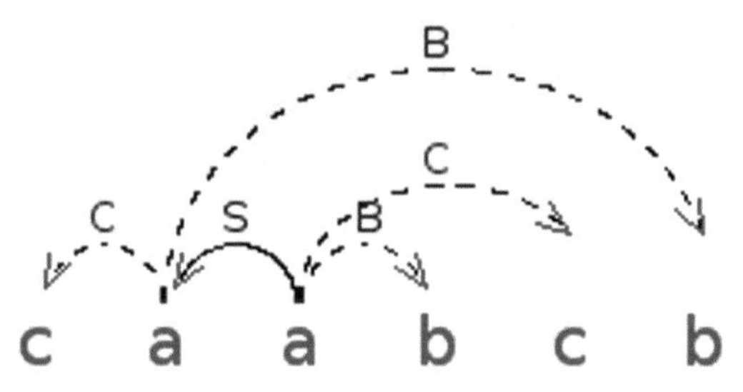

2.2.2 CDG Example: $a^n b^n c^n$

Some mildly context-sensitive languages can also be generated. For instance, the following lexicon defines a CDG that generates $\{a^n b^n c^n, n > 0\}$

a $\mapsto [A]^{\swarrow D}$	b $\mapsto [B \,/\, C]^{\nwarrow D}$	c $\mapsto [C]$
a $\mapsto [A \setminus A]^{\swarrow D}$	b $\mapsto [A \setminus S \,/\, C]^{\nwarrow D}$	c $\mapsto [B \setminus C]$

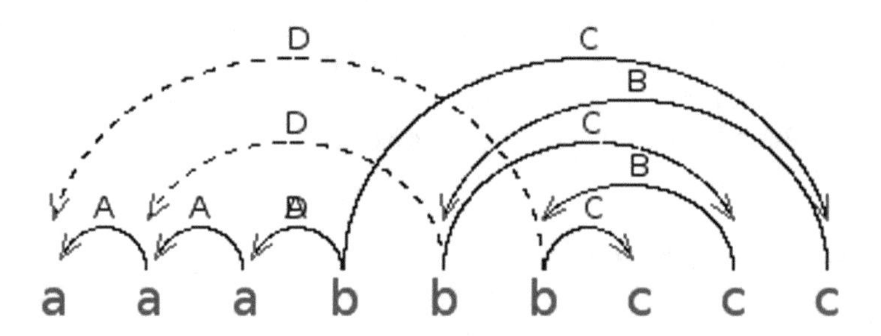

Moreover, it is possible to define a CDG that generates $\{a_1^n a_2^n \cdots a_k^n, n > 0\}$ for any $k > 0$ or languages with equality constraints such as $\{a^n b^m c^n d^m, n > 0, m > 0\}$. The following grammar generates this language:

$a \mapsto [S \,/\, A]^{\nearrow E}$	$a \mapsto [S \,/\, B]^{\nearrow E}$
$a \mapsto [A \,/\, A]^{\nearrow E}$	$a \mapsto [A \,/\, B]^{\nearrow E}$
$b \mapsto [B \,/\, C]^{\nearrow F}$	$b \mapsto [B \,/\, C]^{\nearrow F}$
$c \mapsto [C \,/\, C]^{\searrow E}$	$c \mapsto [C \,/\, D]^{\searrow E}$
$d \mapsto [D \,/\, D]^{\searrow F}$	$d \mapsto [D]^{\searrow F}$

In this case, the basic types modelize the regular language $a^+ b^+ c^+ d^+$ and the polarised valencies ensure that there are the same numbers of *a* and *c* and the same number of *b* and *d*: A non-projective dependency must link each *a* with a *c* and each *b* with a *d*.

3 CDG Analysis

The section presents a parsing algorithm for CDG *CDGAnalyst*, a tool with several parsers *CDGLab* and a greedy parser.

3.1 CDGAnalyst

The parsing algorithm is described in [13]. It is a dynamic programming parsing algorithm based on CYK parsing algorithm. For each item, the remaining basic types and deficits of polarized valencies are computed.

The complexity of CDGAnalyst depends mainly on the number of different polarized valency names but it remains polynomial from the size of the input string.

$$\mathbf{O}(I_G \cdot {a_G}^2 \cdot (\Delta_G \cdot n)^{2p_G} \cdot n^3)$$

n	The length of the input string
I_G	The number of assignments in the lexicon of G
a_G	The maximal number of left or right subtypes in G
Δ_G	The maximal valency deficit in G
p_G	The number of polarized valency names in G

3.2 CDGLab with Filtering

CDGLab [1] contains an implementation of CDGAnalyst. This platform is dedicated to the definition of CDG, the analysis of strings using CDG and the creation of dependency structure corpora (associated to CDG). The parser also allows to reduce the search space. For instance, it is possible to limit the size of the deficits of polarized valencies (this principle is based on the idea that there is a limit on the "complexity" of a natural language: it is not possible to have too many non-projective dependencies above a word in the dependency structure of a sentence).

3.3 Greedy Parser

A transition-based dependency parser [22] has been also implemented. This parser uses three steps: the first step computes the best projective (with anchor) dependency structure. Then a second step adds the left non-projective dependencies and a last step adds the right non-projective dependencies. The parser uses three oracles (one for each step) that are trained using CDG dependency structure corpora. In this case, the model has a linear time complexity.

4 Grammatical Inference

We first describe Gold's learning paradigm [17] in our context. Let $\mathcal{C}$ be a class of grammars that we wish to learn from examples. The issue is to define an algorithm, that when applied to a finite set of examples generated by a grammar in $\mathcal{C}$, yields a grammar in $\mathcal{C}$ that generates the examples; the algorithm is also required to converge.

An *observation set* $\Phi(G)$ of G is associated with every grammar $G \in \mathcal{C}$. This may be the generated language $L(G)$ or an image of structures generated by G. Below we call an enumeration of $\Phi(G)$, a *training sequence* for G.

Definition 4 (Inference Algorithm) An algorithm $\mathcal{A}$ is an *inference algorithm* for $\mathcal{C}$ if, for every grammar $G \in \mathcal{C}$, every *training sequence* σ for G, and every initial subsequence $\sigma[i] = \{s_1, \ldots, s_i\}$ of σ, it returns a grammar $\mathcal{A}(\sigma[i]) \in \mathcal{C}$ (called a *hypothesized grammar*). The algorithm $\mathcal{A}$ *learns* a *target grammar* $G \in \mathcal{C}$ if, on any training sequence σ for G, $\mathcal{A}$ stabilizes on a grammar $\mathcal{A}(\sigma[T])$ generating the same examples as G.[4] The grammar $\lim_{i \to \infty} \mathcal{A}(\sigma[i]) = \mathcal{A}(\sigma[T])$ returned at the stabilization step is the *limit grammar*. The algorithm $\mathcal{A}$ *learns* $\mathcal{C}$ if it learns every grammar in $\mathcal{C}$. $\mathcal{C}$ is *learnable* if there is an inference algorithm learning $\mathcal{C}$.

Learning from **strings** corresponds to sequences σ of examples in $L(G_T)$, while learning from **structures** corresponds to examples in $\Delta(G_T)$, where T is the stabilization step on σ and Δ denotes the set of structures generated by G.

4.1 Learnability Properties

Learnability and unlearnability properties in this Gold's paradigm [17] have been widely studied from a theoretical point of view, and for various classes of grammars. On the one hand, it follows from a theorem of Wright (1989) [25, 29] that if the set of languages of the grammars in a class $\mathcal{C}$ has *finite elasticity* then $\mathcal{C}$ is learnable. On the other hand, if the languages of the grammars in a class $\mathcal{C}$ have a *limit point* then the class $\mathcal{C}$ is unlearnable. For example, informally, for the training sequences $\sigma(n) = (a, ab, aab, aaab, ..., a^n b)$ at any step k we could not choose between two hypothesis: $\{a^n b/0 \leq n \leq k\}$ and $\{a^n b/n \geq 0\}$ (the limit point). We refer to [19] for definitions and details.

Following learning impossibility results for whole classes, subclasses have been proposed. In the case of classical categorial grammars, the notion of k-valued gram-

[4] $\mathcal{A}$ *stabilizes* on σ on step T means that T is the minimal number t for which there is no $t_1 > t$ such that $\mathcal{A}(\sigma[t_1]) \neq \mathcal{A}(\sigma[t])$..

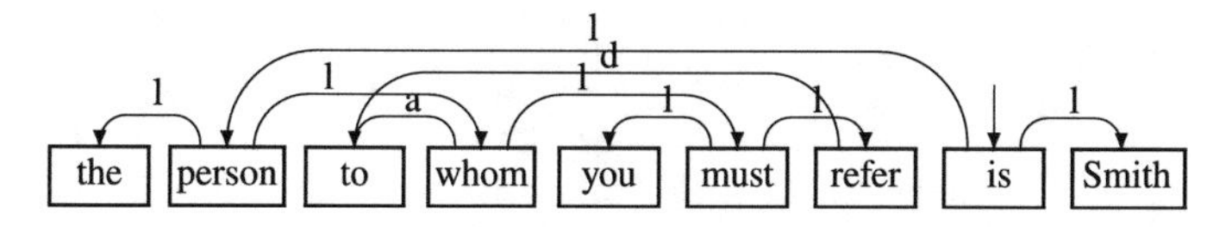

Fig. 1 An untyped dependency net (*d* indicates a discontinuous dependency)

mars subclasses has shown fruitful (where k is the maximum number of types per word) [19], for learning from structures and from strings, relying on finite elasticity properties. However such results extend neither to Lambek-like grammars [7] (categorial grammars based on Lambek calculus or pregroup grammars) nor to CDG (even without potentials) as explained below; such impossibility results are also properties of the given classes of grammars, which also signifies that these latter classes are more expressive (in particular a class of k-valued CDG stricly includes each class of k'-valued classical categorial grammars, for each $k' \leq k$).

4.2 Learnability of K-valued CDG

Below are some key-points on CDG-learnability, depending on a bound k on the number of types per word (we consider k-valued CDG) and on the presence of iterated types (we consider types with or without *). In this section, we consider several variants of structures, reflecting the dependency structure: fully labelled, partially labelled or un-labelled structures, with several kinds of labels.

Fact 1. Each class of k-valued CDG **without** * **iterated types** is learnable from structures and from strings (for each $k \geq 1$); this property was shown in [5] with an algorithm in the rigid case ($k = 1$) to learn from structures called "untyped dependency nets"[5] as in Fig. 1, "finite elasticity" is then shown and applied to export the learnability result from structures to strings.

Variants. In [5] the definition of types with iterative A^* types is extended with repetitive A^+ types and with optional $A^?$ types.[6] The positive learnability result still holds in the presence of A^+ types.

Fact 2. The class of 3rigid CDG **with** * is **not** learnable from strings[7]. A limit point can be seen with this construct:

[5] An untyped dependency net is a list of untyped nodes connected by local (l), anchored (a) and discontinuous (non-projective) (d) dependencies that correspond to a parsing of the sentence. Each nodes contains a word without its type.

[6] With these rules (similar on the right):

- Repetivive $\mathbf{R^l}$ $[C]^P[C^+\backslash\alpha]^Q \vdash [C^*\backslash\alpha]^{PQ}$
- Optional $\mathbf{O^l}$ $[C]^P[C^?\backslash\alpha]^Q \vdash [\alpha]^{PQ}$
  where the Ω rule also holds for optional types: $\Omega^\mathbf{l}$ $[C^?\backslash\beta]^P \vdash [\beta]^P$.

[7] Similar results hold for *optional* ? types.

$$\begin{array}{ll} G'_0 = \{a \mapsto [A], b \mapsto [B], c \mapsto [C'_0]\} & C'_0 = S \\ G'_{n+1} = \{a \mapsto [A], b \mapsto [B], c \mapsto [C'_{n+1}]\} & C'_{n+1} = C'_n \ / \ \mathbf{A}^* \ / \ \mathbf{B}^* \\ G'_* = \{a \mapsto [D], b \mapsto [D], c \mapsto [S \ / \ \mathbf{D}^*]\} & \\ L(G'_n) = \{c(b^*a^*)^k \mid k \leq n\} \text{ and } L(G'_*) = c\{b, a\}^*. & \end{array}$$

In this construction, the number of iterative subtypes (A^*) is not bound.

FA-structures for CDG. *Labelled functor-argument structures* for CDG with iterated types are defined using the name of the application rule (skipping the Ω steps and potentials, and merging I and L), with the dependency name as subscript; *unlabelled functor-argument structures* are defined similarly without dependency name subscripts.

FA-structures

$L^l_{[N]}(John, L^r_{[A]}(L^r_{[A]}(ran, fast), yesterday))$ is a labelled FA-structure generated by the following grammar, for the sentence "John ran fast yesterday":

$$\begin{array}{rl} John & \mapsto [N] \\ ran & \mapsto [N \backslash S \ / \ A^*] \\ fast & \mapsto [A] \\ yesterday & \mapsto [A] \end{array}$$

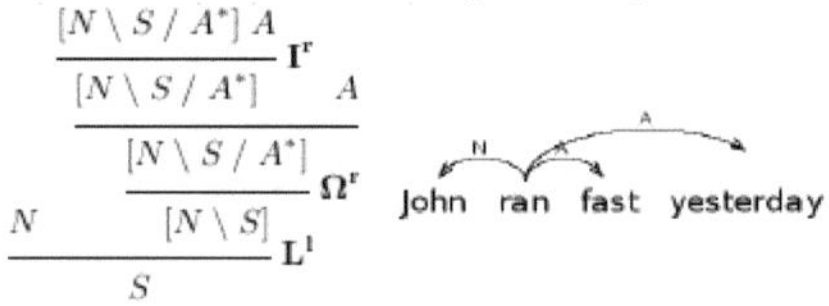

$$\dfrac{N \quad \dfrac{\dfrac{\dfrac{[N \backslash S / A^*] \ A}{[N \backslash S / A^*]}\mathbf{I^r} \quad A}{[N \backslash S / A^*]}}{[N \backslash S]}\Omega^r}{S}\mathbf{L^l}$$

$L^l(John, L^r(L^r(ran, fast), yesterday))$ is an unlabelled FA-structure.

Fact 3. The class of rigid CDG **with** * is not learnable from (unlabelled) functor-argument structures [3]. This result was shown using the same grammars G'_i as for strings, G'_i yielding flat structures such as $L^r(L^r(c, b), a)$.

Fact 4. The paper [3] also provides a limit point for the language of labelled functor-argument structures, the class of CDG considered having no restriction on the number of types per word.

Whether the class of rigid CDG is learnable from labelled FA-structures is an open question.

5 K-star CDG

Traditional unification-based learning algorithms from functor-argument structures thus fail for grammars with iterated types. Other strategies have then been proposed in [3, 4], termed as "Type-Generalize-Expand". These strategies lead to other classes that are learnable as discussed in this section. A recent overview on learning from dependency structures is provided in [11].

Approach \ Parameter	Structured Example	Annotation	Maximum Number (k) of Types per word	Repetition Number (K) for Indiscernibility
Unification-based	functor-argument (FA, proof-tree)	unlabelled (no dep. name)	bound	no bound
Type-Generalize-Expand (TGE, in short)	dependency structure (DS)	labelled (dep. names)	no bound	bound

from [Béchet-Foret, Machine Learning, 2021]

5.1 Learning Approaches with Iterated Types

The following table highlights the main parameters involved in these two learning approaches: unification-based, and type-generalize-expand.

As seen in the previous section, an algorithm based on the first approach cannot converge for CDG with iterated types.

By constrast, the second approach (TGE) will converge for another bound called K-star, that reflects an indiscernability on grammars between "at least K repetitions of a same dependency" (same name and same governor) and its "iterated (star) version".

Vicinity. The TGE method involves a first set of types without iteration, called vicinity, that can be directly obtained from a dependency structure. The *vicinity* $V(w, D)$ of a word w in a (labelled) dependency structure D, is the type

$$V(w, D) = [l_1\backslash \dots \backslash l_k\backslash h/r_m/ \dots /r_1]^P,$$

such that D has:

- the incoming projective dependency or anchor h (or the axiom S),
- the left projective dependencies or anchors $l_k, \dots, l_1$ (in this order),
- the right projective dependencies or anchors $r_1, \dots, r_m$ (in this order),
- the discontinuous dependencies $d_1, \dots, d_n$ with their respective polarities P.[8]

From DS to Vicinity

From this dependency structure D, for a sentence in French[9]:

[8] In details, P is a permutation of $p_1, \dots, p_n$, where p_i are polarized valencies of d_i, in the case of the algorithms, chosen in a standard lexicographical order $<_{lex}$ on d_i (dependency names), compatible with the polarity order $\nwarrow < \searrow < \swarrow < \nearrow$.

[9] meaning "there is also a recent score to recover from the ONPL signed by him."

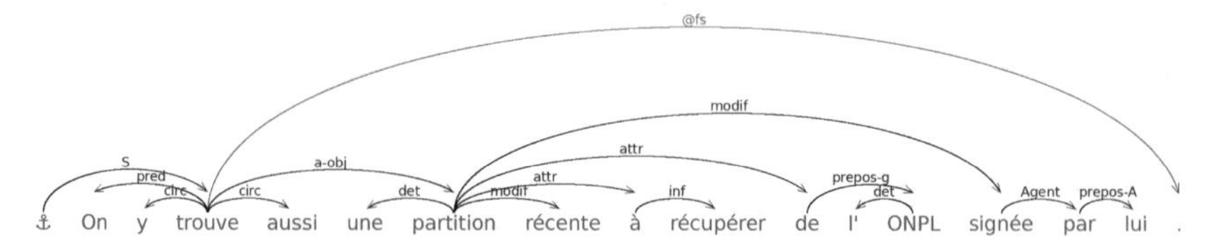

we get these two vicinity types:

$V(partition, D) = [det\backslash a-obj/modif/attr/attr/modif]$

$V(de, D) = [attr/prepos-g]$

TGE-approach. Informally, given a "repetition principle", the vicinity type of a word w $[l_1\backslash \ldots \backslash l_k \backslash h/r_m/\ldots/r_1]^P$ is generalized to a star-typed if appropriate with respect to the repetition principle. This generalized type is then added to w in the target grammar (not the vicinity type).

From Vicinity to CDG-type

If the "2-star repetition principle" applies to $attr$ in the sense that if $attr$ occurs consecutively two times then it can occur consecutively any number of times, we would add this CDG type for Example 5.1: $partition \mapsto [det\backslash a-obj/modif/attr^*/modif]$

Remark. Note that a star-type represents an infinity of vicinities. In the above example, the set of vicinities corresponding to the type of *partition* is

$[det\backslash a-obj/modif/modif]$
$[det\backslash a-obj/modif/attr/modif]$
$[det\backslash a-obj/modif/attr/attr/modif]$
$[det\backslash a-obj/modif/attr/attr/attr/modif]$
...

5.2 TGE-Algorithm

Algorithm TGE$^{(K)}$ (Type-Generalize-Expand):
Input: σ, a training sequence of length N.
Output: CDG **TGE**$^{(K)}(\sigma)$.

```
let G_H = (W_H, C_H, S, λ_H) where W_H := ∅; C_H := {S}; λ_H := ∅;
(loop) for i = 1 to N                        // loop on σ
  let D such that σ[i] = σ[i − 1] · D;       // the i-th DS of σ
  let (X, E) = D;
  (loop) for every w ∈ X            // the order of this loop is not important
     W_H := W_H ∪ {w};
     let t_w = V(w, D)                       // the vicinity of w in D
     (loop) while t_w = [α\l\d\ ··· \d\r\β]^P
        with at least K consecutive occurrences of d, l ≠ d (or α\l not present)
                                           and r ≠ d (or r\β not present)
             t_w := [α\l\d*\r\β]^P
     (loop) while t_w = [α/l/d/ ··· /d/r/β]^P
        with at least K consecutive occurrences of d, l ≠ d (or α/l not present)
                                           and r ≠ d (or r/β not present)
             t_w := [α/l/d*/r/β]^P
     λ_H(w) := λ_H(w) ∪ {t_w};               // lexicon expansion
     end end
return G_H
```

Fact. TGE$^{(K)}$ learns K-star revealing CDG from DS.
This fact is a theorem in [3], to which we refer for the technical definition of K-star revealing grammars. In short, a grammar is K-star revealing iff it is strongly equivalent (generating the same DS) to its K-star generalization, obtained by adding iterated types complying to the K-repetition principle. Importantly, no bound on the number of types is assumed.

The notion of K-star revealing is a complex equivalence property. Two simpler criteria have been proposed later [9, 11] that define learnable classes with same TGE-algorithm. These criteria of syntactic nature are easy to check on a given grammar.

Definition 5 (Simple K-star) Let $K > 1$ be an integer. Let t denote a categorial type and d denote a dependency name. The type t is said to be *simple left K-star on* d if for any successive occurrences $l_1 \backslash l_2 \backslash ...l_p \backslash$ on the left where each l_i is either d or some x^*, there are:

(1) at most $K - 1$ occurrences of d and
(2) no occurrence of d if there exists at least one occurrence of d^*.

The type t is said to be *simple left K-star* if it is simple left K-star on d, for all d. These two notions are defined similarly on the right. The type t is said to be *simple K-star* if it is *simple left K-star* and *simple right K-star*. The CDG G is said to be *simple K-star* whenever all types in its lexicon are simple K-star.

A close variant called global simple K-star is introduced in [11], as another interpretation of the repetition principle, with weaker conditions.

Definition 6 (Global Simple K-star) Let $K > 1$ be an integer. Let t denote a categorial type and d denote a dependency name. The type t is said to be *global simple left K-star on d* if for any successive occurrences $l_1 \backslash l_2 \backslash ...l_p \backslash$ on the left there are:

(1) at most $K - 1$ occurrences of d and
(2') no occurrence of d if there exists at least one occurrence of d^*.

The type t is said to be *global simple left K-star* if it is global simple left K-star on d, for all d. These two notions are defined similarly on the right.
The type t is said to be *global simple K-star* if it is *global simple left K-star* and *global simple right K-star.*
The CDG G is said to be *global simple K-star* whenever all types in its lexicon are global simple K-star.

Notice that given K, the set of global simple K-star is still infinite (consider for example [S / A^* / B^* / A^* / B^*...]). And obviously "global simple K-star" entails "simple K-star". A summary picture is as follows:

K-star revealing languages
$\supseteq$ Simple K-star (syntactic): criterion on types (and grammars) on d:
(1) at most $K - 1$ occurrences of d
and (2) no occurrence of d if there exists at least one occurrence of d^*
in each $l_1 \backslash l_2 \backslash ...l_p \backslash$ where each l_i is either d or some x^*
(similarly on the right)
$\supseteq$ Global Simple K-star:
(1) at most $K - 1$ occurrences of d
and (2') no occurrence of d if there exists at least one occurrence of d^*
(both sides)

Learning Steps

We may consider the following target grammar (Global Simple K-star), generating the DS below on the right:

$$John \mapsto [N] to\_the\_station \mapsto [L]$$
$$ran \mapsto [N\backslash A^*\backslash S/A^*/L/A^*],\ [N\backslash A^*\backslash S/A^*]$$
$$seemingly,\ slowly,\ alone,\ every\_morning \mapsto [A]$$

types (below on the left) are added by Algorithm **TGE**$^{(2)}$ for these DS:

ran $\mapsto [N\backslash S]$ for $(i = 1)$: John ran .

ran $\mapsto [N\backslash S/A]$ for $(i = 2)$: John ran slowly .

ran $\mapsto [N\backslash S/L/A]$ for $(i = 3)$: John ran slowly to_the_station .

ran $\mapsto [N\backslash A\backslash S/\ A^*\ ]$ for $(i = 4)$: seemingly John ran slowly alone . ...

Step 4 reveals one occurrence of A^* in a type for *ran*.

5.3 *Criteria and Readings of the "Repetition Principle"*

We experiments with of repeatable dependency (local with simple K-star condition or global to a type with global simple K-star condition).

Grammar Classes

Let $G_i = a \mapsto [A], b \mapsto [B], c \mapsto t_i$, according to t_i below:

Type t_i	G_i class
$t_0 = [S\ /\ A\ /\ B\ /\ A]$	Simple 2-star, not global simple 2-star
$t_1 = [A^* \setminus S/A^*]$	Global simple 2-star
$t_2 = [A^* \setminus B^* \setminus A^* \setminus S]$	Global simple 2-star
$t_3 = [A^* \setminus B \setminus A^* \setminus S]$	Global simple 2-star
$t_4 = [A^* \setminus A \setminus S]$	Not simple 2-star, not 2-star-revealing
$t_5 = [A^* \setminus B^* \setminus A \setminus S]$	Not simple 2-star, not 2-star-revealing
$t_6 = [A \setminus B^* \setminus A \setminus S]$	Not simple 2-star, not 2-star-revealing

Grammar Classes and Inclusion. From [11]: G'_6 obtained by adding $c \mapsto [A^* \setminus S]$ to G_6 becomes 2-star revealing, but there is no simple 2-star grammar strongly equivalent to G'_6 (generating the same set of dependency structures).

Grammar Classes, TGE-Algorithm, and K-star Generalization. The same TGE^K algorithm can be run to learn the class of simple K-star grammars. It can be adjusted to learn global K-star grammars by adding a final step replacing each type t of the output of TGE^K by its *global simple K-star generalization* $gs^{(K)}(t)$ obtained as follows [11]:

- for each d on the left, if $d \setminus$ occurs at least K times or $d^* \setminus$ is present, then replace each $d \setminus$ with its starred version $d^* \setminus$

– for each d on the right, proceed similarly.

Variants and Extended Types. More flexible interpretations than the strict reading of repeatable optional dependencies as “consecutive repetitions” have been proposed.

– “Dispersed iteration” [4], $\{d_1^*, \ldots, d_p^*\}$ represents the case where the subordinates through a repeatable dependency may occur in any position on the left (respectively, on the right) of the governor.
– Another extension, called “choice iteration” [4], $(d_1|\ldots|d_k)^*$ represents the case where the subordinates through one of several repeatable dependencies may occur in one and the same argument position.

Using a similar approach in the dispersed case, an algorithm TGE^K_{disp} has been shown to learn $K-$star dispersed revealing grammars. A similar learning algorithm TGE^K_{ch} is provided for “choice iteration”.

– CDG with “sequence iteration” have later been proposed in [8] as a generalization of d^*: repeating $/\ d_2\ /\ d_1\ /\ d_2\ /\ d_1$, etc. as $/\ (d_1 \bullet d_2)^*$. An extended CDG-calculus and a TGE-like algorithm for sequences of length 2 is provided in [8][10].

This extension seems relevant for treebanks (see the next section). An inference algorithm $TGE^{(K)}_{J-seq}$ has been proposed in [8] ($TGE^{(K)}_{J-seq}$ where K corresponds to the number of consecutive identical dependencies, and J is for internal length of sequences).

More global variants are still possible: such as having “super-global” dependencies that are either “repeatable everywhere” or “never repeatable”, in the whole grammar.

5.4 *Experiments on Corpora and Prospective Work*

Experiments have been conducted on treebanks in the CONLL UD format.[11] These experiments take into account the CDG-calculus extended with sequence iteration and the inference algorithm $TGE^{(K)}_{J-seq}$ proposed in [8] ($TGE^{(K)}_{J-seq}$ introduces a sequence iteration for each group of at least K consecutive identical sequences of local dependencies, J indicates the maximum internal length of the sequences that are transformed into sequence iterations). We are particularly interested in French and in some under-resourced languages, such as Breton (a celtic language, for which such a treebank has been recently annotated [28]).

Through these experiments, we can produce a CDG-grammar (with variants), where iteration provides compact types; we also look for properties of some dependencies, repeating patterns; we measure them in [8] for example where we get the number of pattern with $K = 2$, $J = 2$ on the left or on the right of a governor. Experiments have been reported in [8] using databases and formal concept analysis tools.

[10] Sequence iteration does not introduce new string languages.

[11] https://universaldependencies.org/format.html.

We conducted new experiments with the graph rewriting tool GREW [18] to test patterns followed by dependencies in a corpus.

Searching a UD Corpus for Iterative Dependencies

To search for 2 repetitions anywhere (on the left or on the right):

```
pattern { e: GOV -> DEP1;  f: GOV -> DEP2;
          e.label = f.label;
          DEP1 << DEP2 }
```

After selecting `e.label`, we get a list of dependencies and statistics; Clicks show the structured sentences where the searched pattern is highlighted.

This can be run for example on a Breton corpus: http://universal.grew.fr/?corpus=UD_Breton-KEB@2.9.

Such a tool and output should help to understand which repetition interpretation is preferable for some dependencies in a language. This tool is also a graph rewriting tool, which could be the support of TGE-like algorithms as described in this article.

6 Open Problems

In this section we discuss general questions related to the CDG framework and point to some open problems.

6.1 CDG-Types and Learning from Structures

We have discussed two models of learning iterated dependencies. These models deal with unlimited iterated types from input dependency structures without marked iteration. The frontier between learnability and un-learnability remains open with respect to several aspects.

Rigid and k-valued grammars. One such open question is the learnability of CDG-grammars (with iteration) from labelled FA or labelled DS (no marked iteration), for the subclass of rigid (or k-rigid for a given k). Linking CDG-grammars with developments on regular expressions might be fruitful in that direction: CDG-types are close to "simple looping" expressions and "simple looping" automata for which inference algorithms and block-alignment techniques are known (see algorithms and Theorem 4 in [16]); these are cases of "one-unambiguous" languages (Theorem 5 in [16]) for which some generalizations are proposed in [12].

Extended CDG-types. The notion of K-star simple grammars has been introduced as a syntactic alternative to the semantic notion of "K-star revealing", also providing learnable subclasses. Whether a grammar is K-star simple can easily be checked.

How this syntactic notion could be adapted to extended CDG-types (flexible iteration, choice iteration or sequence iteration) is an open question. A key point for classical CDG is that a K-star simple grammar is K-star revealing, and learned by the same TGE-algorithm; this property should be preserved for new proposals.

6.2 Membership Properties and Hierarchies

CDG-languages are context-sensitive languages. The rules of the CDG calculus of type aren't directly context sensitive rules because the rules have two parts, one for basic dependency types (type with an empty potential) and the other for potentials. The parts are independent and it is possible to define rules that correspond only on the basic dependency types (with an empty potential) and other for potentials (without the basic type). A second problem comes from the rules $\mathbf{D^l}$ and $\mathbf{D^r}$. They need a potential P between the two dual polarized valencies that must satisfy the FA principle. These rules on potentials can be simulated by several context-sensitive rules: rules when P is empty and other rules that switch polarized valencies except for dual polarized valencies (the simulation is simpler if we can use non-contracting rules rather than just context-sensitive rules but the context-sensitive languages and the non-contracting languages are the same). The lexicon of a CDG can be simulated by a set of context-sensitive rules that choose one of the types for each word. In fact, the simulation must code a complex type by a unique symbol: for a given lexicon, there is a finite number of basic dependency subtypes and there is a finite number of different potentials with a number of polarized valencies less than or equal to the maximum number of polarized valencies by type in the lexicon.

CDG can generate some mildly context-sensitive languages like $\mathbf{a^nb^nc^n}$ or **MIX** (with 3 symbols) that are not context-free languages. CDG can also generate languages similar to $\mathbf{a^nb^nc^n}$ with k symbols and with any set of equality constraints on the number of instances of each symbol. MIX can be generalised to any number of symbols (more than 3 symbols). But, the status of other mildly context-sensitive languages isn't known in general. For instance, we don't know if the **copy** language $\{mm, m \in \{a, b\}^+\}$ can be generated by a CDG. From a theoretical point of view, we don't have an effective tool for showing that a particular language is not a CDG-language like the pumping lemmas for regular languages or context-free languages. The model behind CDG are push-down automata with independent counters [13] but this model has not given us a useful tool so far.

Thus, it is possible to prove that a class of languages is a sub-class of CDG-languages. For instance, non-empty regular languages or non-empty context-free languages are CDG-languages. But we aren't able to prove that some mildly context-sensitive languages aren't CDG-languages. The copy language for instance is generated by a Tree Adjoining Grammar (TAG) but we aren't able to find a CDG for it or prove that it isn't a CDG-language. In the other side, because MIX is a CDG-language but it isn't a TAG language [21], the comparison with the classes of TAG languages is an open problem. Because MIX (with 3 symbols) is a mildly context-sensitive language [26, 27], one may expect that CDG languages are also mildly context-sensitive languages. We also don't known if the CDG-languages are semilinear [13].

Another open problem comes from the hierarchy of CDG-languages when we restrict the number of different non-projective dependency names. We don't know if the hierarchy is strict or not. As it was shown, the complexity of the membership problem of a string to the language of a CDG depends on the number of different non-projective dependency names. Thus we postulate that the hierarchy of languages corresponding to the hierarchy of CDG with a limited number of non-projective dependency names is strict. Otherwise, it would certainly mean that we could find a parsing algorithm with a polynomial limit from both the input string and the CDG.
Remark. An extension of CDG called multimodal categorial dependency grammars (mmCDG) [14] where the first available **FA** is replaced by a more complex principle (for instance, the non-projective dependency names form classes and non-projective dependencies of the same class cannot cross each other) can simulate mildy context-sensitive languages like the copy language. However the membership problem with mmCDG is more complex. It is an NP-complete problem [14].

6.3 *Abstract Family of Languages*

We are interested in closure properties of a family $\mathcal{F}$ of languages, as those of AFL. Before considering such questions for CDG, we give background definitions ([20]).
Homomorphisms. For finite alphabets V_1, V_2: a homomorphism[12] h from V_1^* to V_2^* is ϵ-free if $h(w) = \epsilon$ implies $w = \epsilon$.

A family $\mathcal{F}$ is closed under inverse homorphism if whenever $L \subseteq V_1^*$ is in $\mathcal{F}$ and h is a homomorphism from V_2^* to V_1^*, then $h^{-1}(L)$ is also in $\mathcal{F}$, where:

$$h^{-1}(L) = \{w \in V_2^* \mid h(w) \in L\}$$

Substitutions. A substitution is a mapping f from V_1 to $\mathcal{P}(V_2^*)$, it is naturally extended to strings in V_1^* (by concatenation) and to sets of strings (by union).[13]

A family $\mathcal{F}$ is closed under substitution if whenever $L \in V_1^*$ is in $\mathcal{F}$ and f is a substitution from V_1 such that $f(a) \in \mathcal{F}$ for all $a \in V_1$, then $f(L)$ is also in $\mathcal{F}$.
AFL. $\mathcal{F}$ is an Abstract Family of Languages (AFL) if it is closed under union, concatenation, Kleene plus, ϵ-free homomorphism, inverse homomorphism and intersection with regular sets. A *full AFL* is defined similarly, with Kleene star (not just Kleene plus) and arbitrary homomorphisms (not just ϵ-free).

---

[12] Each character is replaced by a single string https://en.wikipedia.org/wiki/String_operations.

[13] $f(\epsilon) = \{\epsilon\}$, $f(wa) = f(w)f(s)$, $f(L \cup \{w\}) = f(L) \cup f(w)$.

For example, the class of multiple context-free grammars is an AFL [23]. The class of string languages generated by abstract categorial grammars is a substitution-closed full AFL, as shown in [20].
CDG-languages are closed under these AFL-operations:
union ✓, concatenation ✓
ϵ-free homomorphism ✓, inverse homomorphism ✓
and intersection with regular sets ✓
The CDG family is thus a trio (closed under $epsilon$-free homomorphism, inverse homomorphism, and intersection with regular language) and also a semi-AFL (a trio closed under union). However the AFL question is open for CDG-languages as we do not know if they are closed for Kleene plus (the conjecture is "no" in [13]). In [14] it is shown that the mmCDG class, extending CDG with a multimodal rule, defines an AFL.

These closure properties are established for string-languages. In [20] closure properties for ACG tree-languages are also shown. In the case of CDG, such closure questions could be adressed at the level of dependency structures too.

7 Conclusion

In this paper we have given an overview on the family of Categorial Dependency Grammars (CDG) that define dependency grammars beyond context-free languages. The evolutions of CDG come from practical issues to write grammars, theoretical issues to ensure good language properties and to obtain good learning properties.

On the one hand, the original CDG types have been extended with:

- unary operators other than iteration: option $d^?$ and repetition d^+ (where d is a dependency name);
- several kinds of iteration: sequence iteration, choice iteration, dispersed iteration [4]
- with variants for the potential pairing rule: FA (first available) is the standard CDG pairing rule, the FC (first cross) alternative has also been proposed, as sole rule or mixed with FA in the multimodal case (mmCDG).

On the other hand, some measures and contraints on CDG have been proposed, defining subclasses of interest for exact learning: either based on the maximum number of types per word, or on the notion of K-star (a measure of indiscernability between K repetition and star iteration).

On the theoretical side, open questions remain for standard CDG and for their variants. We also believe that CDG, considering different variants, are worth further experiments, as more treebanks become available.

References

1. Alfared, R., Béchet, D., Dikovsky, A.: CDG Lab: a toolbox for dependency grammars and dependency treebanks development. In: Gerdes, K., Hajičová, E., Wanner, L.(eds.) Proceedings of the 1st International Conference on Dependency Linguistics (Depling 2011), Barcelona, Spain, September 5–7 2011 (2011). http://depling.org/proceedingsDepling2011/
2. Béchet, D., Dikovsky, A., Foret, A.: Dependency structure grammar. In: Blache, P., Stabler, E., Busquets, J., Moot, R. (eds.) Logical Aspects of Computational Linguistics, 5th International Conference, LACL 2005, Bordeaux, France, April 28–30, 2005. Proceedings, Lecture Notes in Artificial Intelligence (LNAI), vol. 3492, pp. 18–34. Springer (2005). https://doi.org/10.1007/b136076
3. Béchet, D., Dikovsky, A., Foret, A.: Two models of learning iterated dependencies. In: de Groote, P., Nederhof, M. (eds.) Formal Grammar, 15th and 16th International Conferences, FG 2010, Copenhagen, Denmark, August 2010, FG 2011, Ljubljana, Slovenia, August 2011, Revised Selected Papers, Lecture Notes in Computer Science (LNCS), vol. 7395, pp. 17–32. Springer (2010). https://doi.org/10.1007/978-3-642-32024-8_2
4. Béchet, D., Dikovsky, A., Foret, A.: On dispersed and choice iteration in incrementally learnable dependency types. In: Logical Aspects of Computational Linguistics, 6th International Conference, LACL 2011, Montpellier, France, June 29 – July 1, 2011. Proceedings, Lecture Notes in Computer Science (LNCS), vol. 6736, pp. 80–95 (2011). https://doi.org/10.1007/978-3-642-22221-4_6
5. Béchet, D., Dikovsky, A., Foret, A., Moreau, E.: On learning discontinuous dependencies from positive data. In: Monachesi, P. (ed.) Proceedings of the 9th International Conference on Formal Grammar (FGNancy), Nancy, France, 7-8 August 2004, pp. 1–16. CSLI Publications (2004). http://cslipublications.stanford.edu/FG/2004
6. Béchet, D., Dikovsky, A., Lacroix, O.: "CDG Lab": an integrated environment for categorial dependency grammar and dependency treebank development. In: Gerdes, K., Hajičová, E., Wanner, L. (eds.) Computational Dependency Theory, Frontiers in Artificial Intelligence and Applications, vol. 258, pp. 153–169. IOS Press (2014). https://doi.org/10.3233/978-1-61499-352-0-153
7. Béchet, D., Foret, A.: k-valued non-associative Lambek categorial grammars are not learnable from strings. In: Proceedings of the 41st Annual Meeting of the Association for Computational Linguistics (ACL 2003), Sapporo, Japan, July 2003, pp. 351–358. ACL (2003). https://doi.org/10.3115/1075096.1075141
8. Béchet, D., Foret, A.: Categorial dependency grammars with iterated sequences. In: Logical Aspects of Computational Linguistics. Celebrating 20 Years of LACL (1996-2016) - 9th International Conference, LACL 2016, Nancy, France, December 5-7, 2016, Proceedings, pp. 34–51 (2016). https://doi.org/10.1007/978-3-662-53826-5_3
9. Béchet, D., Foret, A.: Simple k-star categorial dependency grammars and their inference. In: Verwer, S., van Zaanen, M., Smetsers, R. (eds.) Proceedings of the 13th International Conference on Grammatical Inference, ICGI 2016, Delft, The Netherlands, October 5-7, 2016, JMLR Workshop and Conference Proceedings, vol. 57, pp. 3–14. JMLR.org (2016). http://proceedings.mlr.press/v57/bechet16.html
10. Béchet, D., Foret, A.: On Categorial Grammatical Inference and Logical Information Systems. In: Logic and Algorithms in Computational Linguistics 2018, Series: Advances in Intelligent Systems and Computing, Volume 860, pp. 125–153. Springer International Publishing (2019). https://doi.org/10.1007/978-3-030-30077-7_6
11. Béchet, D., Foret, A.: Incremental learning of iterated dependencies. Mach. Learn. (2021). https://doi.org/10.1007/s10994-021-05947-2
12. Caron, P., Mignot, L., Miklarz, C.: On the hierarchy of generalizations of one-unambiguous regular languages. Theor. Comput. Sci. **679**, 95–106 (2017). https://doi.org/10.1016/j.tcs.2016.07.004
13. Dekhtyar, M., Dikovsky, A., Karlov, B.: Categorial dependency grammars. Theor. Comput. Sci. **579**, 33–63 (2015). https://doi.org/10.1016/j.tcs.2015.01.043

14. Dekhtyar, M.I., Dikovsky, A.J., Karlov, B.: Iterated dependencies and kleene iteration. In: de Groote, P., Nederhof, M. (eds.) Formal Grammar—15th and 16th International Conferences, FG 2010, Copenhagen, Denmark, August 2010, FG 2011, Ljubljana, Slovenia, August 2011, Revised Selected Papers, Lecture Notes in Computer Science, vol. 7395, pp. 66–81. Springer (2010). https://doi.org/10.1007/978-3-642-32024-8_5
15. Dikovsky, A.: Multimodal categorial dependency grammars. In: Proceedings of the 12th Conference on Formal Grammar, pp. 1–12. Dublin, Ireland (2007)
16. Fernau, H.: Algorithms for learning regular expressions from positive data. Inf. Comput. **207**(4), 521–541 (2009). https://doi.org/10.1016/j.ic.2008.12.008
17. Gold, E.M.: Language identification in the limit. Inf. Control **10**, 447–474 (1967)
18. Guillaume, B.: Graph Matching and Graph Rewriting: GREW tools for corpus exploration, maintenance and conversion. In: EACL 2021—16th conference of the European Chapter of the Association for Computational Linguistics. Kiev/Online, Ukraine (2021). https://hal.inria.fr/hal-03177701
19. Kanazawa, M.: Learnable classes of categorial grammars. Studies in Logic, Language and Information. FoLLI & CSLI (1998)
20. Kanazawa, M.: Abstract families of abstract categorial languages. Electron. Notes Theor. Comput. Sci. **165**, 65–80 (2006). https://doi.org/10.1016/j.entcs.2006.05.037
21. Kanazawa, M., Salvati, S.: MIX is not a tree-adjoining language. In: The 50th Annual Meeting of the Association for Computational Linguistics, Proceedings of the Conference, July 8-14, 2012, Jeju Island, Korea - Volume 1: Long Papers, pp. 666–674. The Association for Computer Linguistics (2012). https://aclanthology.org/P12-1070/
22. Lacroix, O., Béchet, D.: A three-step transition-based system for non-projective dependency parsing. In: COLING 2014, 25th International Conference on Computational Linguistics, Proceedings of the Conference: Technical Papers, August 23-29, 2014, Dublin, Ireland, pp. 224–232 (2014). http://aclweb.org/anthology/C14-1023/
23. Matsumura, T., Seki, H., Fujii, M., Kasami, T.: The generative power of multiple context-free grammars and head grammars. Syst. Comput. Jpn. **22**(4), 41–56 (1991). https://doi.org/10.1002/scj.4690220405
24. Mel'čuk, I.: Dependency Syntax. SUNY Press, Albany, NY (1988)
25. Motoki, T., Shinohara, T., Wright, K.: The correct definition of finite elasticity: Corrigendum to identification of unions. In: The fourth Annual Workshop on Computational Learning Theory, p. 375. San Mateo, Calif. (1991)
26. Nederhof, M.J.: A short proof that o_2 is an MCFL. In: Proceedings of the 54th Annual Meeting of the Association for Computational Linguistics (Volume 1: Long Papers), pp. 1117–1126. Association for Computational Linguistics, Berlin, Germany (2016). https://doi.org/10.18653/v1/P16-1106
27. Salvati, S.: MIX is a 2-MCFL and the word problem in z^2 is captured by the IO and the OI hierarchies. J. Comput. Syst. Sci. **81**(7), 1252–1277 (2015). https://doi.org/10.1016/j.jcss.2015.03.004
28. Tyers, F.M., Ravishankar, V.: A prototype dependency treebank for breton. In: Sébillot, P., Claveau, V. (eds.) Actes de la Conférence TALN. CORIA-TALN-RJC 2018 - Volume 1 - Articles longs, articles courts de TALN, Rennes, France, May 14-18, 2018, pp. 197–204. ATALA (2018). https://aclanthology.org/2018.jeptalnrecital-court.1/
29. Wright, K.: Identification of unions of languages drawn from an identifiable class. In: Rivest, R.L., Haussler, D., Warmuth, M.K. (eds.) Proceedings of the Second Annual Workshop on Computational Learning Theory, COLT 1989, Santa Cruz, CA, USA, July 31–August 2, 1989., pp. 328–333. Morgan Kaufmann (1989)

Diamonds Are Forever

Theoretical and Empirical Support for a Dependency-Enhanced Type Logic

Michael Moortgat, Konstantinos Kogkalidis, and Gijs Wijnholds

Abstract Extended Lambek calculi enlarge the type language with adjoint pairs of unary modalities. In previous work, modalities have been used as licensors for controlled forms of restructuring, reordering and copying. Here, we study a complementary use of the modalities as dependency features coding for grammatical roles. The result is a multidimensional type logic simultaneously inducing dependency and function argument structure on the linguistic material. We discuss the new perspective on constituent structure suggested by the dependency-enhanced type logic, and we experimentally evaluate how well a neural language model like BERT can deal with the subtle interplay between logical and structural reasoning that this type logic gives rise to.

Keywords Lambek calculus · Typelogical grammar · Dependency modalities · Neural language models · Probing

1 Introduction

Lambek's Syntactic Calculus [19, 20] is an early representative of substructural logic. The original versions of this syntactic calculus lack the required expressivity to serve as a tool for realistic grammar development. The extended Lambek calculi introduced in the 1990ies enrich the type language with modalities for structural control. These categorial modalities have found two distinct uses as argued in [17]. On the one hand, they can act as *licenses* granting modally marked formulas access

---

M. Moortgat (✉) · K. Kogkalidis · G. Wijnholds
UiL-OTS Utrecht University, Utrecht, The Netherlands
e-mail: m.j.moortgat@uu.nl

K. Kogkalidis
e-mail: k.kogkalidis@uu.nl

G. Wijnholds
e-mail: g.j.wijnholds@uu.nl

R. Loukanova et al. (eds.), *Logic and Algorithms in Computational Linguistics 2021 (LACompLing2021)*, Studies in Computational Intelligence 1081,
https://doi.org/10.1007/978-3-031-21780-7_3

to structural operations that by default would not be permitted. On the other hand, modalities can be used to *block* structural rules that otherwise would be available.

Examples of modalities as licensors relate to various aspects of grammatical resource management: multiplicity, order and structure. As for multiplicity, under the control of modalities limited forms of copying can be introduced in grammar logics that overall are resource-sensitive systems, see [3, 11, 29] for some recent examples. As for order and structure, modalities may be used to license changes of word order and/or constituent structure that leave the form-meaning correspondence intact, as illustrated e.g., in [23, 26].

In this chapter, we concentrate on the complementary use of modalities as demarcations of locality domains and study a modally-extended type language designed to simultaneously account for function-argument structure and *dependency structure*. For function-argument structure the key opposition is between a *function* type A/B (or $B\backslash A$) and the *argument* B that it combines with to produce an A. Dependency structures [21] on the other hand are based on the opposition between a *head* and its *dependents*; these dependents can either be *complements* selected by the head, or *adjuncts* modifying the head. A dependency-enhanced type lexicon of the sort described here is at the core of the supertagger and neural parser proposed in [12–14]. Here we focus on the logical foundations of the underlying type logic.

The chapter is structured as follows. Section 1.1 sets the stage and introduces the type logics for syntax and semantics that constitute the Lambek hierarchy. We present Natural Deduction proof systems for these logics and discuss Montague's view of compositional interpretation as a structure-preserving map relating types and proofs of a syntactic source logic to their counterparts in a semantic target logic. Section 1.2 briefly recaps the motivation for extending the type language with modalities, and how they can be used to license restricted forms of structural reasoning. Section 2 introduces the dependency-enhanced type system. Section 2.1 discusses how this system invites us to rethink categorial notions of constituency; the Dutch crossed dependencies presented in Sect. 2.2 offer a worked-out case study of the new perspective. Section 3 presents experimental evaluation of how well a neural language model like BERT can deal with the subtle interplay between logical and structural reasoning that the dependency-enhanced type logic gives rise to.

1.1 The Lambek Hierarchy

A categorial grammar consists of a language-specific *lexicon* assigning a finite number of types to the words of the language together with a universal *type logic* for the combinatorics of these types. In his [19, 20] papers, Lambek presents two views on a type calculus for natural language syntax. The most elementary one is the system known as **NL**, i.e., the Non-associative Lambek Calculus of [20]. The original presentation is in terms of statements $A \longrightarrow B$, i.e., derivability is modelled as a relation holding between types. The derivability relation is reflexive and transitive:

$$A \longrightarrow A \qquad \frac{A \longrightarrow B \quad B \longrightarrow C}{A \longrightarrow C} \tag{1}$$

and one considers three type-forming operations, a product $\bullet$ and two division operations $/$ and $\backslash$, together constituting a residuated triple:

$$B \longrightarrow A\backslash C \quad \text{iff} \quad A \bullet B \longrightarrow C \quad \text{iff} \quad A \longrightarrow C/B \tag{2}$$

A product type $A \bullet B$ here stands for the *merger* of a phrase A followed by a phrase B. The slashes are for *incomplete expressions*: C/B combines with a B to its right to merge into C, $A\backslash C$ wants a phrase A to its left to merge into a C.

The pure residuation logic **NL** can be extended by attributing extra properties to the product. The system **L** of [19] turns the product into an associative operation. Whereas **NL** assigns types to phrases (bracketed strings), **L** assigns types to strings, ignoring hierarchical phrase structure. Adding product commutativity, one obtains the system **LP**, studied by van Benthem [2] as the calculus for semantic composition, and also known as MILL, Multiplicative Intuitionistic Linear Logic. For each of these systems, one can further consider a multiplicative unit I. The rules for associativity, commutativity and the unit, respectively, are as follows.

$$\begin{array}{c} A \bullet (B \bullet C) \longleftrightarrow (A \bullet B) \bullet C \\ A \bullet B \longrightarrow B \bullet A \\ I \bullet A \longleftrightarrow A \longleftrightarrow A \bullet I \end{array} \tag{3}$$

Natural Deduction We will represent the sample derivations in this chapter in sequent-style Natural Deduction (N.D.) format. Basic judgements then are statements, called *sequents*, of the form $\Gamma \vdash A$, with A a type formula, and Γ a structure. Structures are binary trees with formulas at the leaves: $\Gamma, \Delta ::= A \mid \Gamma \cdot \Delta$, where the punctuation '$\cdot$' is the structural counterpart of $\bullet$.

For the $/, \backslash$ implicational fragment, we have Axioms $A \vdash A$, and for each of the type-forming operations an Elimination and an Introduction rule.

$$\frac{\Gamma \vdash A \quad \Delta \vdash A\backslash B}{\Gamma \cdot \Delta \vdash B} \backslash E \quad \frac{A \cdot \Gamma \vdash B}{\Gamma \vdash A\backslash B} \backslash I \quad \frac{\Gamma \vdash B/A \quad \Delta \vdash A}{\Gamma \cdot \Delta \vdash B} /E \quad \frac{\Gamma \cdot A \vdash B}{\Gamma \vdash B/A} /I \tag{4}$$

Structural rules, if present, are explicitly represented, e.g., the restructuring under associativity that distinguishes **L** from **NL**, or product commutativity for **LP**.[1]

$$\frac{\Gamma[\Delta \cdot (\Delta' \cdot \Delta'')] \vdash A}{\Gamma[(\Delta \cdot \Delta') \cdot \Delta''] \vdash A} \mathsf{A}^r \quad \frac{\Gamma[(\Delta \cdot \Delta') \cdot \Delta''] \vdash A}{\Gamma[\Delta \cdot (\Delta' \cdot \Delta'')] \vdash A} \mathsf{A}^l \quad \frac{\Gamma[\Delta' \cdot \Delta] \vdash A}{\Gamma[\Delta \cdot \Delta'] \vdash A} \mathsf{C} \tag{5}$$

As an example, consider the relative clause 'paper that Bob rejected'. A derivation in **L** is presented in Fig. 1, with associative restructuring to make the direct object hypothesis extractable. For readability, left of the turnstile we use words instead of

[1] The notation $\Gamma[\ ]$ is for a *context*: a structure with a hole.

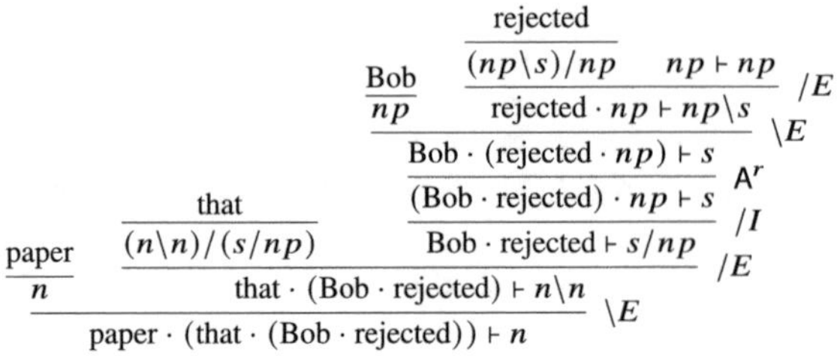

Fig. 1 Relative clause derivation in **L**

$$\dfrac{\dfrac{\text{paper}}{n} \qquad \dfrac{\dfrac{\text{that}}{(n\backslash n)/(s/np)} \qquad \dfrac{\dfrac{\dfrac{\dfrac{\text{Bob}}{np} \qquad \dfrac{\dfrac{\text{rejected}}{(np\backslash s)/np} \qquad np \vdash np}{\text{rejected} \cdot np \vdash np\backslash s}\,/E}{\text{Bob} \cdot (\text{rejected} \cdot np) \vdash s}\,\backslash E}{(\text{Bob} \cdot \text{rejected}) \cdot np \vdash s}\,\mathsf{A}^r}{\text{Bob} \cdot \text{rejected} \vdash s/np}\,/I}{\text{that} \cdot (\text{Bob} \cdot \text{rejected}) \vdash n\backslash n}\,/E}{\text{paper} \cdot (\text{that} \cdot (\text{Bob} \cdot \text{rejected})) \vdash n}\,\backslash E$$

their types where possible, and write these words above their required lexical type assignment in the case of axioms.

Compositional Interpretation(s)

We review Montague's [22] view on compositionality as a homomorphism, a structure-preserving map respecting types and proofs/derivations.

Proofs and Terms The Curry–Howard correspondence establishes an *isomorphism* between proofs and a term language encoding them. N.D. derivations are seen as *typing judgements*, checking whether a program (=term) is well-typed. Sequents $\Gamma \vdash B$ with $A_1, \ldots, A_n$ the yield of the antecedent structure Γ now become statements of the form $x_1 : A_1, \ldots, x_n : A_n \vdash M : B$. These statements mean that program M is a well-formed expression of type B given type declarations $x_1 : A_1, \ldots, x_n : A_n$ (a *typing environment*).

The proofs/terms correspondence for a hierarchy of sublogics of Intuitionistic Propositional Logic has been investigated in [36], starting with Lambek's Syntactic Calculus. Let us compare the implicational fragments of **(N)L** and of the semantic type calculus **LP**, with non-directional implication $A \multimap B$, in the light of the Curry–Howard correspondence.

For the $/, \backslash$ fragment of the syntactic source **L**, assume types $A, B ::= s \mid np \mid n \mid A\backslash B \mid B/A$. To encode proofs, the term language makes a distinction between left versus right application/abstraction: $M, N ::= x \mid \lambda^r x.M \mid \lambda^l x.M \mid (M \ltimes N) \mid (N \rtimes M)$. Sequents $\Gamma \vdash M : B$ where the sequence $x_1 : A_1, \ldots, x_n : A_n$ is the yield of the antecedent structure Γ require the antecedent variables to be distinct. The typing rules start from Axiom $x : A \vdash x : A$, with the following term-decorated inference rules for the $/, \backslash$ Introduction and Elimination steps. The $/$ ($\backslash$) Introduction rule requires x to occur free in M exactly once as the rightmost (leftmost) free variable. The $/, \backslash$ Elimination rules require the sets of free variables of Γ and Δ to be disjoint. The Associativity structural rules have no effect on the constructed term.

$$\frac{\Gamma \cdot x : A \vdash M : B}{\Gamma \vdash \lambda^r x.M : B/A} \; I/ \qquad \frac{x : A \cdot \Gamma \vdash M : B}{\Gamma \vdash \lambda^l x.M : A\backslash B} \; I\backslash$$

$$\frac{\Gamma \vdash M : B/A \quad \Delta \vdash N : A}{\Gamma \cdot \Delta \vdash (M \ltimes N) : B} \; E/ \quad \frac{\Gamma \vdash N : A \quad \Delta \vdash M : A\backslash B}{\Gamma \cdot \Delta \vdash (N \rtimes M) : B} \; E\backslash \tag{6}$$

For derivations of the non-associative calculus **NL** we will use the same term language as for **L**, contenting ourselves with a proof-term correspondence, rather than a proper isomorphism. The language of **NL** terms then is the sublanguage of **L** terms that do not depend on the Associativity structural rules.

The target logic **LP** extends **L** with product commutativity. **LP** is a.k.a. MILL, the multiplicative intuitionistic fragment of Girard's Linear Logic [6]. In MILL$_{\multimap}$, the slashes $/, \backslash$ collapse to linear implication $\multimap$. For the usual set-theoretic interpretation, the target signature can be built from atomic types e, t, for entities and truth values. The full set of types then is $A, B ::= e \mid t \mid A \multimap B$; The term language has abstraction and application: $M, N ::= x \mid \lambda x.M \mid M\ N$. The typing rules start from Axioms $x : A \vdash x : A$ and are subject to the same linearity requirements as those of **L**, except that now the $\multimap$ Introduction rule simply requires x to occur free in M exactly once: in the presence of Commutativity the rightmost/leftmost restriction can be dropped.

$$\frac{\Gamma, x : A \vdash M : B}{\Gamma \vdash \lambda x.M : A \multimap B} \; (\multimap I) \qquad \frac{\Gamma \vdash M : A \multimap B \quad \Delta \vdash N : A}{\Gamma, \Delta \vdash M\ N : B} \; (\multimap E) \tag{7}$$

A homomorphism $\lceil\cdot\rceil$ realizing a compositional interpretation maps types and derivations of the syntactic source logic to their counterparts in the **LP** target logic.

$$\textbf{(N)L}^{s,np,n}_{/,\backslash} \quad \xrightarrow{\lceil\cdot\rceil} \quad \textbf{LP}/\text{MILL}^{e,t}_{\multimap} \tag{8}$$

For the action of $\lceil\cdot\rceil$ on *types*, we can assume $\lceil s \rceil = t$, $\lceil np \rceil = e$, $\lceil n \rceil = e \multimap t$ for the atomic syntactic types, and $\lceil A\backslash B \rceil = \lceil B/A \rceil = \lceil A \rceil \multimap \lceil B \rceil$. For the translation of (terms encoding) proofs, $\lceil x \rceil = \widetilde{x}$ translates Axioms, where we write $\widetilde{x}$ for the target variable corresponding to source variable x; the directional abstractions/applications of the source logic are sent to their non-directional target counterparts.

$$\begin{array}{rcl} \lceil \lambda^l x.M \rceil = \lceil \lambda^r x.M \rceil & = & \lambda \widetilde{x}.\lceil M \rceil \\ \lceil N \rtimes M \rceil = \lceil M \ltimes N \rceil & = & \lceil M \rceil\ \lceil N \rceil \end{array} \tag{9}$$

Illustration To see the interpretation homomorphism in action, let us return to the relative clause example 'paper that Bob rejected' of Fig. 1.

To understand the effect of $\lceil\cdot\rceil$, it is useful to make a distinction between *derivational* and *lexical* semantics, where the former treats word meanings as black boxes, and the latter further spells out the word-internal computations. The derivational semantics $\lceil M \rceil$ of source term M encoding the derivation of 'paper that Bob rejected' stops short of specifying the translation of the lexical source constants.

Table 1 Syntactic types, meaning recipes and semantic types for our running example

WORD	SYN TYPE	$\lceil\cdot\rceil$	SEM TYPE
Paper	n	PAPER	$e \multimap t$
that	$(n\backslash n)/(s/np)$	$\lambda x \lambda y \lambda^! z.((y\ z) \wedge (x\ z))$	$(e \multimap t) \multimap (e \multimap t) \multimap (!e \multimap t)$
Bob	np	BOB	e
rejected	$(np\backslash s)/np$	REJECTED	$e \multimap e \multimap t$

$$\begin{array}{rcl} M & = & \text{paper} \rtimes (\text{that} \ltimes \lambda^r x.(\text{Bob} \rtimes (\text{rejected} \ltimes x))) : n \\ \lceil M \rceil & = & ((\lceil \text{that} \rceil\ \lambda x.((\lceil \text{rejected} \rceil\ x)\ \lceil \text{Bob} \rceil))\ \lceil \text{paper} \rceil) : e \multimap t \end{array} \tag{10}$$

Moving on to the lexical semantics, we can *unpack* the black box word meanings as per Table 1, assuming the target signature includes constants $\text{PAPER}^{e \multimap t}$, BOB^{e}, $\text{REJECTED}^{e \multimap e \multimap t}$, $\wedge^{t \multimap t \multimap t}$. Whereas the derivational semantics is captured by a linear lambda term, specification of the lexical semantics may require expressivity beyond linear. The translation of the relative pronoun in our example is a non-linear meaning recipe, that can be expressed in an extension of MILL with a ! exponential licensing copying of the z parameter. The jury is out as to what an appropriate notion of copying would be for natural language semantics. Moot and Retoré [27] argue that the full power of the Linear Logic ! exponential is too much, since all lambda term meanings used in formal semantics are already expressible with the restricted copying of Lafont's Soft Linear Logic [18] and the soft lambda calculus going with it [1]. We don't take a position on this issue, leaving this as a topic for further research. Here and below we simply write $\lambda^!$ to alert the reader to the fact that the abstraction is over a reusable variable.

Substituting these lexical translations in $\lceil M \rceil$ and simplifying by means of β reduction[2] produces the final result below, where the meaning recipe for 'that' computes the intersection between the set of papers and the set of things rejected by Bob.

$$\lambda^! x.((\text{PAPER}\ x) \wedge ((\text{REJECTED}\ x)\ \text{BOB})) : !e \multimap t \tag{11}$$

Concluding this illustration of compositional interpretation, here are some general points to keep in mind for the remainder of this paper.

- On the type level, the $\lceil\cdot\rceil$ homomorphism sends source atoms to target *types*, not necessarily atomic. The mapping $\lceil n \rceil = e \multimap t$ for common nouns is a case in point, and has the effect of interpreting both n and verb phrases $np\backslash s$ as (characteristic functions of) sets of individuals. The framework of Abstract Categorial Grammars [7] makes good use of this possibility to characterize the string languages of extended context-free formalisms in terms of type homomorphisms of increasing complexity, as we will see in Sect. 2.2.
- On the term level, similarly, the $\lceil\cdot\rceil$ homomorphism sends source constants of type A to *terms* of type $\lceil A \rceil$, where the target term for a source constant can be

[2] The application $(\lambda x.M)N$ β-reduces to $M[x/N]$ which replaces the occurrences of x in M by N.

complex, as long as it respects the typing. The relative pronoun translation above is an example. Notice that the target signature can have constants of *lower* complexity than the corresponding source constants. An intersective adjective such as 'red' in 'red car' will be a source constant of type n/n. At the target end, one may want to set $\lceil \text{red} \rceil = \lambda x \lambda^! y.((\text{RED } y) \wedge (x\ y))$, a term of type $(e \multimap t) \multimap (!\,e \multimap t)$, specified in terms of a target constant RED of type $e \multimap t$ and boolean conjunction.

- Finally, the compositional setup readily accommodates a wide variety of concrete interpretations, depending on the models one has in mind for the target signature. The e (entity), t (truth value) signature of our example leads to the set-theoretic interpretation familiar from formal semantics; the ACG target signature represents strings as linear λ terms; compositional vector-based semantics in terms of linear λ calculus can be found in [32, 38].

1.2 Modalities for Structural Control

The **L** derivation of our running relative clause example 'paper that Bob rejected' relies on a *global* associativity rule to extract the *np* gap hypothesis out of the verb phrase 'rejected *np*'. Associativity, unfortunalely, is not enough to properly characterize the structural positions that are accessible for extraction: for a relative clause body typed as s/np it restricts extraction to *peripheral* positions at the right edge of the relative clause. Extending the example with an adverbial modifier to 'paper that Bob rejected __ immediately' shows that also phrase internal positions are accessible. Adding a global commutativity option, clearly, is not an option.

Let us consider then an extension of the type language with a pair of unary connectives $\Diamond, \Box$ satisfying

$$A \longrightarrow \Box B \quad \text{iff} \quad \Diamond A \longrightarrow B \tag{12}$$

On the logical level, $\Diamond, \Box$ form a residuated pair, which means they are monotonic operations (from $A \longrightarrow B$ infer $\Diamond A \longrightarrow \Diamond B$ and $\Box A \longrightarrow \Box B$), and one obtains an interior operation from the composition $\Diamond\Box(\cdot)$ and a closure operation from the composition $\Box\Diamond(\cdot)$, i.e., $\Diamond\Box A \longrightarrow A$ and $A \longrightarrow \Box\Diamond A$.

In addition to the purely logical properties of (12), the modalities can be associated with extra structural properties. In particular, modalities make it possible to replace *global* structural rules by *restricted* versions as in the examples below.[3] Postulate $A^\diamond$ is a restricted form of (right) associativity licensed by the $\Diamond$ marking on the formula C. Postulate $C^\diamond$ allows reordering across a phrase boundary but again only for $\Diamond$ marked formulas.

$$\begin{aligned} A^\diamond : (A \bullet B) \bullet \Diamond C &\longrightarrow A \bullet (B \bullet \Diamond C) \\ C^\diamond : (A \bullet B) \bullet \Diamond C &\longrightarrow (A \bullet \Diamond C) \bullet B \end{aligned} \tag{13}$$

[3] In practice, one will work with indexed families $\{\Diamond_i, \Box_i\}_{i \in I}$ so as to parameterize type assignment to particular structural choices. To avoid clutter, we avoid indexing in this section, specifying the intended structural packages in the text.

Control Modalities: N.D. Rules For N.D. presentation, the language of structures is extended with unary punctuation $\langle\cdot\rangle$ as counterpart of the $\Diamond$ operation. Structures are defined then as $\Gamma, \Delta ::= A \mid \langle\Gamma\rangle \mid \Gamma \cdot \Delta$.

The term language, both for the directional calculi **(N)L** and for the non-directional **LP**, is extended with operations capturing the $\Diamond, \Box$ Introduction and Elimination rules: if M is a term, then so are $\triangledown M$, $\vartriangle M$, $\blacktriangledown M$, $\blacktriangle M$. The typing rules then become

$$\frac{\langle\Gamma\rangle \vdash M : A}{\Gamma \vdash \blacktriangle M : \Box A}\ \Box I \qquad \frac{\Gamma \vdash M : \Box A}{\langle\Gamma\rangle \vdash \blacktriangledown M : A}\ \Box E$$
$$\frac{\Gamma \vdash M : A}{\langle\Gamma\rangle \vdash \vartriangle M : \Diamond A}\ \Diamond I \qquad \frac{\Delta \vdash M : \Diamond A \quad \Gamma[\langle x : A\rangle] \vdash N : B}{\Gamma[\Delta] \vdash N[\triangledown M/x] : B}\ \Diamond E \tag{14}$$

The term reductions for the modalities are analogous to the β, η reductions for the binary operations. Specifically, a unary Introduction step immediately followed by its Elimination gives rise to the reductions $\blacktriangledown \blacktriangle M \rightsquigarrow_\beta M$ and $\triangledown \vartriangle M \rightsquigarrow_\beta M$; a unary Elimination step immediately followed by its Introduction simplifies by the reductions $\blacktriangle \blacktriangledown M \rightsquigarrow_\eta M$ and $\vartriangle \triangledown M \rightsquigarrow_\eta M$.

The N.D. structural rules corresponding to the postulates in (13) do not affect the term assignment.

$$\frac{\Gamma[\Delta \cdot (\Delta' \cdot \langle\Delta''\rangle)] \vdash M : A}{\Gamma[(\Delta \cdot \Delta') \cdot \langle\Delta''\rangle] \vdash M : A}\ A^\Diamond \qquad \frac{\Gamma[(\Delta \cdot \langle\Delta''\rangle) \cdot \Delta'] \vdash M : A}{\Gamma[(\Delta \cdot \Delta') \cdot \langle\Delta''\rangle] \vdash M : A}\ C^\Diamond \tag{15}$$

Illustration In Fig. 2 below we return to our relative clause example, this time with a non-peripheral gap. Following the analysis of [23], the type assignment for the relative pronoun 'that' is refined to $(n\backslash n)/(s/\Diamond\Box np)$ and we assume the controlled structural rules (15). This makes the $\Diamond\Box np$ hypothesis into a 'movable' np. Thanks to the $A^\Diamond$, $C^\Diamond$ rules $\Diamond\Box np$ can find its place next to the transitive verb 'rejected' where it is then used as a regular np because of the contraction $\Diamond\Box np \vdash np$.[4] In the derivation, we use $(\Diamond E')$ for the case where the left premise of $(\Diamond E)$ is an Axiom.

The proof term M for the syntactic derivation abstracts over a gap variable x of type $\Diamond\Box np$; x is 'lowered' to a type np term ($\blacktriangledown \triangledown x$) to serve as the direct object argument of 'rejected'; the operations $\blacktriangledown \triangledown$ are the footprint of the $\Diamond$ and $\Box$ Elimination steps behind the contraction $\Diamond\Box np \vdash np$.

For the target image $\lceil M \rceil$ the easy way is to assume that the role of the control modalities is restricted to syntax, setting $\lceil \Diamond A \rceil = \lceil \Box A \rceil = \lceil A \rceil$. The more challenging alternative is to transmit the modal type and term information unchanged in the transition to the target language. To obtain the final property intersection interpre-

[4] The analysis of non-peripheral extraction presented here is quite similar in spirit to the 'subexponentials' $!^i$ used by [3] for controlled structural rules in a non-associative, non-commutative setting. As a matter of fact, one can see the combination $\Diamond\Box(-)$ as the decomposition of an exponential $!(-)$ into an adjoint pair, with the $\Diamond$ part licensing structural reasoning by means of (15) and the ! Dereliction rule turned into a theorem $\Diamond\Box A \vdash A$. We leave a detailed comparison for another occasion.

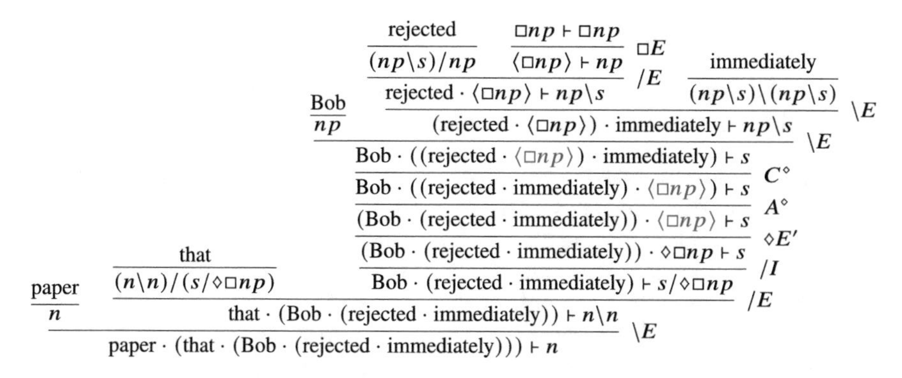

Fig. 2 Extraction of a non-peripheral gap in **NL** extended with $\Diamond$, $\Box$ for structural control

tation of (11), (17) gives an adjusted lexical recipe for the relative pronoun. The abstracted variable v here is of type $\lceil s/\Diamond\Box np \rceil = \Diamond\Box e \multimap t$ whereas w is of type $\lceil n \rceil = e \multimap t$. The final abstraction provides a reusable z variable of type $!\Diamond\Box e$ which is distributed over the $\wedge$ conjuncts.

$$\begin{aligned} M &= \text{paper} \rtimes (\text{that} \ltimes \lambda^r x.(\text{Bob} \rtimes ((\text{rejected} \ltimes (\blacktriangledown \triangledown x))) \rtimes \text{immediately})) : n \\ \lceil M \rceil &= (\lceil \text{that} \rceil\ \lambda x.((\lceil \text{immediately} \rceil\ (\lceil \text{rejected} \rceil\ (\blacktriangledown \triangledown x)))\ \lceil \text{Bob} \rceil))\ \lceil \text{paper} \rceil : e \multimap t \\ &= \lambda z^!.((\text{PAPER}\ (\blacktriangledown \triangledown z)) \wedge ((\text{IMMEDIATELY}\ (\text{REJECTED}\ (\blacktriangledown \triangledown z)))\ \text{BOB})) : !\Diamond\Box e \multimap t \end{aligned} \tag{16}$$

$$\lceil \text{that} \rceil = \lambda v \lambda w \lambda^! z.((w\ (\blacktriangledown \triangledown z)) \wedge (v\ z)) \tag{17}$$

In the example of Fig. 2, the function of the control modalities is to *license* structural rules that otherwise would be forbidden. In a complementary role, modalities can also be used to impose island constraints, thus acting as *obstacles* to structural rules that otherwise would be available. Modalities as structural inhibitors have been advocated by Morrill [28] to block overgeneration of (global or controlled) associativity. The dependency-enhanced type logic to which we now turn generalizes and refines this use of modalities as demarcations of syntactic/semantic locality domains.

2 Dependency-Enhanced Types

Dependency structure articulates the linguistic material on the basis of an opposition between a *head* and its *dependents*, where elements dependent on a head can be further subcategorized into *complements* and *adjuncts*. Examples:

- head—*complement* relations
  - verbal domain: subj, (in)direct object, ...
  - nominal domain: prepositional object, ...

- *adjunct*—head relations
  - verbal domain: (time, manner, ...) adverbial
  - nominal domain: adjectival, numeral, determiner, ...

Function-argument structure and dependency structure are not always aligned. The category of determiners is a case in point. Semantically, we want to assign a function type to words like 'some', 'all', 'no'—the type of the characteristic function of a relation between the sets denoted by the noun they combine with and by the verb phrase. From a morphosyntactic point of view, there are good reasons to view the noun as the head of a noun phrase with the determiner as the dependent element, agreeing with the head noun in number, case, etc. What we need then is a *multidimensional* type logic that treats dependency and function-argument information as autonomous perspectives on grammatical structure.

A first step in this direction is the bimodal syntactic calculus **DNL** of [25]. In this (unpublished) paper, the product $\bullet$ is split in a left-headed $\bullet_l$ and a right-headed $\bullet_r$ version. The familiar residuation laws relate the products $\bullet_l$ and $\bullet_r$ to the concomitant slashes (Fig. 3). In this type system, one can make the distinction between function types that play the role of head ($C/_l B$, $A\backslash_r C$) versus function types that are dependents ($C/_r B$, $A\backslash_l C$). The type system by itself does not commit one to a particular interpretation of the head-dependent asymmetry. As an alternative to the interpretation in terms of grammatical roles that we are interested in here, Hendriks [8] presents a typelogical modeling of *intonation* using **DNL**; in this context, the notion of head is to be understood in terms of prosodic prominence.

A modally-enhanced type system makes it possible to avoid the proliferation of product operations by defining different forms of syntactic combination as compositions of the regular concatenation product $\bullet$ and modal marking. For the dependency products [17] propose a deconstruction that marks the head subtype with $\Diamond$. Here we make the opposite choice and mark the dependent daughter as this opens the possibility of further differentiating dependents according to the specific grammatical role they play:

$$A \bullet_l B := A \bullet \Diamond B \qquad A \bullet_r B := \Diamond A \bullet B \tag{18}$$

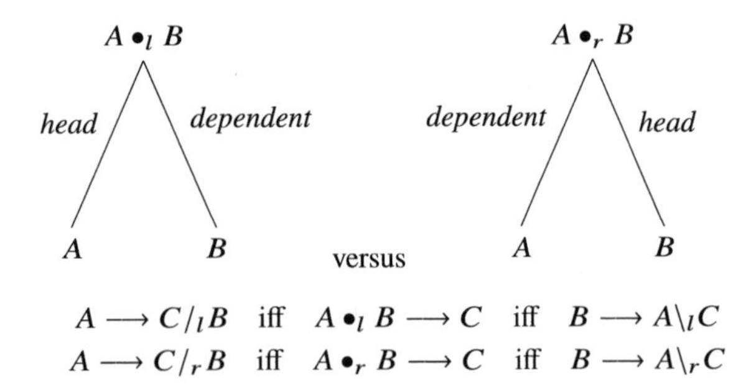

Fig. 3 The **DNL** distinction between a left- and a right-headed tree

Given the above definition of the headed products $\bullet_l$ and $\bullet_r$, the translation of the slashes is obtained by applying the residuation inferences $\Diamond A \longrightarrow B$ iff $A \longrightarrow \Box B$.

$$\dfrac{\dfrac{A \longrightarrow C/_l B}{A \bullet_l B \longrightarrow C}}{B \longrightarrow A\backslash_l C} \rightsquigarrow \dfrac{\dfrac{\dfrac{A \longrightarrow C/\Diamond B}{A \bullet \Diamond B \longrightarrow C}}{\Diamond B \longrightarrow A\backslash C}}{B \longrightarrow \Box(A\backslash C)} \quad \dfrac{\dfrac{A \longrightarrow C/_r B}{A \bullet_r B \longrightarrow C}}{B \longrightarrow A\backslash_r C} \rightsquigarrow \dfrac{\dfrac{\dfrac{A \longrightarrow \Box(C/B)}{\Diamond A \longrightarrow C/B}}{\Diamond A \bullet B \longrightarrow C}}{B \longrightarrow \Diamond A\backslash C} \tag{19}$$

The multimodal generalization to families $\{\Diamond_d, \Box_d\}_{d \in \text{DepLabel}}$, with indices d taken from a set of dependency labels, is straightforward. The translation of the slash types then leads to a distinction between types of the form $\Diamond_d A\backslash C$ or $C/\Diamond_d B$ for *heads* assigning dependency role d to their complement versus types of the form $\Box_d(A\backslash C)$ or $\Box_d(C/B)$ for *dependents* projecting their adjunct role d.

The derivation of Fig. 4 shows the dependency-enhanced type system at work. The intransitive verb 'twitter' and the preposition 'in' are head functors, triggering the $\Diamond$ Introduction steps that assign the dependency roles of subject resp. prepositional object to their complements. The determiner 'the' and the complete prepositional phrase 'in the skies' are typed as dependent functors, projecting the adjunct roles of determiner resp. adverbial modifier via the $\Box$ Elimination steps.

The semantic term associated with the derivation has function application for the $/$ and $\backslash$ Elimination steps, and $\blacktriangledown$, $\vartriangle$ for the $\Box$ Elimination resp. $\Diamond$ Introduction inferences.

$$(\blacktriangledown^{amod}(\text{IN } \vartriangle^{pobj}(\blacktriangledown^{det}\text{THE SKIES}))\ (\text{TWITTER } \vartriangle^{su} \text{SWALLOWS})) \tag{20}$$

The induced dependency structure is computed by localizing the head within each dependency domain (i.e., the terminal element that doesn't carry modal decoration), and drawing outgoing arcs from the head to (the head of) its dependents, as depicted in Fig. 5.

$$\dfrac{\dfrac{\dfrac{\dfrac{\text{swallows}}{np}}{\langle\text{swallows}\rangle^{su} \vdash \Diamond_{su} np}\,\Diamond I \quad \dfrac{\text{twitter}}{\Diamond_{su} np\backslash s}}{\langle\text{swallows}\rangle^{su} \cdot \text{twitter} \vdash s}\,\backslash E \quad \dfrac{\dfrac{\dfrac{\text{in}}{\Box_{amod}(s\backslash s)/\Diamond_{pobj} np} \quad \dfrac{\dfrac{\dfrac{\dfrac{\text{the}}{\Box_{det}(np/n)}}{\langle\text{the}\rangle^{det} \vdash np/n}\,\Box E \quad \dfrac{\text{skies}}{n}}{\langle\text{the}\rangle^{det} \cdot \text{skies} \vdash np}\,/E}{\langle\langle\text{the}\rangle^{det} \cdot \text{skies}\rangle^{pobj} \vdash \Diamond_{pobj} np}\,\Diamond I}{\text{in} \cdot \langle\langle\text{the}\rangle^{det} \cdot \text{skies}\rangle^{pobj} \vdash \Box_{amod}(s\backslash s)}\,/E}{\langle\text{in} \cdot \langle\langle\text{the}\rangle^{det} \cdot \text{skies}\rangle^{pobj}\rangle^{amod} \vdash s\backslash s}\,\Box E}{(\langle\text{swallows}\rangle^{su} \cdot \text{twitter}) \cdot \langle\text{in} \cdot \langle\langle\text{the}\rangle^{det} \cdot \text{skies}\rangle^{pobj}\rangle^{amod} \vdash s}\,\backslash E$$

Fig. 4 An example of a derivation in **NL** extended with $\Diamond$, $\Box$ for dependency demarcation

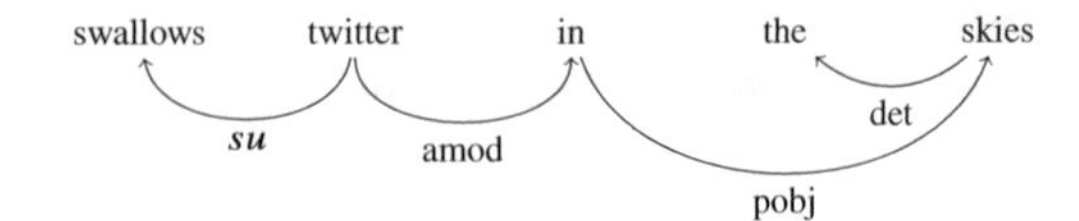

Fig. 5 Shallow dependency structure obtained by traversal of (20)

2.1 Rethinking Constituency

The move to a dependency-enhanced type system invites us to rethink the notion of constituency. In Sect. 1 the default logic was taken to be the non-associative Lambek calculus **NL**. In that system, constituent structure is determined by the non-commutative, non-associative binary product operation, and any change to that structure affecting linear order or grouping has to be explicitly licensed by modalities. In the dependency-enhanced type system **DL**, constituents are determined by the unary dependency modalities; in this setting, the binary product can be seen as an associative operation, leading to a 'flat' structure where a head together with its dependents constitute a dependency domain.

We illustrate the perspective shift with relative clause extraction. Consider first the $\mathbf{NL}_\diamond$ analysis. As shown in [26], in the case of Dutch, where the verb in subordinate clauses is clause-final, material can be extracted from left branches of the constituent structure tree via the controlled associativity and commutativity postulates of (21) below. These postulates are the mirror images of the right branch extraction postulates of the English relativization example we discussed in Sect. 1. We index the modalities for extraction with x to distinguish them from the dependency modalities; also we use $!_x\, A$ as an abbreviation for $\diamond_x \Box_x A$.

$$\begin{array}{ll} (A_\diamond)\ \diamond_x A \bullet (B \bullet C) \longrightarrow (\diamond_x A \bullet B) \bullet C \\ (C_\diamond)\ \diamond_x A \bullet (B \bullet C) \longrightarrow B \bullet (\diamond_x A \bullet C) \end{array} \tag{21}$$

The derived inference rule below (Δ non-empty) telescopes a sequence of structural steps flanked by \ Introduction and $\diamond$, $\Box$ Elimination into a one-step inference rule.

$$\frac{\Gamma[A \cdot \Delta] \vdash B}{\Gamma[\Delta] \vdash !_x\, A \backslash B}\ xleft \tag{22}$$

The typing *die* $::(n\backslash n)/(!_x\, np\backslash s)$ for the Dutch relative pronoun now causes a derivational ambiguity between subject and object relativisation as shown by the English translation of (23) (a) and (b).

$$\begin{array}{lll} (a)\ \text{mannen die \_ vrouwen haten} & \rightsquigarrow & \text{men who hate women} \\ (b)\ \text{mannen die vrouwen \_ haten} & \rightsquigarrow & \text{men who(m) women hate} \end{array} \tag{23}$$

The subject relativisation (23a) doesn't require structural reasoning; the type for the hypothesis $\diamond_x \Box_x np$ can immediately reduce to np, satisfying the subject argument of the transitive verb *haten* $::\ np\backslash(np\backslash s)$. For the object relativisation case (23b), the

Fig. 6 Standard example of crossing dependencies in Dutch

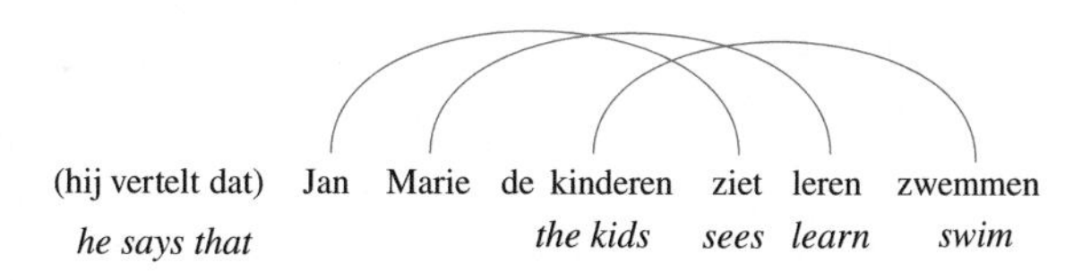

$\Diamond_x$ marking on the hypothesis subtype $\Diamond_x\Box_x np$ licenses application of the controlled commutativity rule, allowing the hypothesis to 'jump over' the subject 'vrouwen', and then reduce to np when it finds itself in the direct object position next to the transitive verb 'haten'.

In the dependency-enhanced type system **DL**, the transitive verb type becomes *haten* :: $\Diamond_{obj}\, np\backslash(\Diamond_{subj}\, np\backslash s)$. The relative pronoun type, accordingly, can be refined to *die* :: $\Box_{mod}(n\backslash n)/\Diamond_{body}(\Diamond_{subj}\, np\backslash s)$ or $\Box_{mod}(n\backslash n)/\Diamond_{body}(\Diamond_{obj}\, np\backslash s)$, making the distinction between subject and object relativisation explicit at the type level. The effect is that the *derivational* ambiguity of relativisation in $\mathbf{NL}_\Diamond$ is turned into a *lexical* ambiguity that is to be resolved in the supertagging phase.

The typings $\Box_{mod}(n\backslash n)/\Diamond_{body}(\Diamond_{subj}\, np\backslash s)$ or $\Box_{mod}(n\backslash n)/\Diamond_{body}(\Diamond_{obj}\, np\backslash s)$ allow for *local* cases of extraction, where the $\Diamond_{subj}$ or $\Diamond_{obj}$ hypothesis is an immediate dependent within the s relative clause body. For non-local cases, where the hypothesis has to be extracted from a more deeply embedded dependency domain, we need to prefix the $\Diamond_{subj}$ or $\Diamond_{obj}$ hypothesis subtype with $!_x$ to license extraction as in the $\mathbf{NL}_\Diamond$ analysis. In addition to associativity and controlled commutativity postulates, **DL** requires extra postulates allowing $\Diamond_x$ to commute with dependency modalities $\Diamond_d$ for $d \in$ DepLabel:

$$(D_\Diamond) \qquad \Diamond_x A \bullet \Diamond_d B \longrightarrow \Diamond_d(\Diamond_x A \bullet B) \tag{24}$$

There is a subtle difference in the way the *xleft* derived inference rule has to be understood in **DL**. In **NL**, the notation $\Gamma[\Delta]$ zooms in on a substructure Δ within a binary product context Γ. In **DL**, the context is built in terms of the binary product structure $(\_ \cdot \_)$ *and* the unary $\langle\_\rangle^d$ structure imposed by the dependency modalities. In its unmodified form, *xleft* allows a hypothesis to be extracted from arbitrarily deeply nested dependency domains. A refined form of *xleft* could then specify which dependency domains d are in fact impenetrable.

2.2 *Case Study: Dutch Verb Clusters*

The verb clusters arising in Dutch embedded clauses form a nice playground to study the interplay between logical and structural reasoning at the syntax-semantics interface. Consider the example of Fig. 6, where the arcs informally indicate 'who does what'.

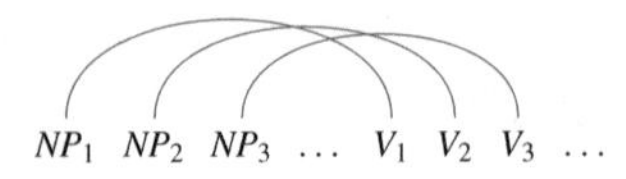

Fig. 7 Templatic pattern generating the example of Fig. 6

As shown by [9, 34], handling these crossing dependencies requires expressivity beyond context-free. Abstractly, the pattern we find here is that of the copy language w^2, a well-known representative of mild context-sensitivity, as discussed in [10] (Fig. 7).

In what follows, we look at Dutch verb clusters from three perspectives. For a start, in Sect. 2.2.1 we present a Multiple Context-Free Grammar (MCFG) for the construction. MCFG [33] is an insightful formalism for mildly context-sensitive phenomena, allowing for a very direct representation of the characteristic crossing dependencies. Then in Sects. 2.2.2–2.2.3, we compare two strategies for obtaining the crossing dependencies in a typelogical setting. Section 2.2.2 illustrates the ACG method that computes the surface word order as the homomorphic image of an 'abstract syntax' source derivation. In Sect. 2.2.3, the sequent structure that forms an appropriate input for semantic composition is transformed into the required surface structure by means of explicit structural rules.

2.2.1 MCFG

The framework of MCFG is a straightforward generalization of context-free rewriting grammars. In an MCFG, non-terminals range over string *tuples* rather than simple strings; for a k-dimensional MCFG, the maximal tuple size is k. To model the Dutch verbal clusters, we can use a 2-MCFG. Consider first the examples below, together with the MCFG rules of (26). For further reference, we give the rules an identifier; for rules introducing lexical items a gloss is provided. Rules are presented in the convenient (simple) Range Concatenation Grammar format of [4] which depicts non-terminals ranging over an n-tuple as n-ary predicate symbols.

iets willen zeggen	*something want say*	= want to say sth	(25)
haar iets laten zeggen	*her something let say*	= let her say sth	

$$\begin{array}{llll}
f_0 & INFP(\epsilon, x) \leftarrow & INF_0(x) & \\
f_1 & INFP(x, y) \leftarrow & NP(x)\ INF_1(y) & \\
g_0 & INFP(y, x \cdot z) \leftarrow & INFP(y, z)\ IVR_0(x) & \\
g_1 & INFP(y \cdot z, x \cdot w) \leftarrow & NP(y)\ INFP(z, w)\ IVR_1(x) & \\
d_0 & INF_0(\textsf{vertrekken}) \leftarrow & & \textit{leave} \\
d_1 & INF_1(\textsf{zeggen}) \leftarrow & & \textit{say} \\
d_2 & IVR_0(\textsf{willen}) \leftarrow & & \textit{want} \\
d_3 & IVR_1(\textsf{laten}) \leftarrow & & \textit{let}
\end{array} \quad (26)$$

Infinitival phrases *INFP* are interpreted as *pairs* of strings (tuples of size 2). The second component of these pairs holds the verb (or verbal cluster), the first component the rest of the phrase. Rules g_i introduce the lexical items responsible for cluster formation IVR_i, the so-called verb raising (VR) triggers. Instead of combining with their *INFP* complement by means of concatenation, they *infix* themselves into that complement, left-adjoining to its verbal head. IVR_0 is for modal and temporal auxiliaries; IVR_1 covers causatives and verbs of perception. Apart from their *INFP* complement, the latter also select a direct object noun phrase, which acts as the understood subject of the *INFP* complement. The preterminal rules d_i give some representative lexical items.

For good measure, a main clause rule r_0 is added for the start symbol S; our focus here is on the structure of infinitival phrases, so we simply start main clauses with a fixed prefix 'hij zal' (he will). The effect of the S rule is to concatenate the strings that constitute the pair interpreting non-terminal *INFP*.

$$r_0 \quad S(\textsf{hij}\cdot\textsf{zal}\cdot x\cdot y) \;\leftarrow\; INFP(x, y) \tag{27}$$

whereas IVR_0 and IVR_1 select a bare infinitival complement, some VR triggers require a *TIP* complement, i.e., an infinitival complement which prefixes the particle 'te' to its verbal head. These 'te'-infinitives also appear in *extraposition* constructions, which realize the complete *TIP* complement intact at the end of the clause. Interestingly, some verbs, e.g., 'proberen' (try), allow the two realizations as shown below.

iets proberen te zeggen (*something try to say*)	VR	(28)
proberen iets te zeggen (*try something to say*)	extraposition	

The relevant MCFG rules are given below; 'proberen' is categorized both as VR trigger IVR_2, and as INF_2, where the latter gives rise to the extraposition construction. The non-terminal $INFP_x$ stands for an infinival phrase missing a *TIP* complement. Rule x_0 introduces $INFP_x$. The lack of the *TIP* complement is then transmitted through the derivation by means of rules x_1, x_2; these are variants of g_1, g_1 with a $INFP_x$ complement rather than a complete *INFP*. Rule x_3 is the extraposition variant of the main clause S rule r_0; the rule satisfies the outstanding request for the *TIP* complement, with a complete sentence as a result.

$$\begin{array}{lrcll}
h_0 & INFP(y, x\cdot z) & \leftarrow & TIP(y, z)\; IVR_2(x) & \\
h_1 & TIP(x, \textsf{te}\cdot y) & \leftarrow & INFP(x, y) & \\
d_4 & IVR_2(\textsf{proberen}) & \leftarrow & & \textit{try} \\
x_0 & INFP_x(\epsilon, x) & \leftarrow & INF_2(x) & \\
x_1 & INFP_x(y, x\cdot z) & \leftarrow & INFP_x(y, z)\; IVR_0(x) & \\
x_2 & INFP_x(x\cdot z, y\cdot w) & \leftarrow & NP(x)\; INFP_x(z, w)\; IVR_1(y) & \\
x_3 & S(\textsf{hij}\cdot\textsf{zal}\cdot x\cdot y\cdot z\cdot w) & \leftarrow & INFP_x(x, y)\; TIP(z, w) &
\end{array} \tag{29}$$

We now turn to available strategies for obtaining the crossing dependencies of (26) in typelogical grammar.

2.2.2 From Abstract Syntax to String Semantics

As we saw in Sect. 1, the Abstract Categorial Grammar method obtains the string languages of (extended) context-free formalisms through a homomorphism relating an 'abstract syntax' source to a target interpretation for the surface string representation. Both source and target are expressed in terms of $\text{ILL}_{\multimap}$, i.e., simply typed Intuitionistic Linear Logic, **LP** in categorial terminology.

The ACG construction for MCFG presented in [7] has a source signature with the MCFG non-terminals as atomic types, and the rule identifiers as abstract constants; a rule $r : A \longleftarrow B_1 \ldots B_n$ is translated into a type assignment $B_1 \multimap \cdots \multimap B_n \multimap A$ to rule constant r. To model strings, the target signature uses a function type $* \multimap *$, where $*$ is some arbitrary atomic type; concatenation is represented by function composition[5] and the empty string by the identity function $\lambda i.i$ of type $* \multimap *$. Below we abbreviate $* \multimap *$ as σ. Corresponding to the MCFG terminals, the target signature has constants of type σ. The interpretation homomorphism models string *tuples* as higher-order types. Concretely for the 2-MCFG rules of (26), one sets $\lceil INFP \rceil = (\sigma \multimap \sigma \multimap \sigma) \multimap \sigma$ to obtain the interpretation of *INFP* as a *pair* of strings; for the other non-terminals, $\lceil NT \rceil = \sigma$.

For the verb-raising fragment, the translations of the abstract rule constants below are target terms that respect the typing. We abbreviate $(\sigma \multimap \sigma \multimap \sigma) \multimap \sigma$ as $\sigma^{(2)}$. For the preterminal rules d_i, the translation is simply the string constant introduced by these rules.

$$\begin{array}{lcll} \lceil f_1 \rceil & = & \lambda x \lambda y \lambda f.(f\ x\ y) & ::\ \sigma \multimap \sigma \multimap \sigma^{(2)} \\ \lceil g_0 \rceil & = & \lambda q \lambda x \lambda f.(q\ \lambda y \lambda z.(f\ y\ x + z)) & ::\ \sigma^{(2)} \multimap \sigma \multimap \sigma^{(2)} \\ \lceil g_1 \rceil & = & \lambda x \lambda q \lambda y \lambda f.(q\ \lambda z \lambda w.(f\ x + z\ y + w)) & ::\ \sigma \multimap \sigma^{(2)} \multimap \sigma \multimap \sigma^{(2)} \end{array} \tag{30}$$

As an example, consider the abstract source term for the infinitival phrase 'haar iets willen laten zeggen' and its β-normalized target translation as a pair of strings 'haar iets' and 'willen laten zeggen'.

$$\lceil g_0\ (g_1\ d_5\ (f_1\ d_6\ d_1)\ d_3)\ d_2 \rceil \quad = \quad \lambda f.(f\ \mathsf{haar} + \mathsf{iets}\ \ \mathsf{willen} + \mathsf{laten} + \mathsf{zeggen}) \tag{31}$$

The translation of the main clause rule r_0 then concatenates the element of the string pair by providing the concatenation combinator for the abstraction over f, a variable of type $\sigma \multimap \sigma \multimap \sigma$.

$$\lceil r_0 \rceil \quad = \quad \lambda q.\mathsf{hij} + \mathsf{zal} + (q\ \lambda x \lambda y.(x + y)) \ :: \ \sigma^{(2)} \multimap \sigma \tag{32}$$

As said, ACG uses **LP**, with non-directional $\multimap$ implicational types, both for the source and the target logic. But the ACG method is easily adapted to a **NL** source. Instead of abstract rule constants, here we would use *words* as abstract source constants. A MCFG rule $B \longleftarrow A_1 \ldots A_n\ C$, with non-terminal C providing the lexical

[5] Infix '+', defined as $\lambda x \lambda y \lambda i.(x\ (y\ i))$, represents concatenation.

$$\dfrac{\dfrac{\dfrac{\text{haar}}{NP} \quad \dfrac{\dfrac{\dfrac{\text{iets}}{NP} \quad \dfrac{\text{zeggen}}{NP\backslash INFP}}{\text{iets}\cdot\text{zeggen} \vdash INFP}\,\backslash E \quad \dfrac{\text{laten}}{INFP\backslash(NP\backslash INFP)}}{(\text{iets}\cdot\text{zeggen})\cdot\text{laten}\vdash NP\backslash INFP}\,\backslash E}{\text{haar}\cdot((\text{iets}\cdot\text{zeggen})\cdot\text{laten})\vdash INFP}\,\backslash E \quad \dfrac{\text{willen}}{INFP\backslash INFP}}{\dagger \quad (\text{haar}\cdot((\text{iets}\cdot\text{zeggen})\cdot\text{laten}))\cdot\text{willen}\vdash INFP}\,\backslash E$$

Fig. 8 Source syntactic derivation in **NL**, employing purely functional types

anchoring provided by the preterminal rules d_i is turned into a **NL** type assignment $A_n\backslash\cdots\backslash A_1\backslash B$ to that lexical anchor.

Translation of the abstract lexical constants now takes the form below.

$$\begin{array}{rcll} \lceil\text{zeggen}\rceil^{string} & = & \lambda x\lambda f.(f\ x\ \mathsf{zeggen}) & ::\ \sigma \multimap \sigma^{(2)} \\ \lceil\text{willen}\rceil^{string} & = & \lambda q\lambda f.(q\ \lambda y\lambda z.(f\ y\ \mathsf{willen} + z)) & ::\ \sigma^{(2)} \multimap \sigma^{(2)} \\ \lceil\text{laten}\rceil^{string} & = & \lambda q\lambda x\lambda f.(q\ \lambda z\lambda w.(f\ x + z\ \mathsf{laten} + w)) & ::\ \sigma^{(2)} \multimap \sigma \multimap \sigma^{(2)} \end{array} \tag{33}$$

The abstract syntax derivation for the *INFP* 'haar iets willen laten zeggen' directly reflects the scopal relations for semantic composition, which has the elements of the verbal cluster in the *reverse* order with respect to their appearance in the surface string. The string image of the derivation agrees with the translation of the ACG abstract syntax term: $\lceil\dagger\rceil^{string} = \lambda f.(f\ \ \mathsf{haar} + \mathsf{iets}\ \ \mathsf{willen} + \mathsf{laten} + \mathsf{zeggen})$. A second homomorphism $\lceil\cdot\rceil^{sem}$ produces the term for meaning assembly. On the type level, directional implications are sent to non-directional $\multimap$. Abstract source constants translate as target constants (small caps) of the appropriate type: $\lceil\dagger\rceil^{sem} = \textsc{want}\ (\textsc{let}\ (\textsc{say}\ \textsc{something})\ \textsc{her})$.

The semantic target types of Fig. 8 are purely functional. To add the dependency information, function types $A_n\backslash\ldots\backslash A_1\backslash B$ are are turned into $\Diamond_{d_n} A_n\backslash\ldots\backslash\Diamond_{d_1} A_1\backslash B$, assigning complements A_i the dependency roles d_i, cf the derivation below where the dependency label *vc* stands for verbal complement (Fig. 9). The $\lceil\cdot\rceil^{sem}$ homomorphism then records the $\Diamond$ Introduction steps:

$$\lceil\dagger\rceil^{sem} = (\textsc{want}\ \ \triangle^{vc}((\textsc{let}\ \ \triangle^{vc}(\textsc{say}\ \ \triangle^{obj}\ \textsc{something}))\ \ \triangle^{obj}\ \textsc{her})) \tag{34}$$

The $\lceil\cdot\rceil^{string}$ homomorphism ignores the dependency marking, setting $\lceil\Diamond_d\ A\rceil^{string} = \lceil\Box_d\ A\rceil^{string} = \lceil A\rceil^{string}$.

2.2.3 Alternating Logical and Structural Reasoning

An alternative typelogical approach more in the spirit of Sect. 1 reads off the syntactic surface string as the yield of the antecedent structure term Γ in a derivation $\Gamma \vdash A$. For our verb raising fragment, a derivation then starts with a purely logical phase with Introduction and Elimination steps for the typeforming operations followed by a structural phase that transforms the sequent antecedent into a structure with the desired surface string as its yield. Schematically, writing $\mid A_1 \ldots A_n \mid$ for the yield

$$
\dfrac{
  \dfrac{\dfrac{\dfrac{\text{haar}}{NP}}{\langle \text{haar}\rangle^{obj} \vdash \Diamond_{obj} NP}\,\Diamond I
  \qquad
  \dfrac{
    \dfrac{\dfrac{\dfrac{\dfrac{\dfrac{\text{iets}}{NP}}{\langle \text{iets}\rangle^{obj} \vdash \Diamond_{obj} NP}\,\Diamond I \quad \dfrac{\text{zeggen}}{\Diamond_{obj} NP\backslash INFP}}{\langle \text{iets}\rangle^{obj} \cdot \text{zeggen} \vdash INFP}\,\backslash E}{\langle\langle \text{iets}\rangle^{obj} \cdot \text{zeggen}\rangle^{vc} \vdash \Diamond_{vc} INFP}\,\Diamond I}{}
    \quad
    \dfrac{\text{laten}}{\Diamond_{vc} INFP\backslash(\Diamond_{obj} NP\backslash INFP)}
  }{\langle\langle \text{iets}\rangle^{obj} \cdot \text{zeggen}\rangle^{vc} \cdot \text{laten} \vdash \Diamond_{obj} NP\backslash INFP}\,\backslash E
  }{\dfrac{\langle \text{haar}\rangle^{obj} \cdot (\langle\langle \text{iets}\rangle^{obj} \cdot \text{zeggen}\rangle^{vc} \cdot \text{laten}) \vdash INFP}{\langle\langle \text{haar}\rangle^{obj} \cdot (\langle\langle \text{iets}\rangle^{obj} \cdot \text{zeggen}\rangle^{vc} \cdot \text{laten})\rangle^{vc} \vdash \Diamond_{vc} INFP}\,\Diamond I}\,\backslash E
  \quad
  \dfrac{\text{willen}}{\Diamond_{vc} INFP\backslash INFP}
}{\langle\langle \text{haar}\rangle^{obj} \cdot (\langle\langle \text{iets}\rangle^{obj} \cdot \text{zeggen}\rangle^{vc} \cdot \text{laten})\rangle^{vc} \cdot \text{willen} \vdash INFP}\,\backslash E
$$

Fig. 9 The derivation of Fig. 8, expanded with dependency modalities

of an antecedent structure, we have the following situation for our running example 'haar iets willen laten zeggen':

$$
\dfrac{\dfrac{\vdots}{\mid \text{haar iets zeggen laten willen} \mid \vdash INFP}\ \textit{logical phase}}{\dfrac{?}{\mid \text{haar iets willen laten zeggen} \mid \vdash INFP}\ \textit{structural phase}} \tag{35}
$$

The challenge here is that to reach the end sequent, the verb raising triggers (the categories IVR_0, IVR_1 of the MCFG rules in (26) have to break into the dependency domain of their infinitival complements. To properly control this process, we need *head marking*, as in Head Grammars [31]. The adjusted type assignments below use $\Box_{hd}$ to mark plain infinitives, and $\Box_{vr}$ to identify the verb raising triggers.

$$
\begin{array}{rcl}
\text{zeggen} :: \Diamond_{obj} NP\backslash INFP & \rightsquigarrow & \Box_{hd}(\Diamond_{obj} NP\backslash INFP) \\
\text{willen} :: \Diamond_{vc} INFP\backslash INFP & \rightsquigarrow & \Box_{vr}(\Diamond_{vc} INFP\backslash INFP) \\
\text{laten} :: \Diamond_{vc} INFP\backslash(\Diamond_{obj} NP\backslash INFP) & \rightsquigarrow & \Box_{vr}(\Diamond_{vc} INFP\backslash(\Diamond_{obj} NP\backslash INFP))
\end{array} \tag{36}
$$

To compute the order reversal of the verbal cluster, the two structural transformations of Fig. 10 are required. Viewed in forward chaining fashion (from lexical axioms to conclusion), the first one causes a verb raising trigger vr in construction with the head hd of its verbal complement vc to merge into a complex head. The second one is a form of controlled associativity, reassociating a complex verbal complement across the vc phrasal boundary. The joint effect of these structural transformations is to left-adjoin the verb raising triggers to the heads of their verbal complement (Fig. 11).

The astute reader will have noticed that the intermediate stages between the output of the logical phase and the endsequent that has the verb cluster in the desired order all qualify as well-formed *INFP* phrases: the proposed structural rules *allow* the

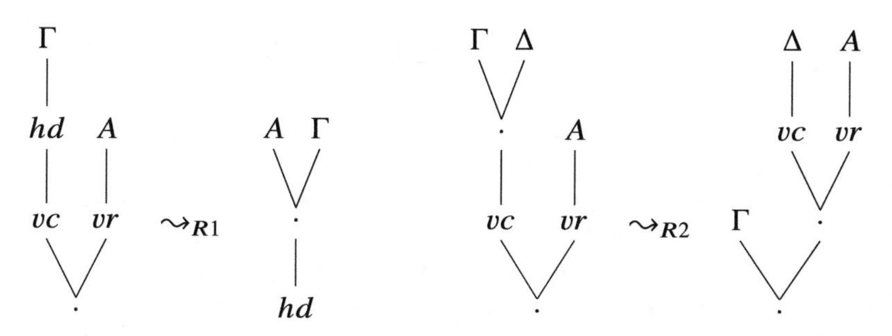

Fig. 10 Structural transformations required to turn the logical judgement of (36) into its surface representation

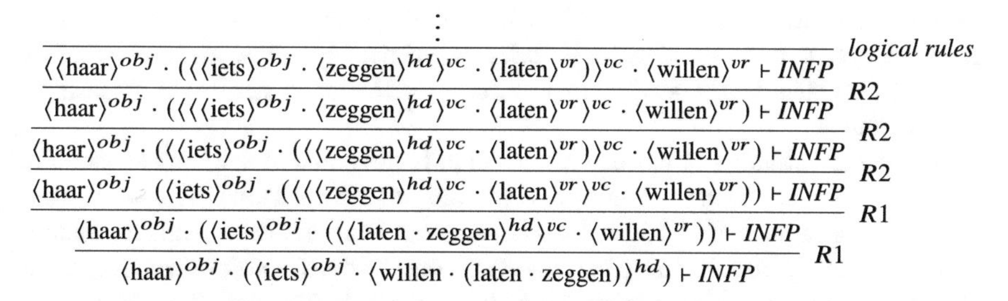

Fig. 11 Applying the rules of Fig. 10 to the logical judgement of (36)

verbal cluster to be derived, but they don't *force* this to happen. To achieve the latter, one can change the goal type of the end sequent to $\Box_{hd}$ *INFP*, 'abstracting' the head feature via a distributivity postulate $\Diamond_{hd}(A \bullet B) \longrightarrow A \bullet \Diamond_{hd} B$ as shown in [24].

In general, the issue of characterizing all-and-*only* well-formed phrases as logically derivable deserves careful attention in refined typelogical systems of the kind discussed here. If one wants to use the type logic in a 'generative' mode, overgeneration will be hard to avoid, but can give an impetus to further refinement. If, as we do here, one sees the parsing problem as the task of associating well-formed input with the correct semantic analysis one can be tolerant of a modest degree of overgeneration.

Comparing the typelogical approaches of Sects. 2.2.2–2.2.3, we see that the ACG abstract syntax method has a very simple combinatorics, with a purely applicative derivation. The full burden of this approach is put on the type homomorphism that inflates a source atom to a third order target type: $\lceil INFP \rceil = (\sigma \multimap \sigma \multimap \sigma) \multimap \sigma$. The alternative approach rests on a division of labour between logical and structural reasoning. The logical phase of a derivation transparently reflects semantic composition; the structural phase consists of meaning-preserving transformations that ultimately produce the surface form.

3 Dependencies and Neural Language Models

The theoretical modelling presented above allows for a crisp and tractable formal description of verbal cluster formation in Dutch. Implementing such a pipeline requires in essence two key components: a parser capable of extracting the derivational structure described [15], and a plug-and-play lexicon of semantic recipes to deal with word-internal function/argument structures. In stark contrast to the above, the dominant strain of applied NLP research commonly employs an overparameterized end-to-end language model trained with self supervision on massive raw text corpora. A natural question then is whether indeed such language models have the capacity of automatically "understanding" the kind of syntactic and semantic subtleties discussed, when having compiled away any kind of explicit structure manipulation. In this section, we describe an experiment that allows us to test whether and to what extent this is the case for a variety of different cluster forming constructions in Dutch.

The overall strategy of our experiment is to target a neural language model like BERT [5]—which in our experiments will be the Dutch incarnation [35]—and investigate whether such a model (a) is capable of recognizing subtle semantic dependencies arising in various clustering/extraposition constructions, and (b) whether specific constructions are more easy to recognize than others. In this, we follow up on previous work [16, 37], and generate a variety of instances of clustering/extraposition constructions that are then generated by considering lexical variation over the word categories in the sentences. The way we evaluate is by training a probe classifier on top of a BERTje instance to predict subjecthood for each verb in a sentence.

The novelty of the current experiment here is in the process of generating the base examples that we want to analyze. The ACG method of Sect. 2.2.2 lends itself well for the type of empirical investigation we want to carry out as it allows us to generate many different examples on the spot, but also to classify sentences based on their semantic properties. The example of verb raising versus extraposition (28) illustrates how non-obligatory verb raisers may occupy a different position in the sentence while preserving the meaning. Both sentences ('iets proberen te zeggen' and 'proberen iets te zeggen') have identical meaning, but it is not said that a language model like BERT would be automatically inclined to recognize this.

The process involves three steps. First, we implement an ACG-style grammar to capture the tectogrammatic function-argument and dependency structures present in the sentences of interest. The grammar will find use in generating a collection of unique abstract syntactic representations, each one endowed with the means to be converted to a concrete sentence (read: string) and meaning representation (read: semantic λ-term) by virtue of type- and term-level morphisms defined upon the grammar's constants. Using the generated abstract samples, we will use a lexicon, i.e., a mapping from grammatical types to lexical inhabitants of these types, to automatically populate an annotated dataset of the phenomena addressed. Finally, we are going to throw the lexically varied samples at the neural model, and use a tiny probe attached to it in order to extract its comprehension of the input; combining the probe's output with the syntactic and semantic grounding of the samples will allow us to measure and quantify the model's performance, in the process giving us concrete evidence to soothe or aggravate our apprehensions.

Our code is available online at github.com/gijswijnholds/malin_2022.

3.1 Experimental Design

Abstract Generation The first step of the experimental setup involves the abstract generation phase. Here, the ACG style modelling of Sect. 2.2.2 uses rule constants that encode the rules of the MCFG of interest; out of that we define the usual two homomorphisms to realize the string interpretation (surface form) and the semantic interpretation. Next to that, we define a third homomorphism[6] that relates each occurrence of a verb with its (understood) subject. For the example sentence 'hij zal haar iets willen laten zeggen' the result of the three maps is shown below:

$$\begin{aligned} \lceil \dagger \rceil^{string} &= \textsf{hij} + \textsf{zal} + \textsf{haar} + \textsf{iets} + \textsf{willen} + \textsf{laten} + \textsf{zeggen} \\ \lceil \dagger \rceil^{sem} &= \text{WILL } M \; (\Delta^{subj} \text{ HE}) \\ M &= \Delta^{vc}(\text{WANT } \Delta^{vc}((\text{LET } \Delta^{vc}(\text{SAY } \Delta^{obj} \text{ SOMETHING})) \; \Delta^{obj} \text{ HER})) \\ \lceil \dagger \rceil^{pair} &= [(\textsf{zal}, \textsf{hij}), (\textsf{willen}, \textsf{hij}), (\textsf{laten}, \textsf{hij}), (\textsf{zeggen}, \textsf{haar})] \end{aligned} \tag{37}$$

The first interpretation gives the surface string as a prototypical sentence illustrating the syntactic pattern encoded by the underlying abstract syntax tree (AST). A second mapping translates the AST into a semantic term, made of binary function-argument structures and unary structures demarcating specific dependency domains, like *su* for subject, *vc* for verbal complement and so forth. Finally, a third translation gives us the mapping between verbs occurring in the sentence and their respective subject nouns.

The MCFG in ACG style, together with the homomorphic interpretation maps are implemented in Prolog, generating a file with 298 unique abstract syntax trees together with their example sentences, semantic terms, and list of verb-noun pairs encoding subject-relationships.

Lexical Generation In order to instantiate the list of items generated in the abstract generation phrase, we substitute each sample word by its corresponding lexical category, and then effectively sample 10 concrete instantiations based on a manually curated lexicon, depicted in Table 2.

For nouns, we distinguish between inanimate nouns, for which we carefully selected a list of 15 nouns, and animate nouns which we fill from a list of common Dutch first names (1 184 male, 1 493 female, all with a minimum occurrence of 500

[6] This homomorphism can be defined using standard tools from λ calculus for modelling tuples and lists.

Table 2 Word categories with their description, and examples

Category	Description	Examples
INF0	intransitive infinitive	vertrekken, stemmen, verliezen, ...
INF1	transitive infinitive with inanimate object	zeggen, begrijpen, merken, ...
INF1A	transitive infinitive, animate object	ontmoeten, bedanken, kennen, ...
IVR0	obligatory verb raiser	willen, zullen, moeten, ...
IVR1	obligatory verb raiser, subject flipper	laten, doen
IVR2	non-obligatory verb raiser	proberen, weigeren, trachten, ...
INF2	extraposition	proberen, weigeren, trachten, ...
INF3	extraposition, object control	verzoeken, dwingen, verplichten, ...
INF4	extraposition, subject control	beloven, verzekeren, zweren, ...
OBJ1A	animate direct object	Karin, Wouter, ...
OBJ1I	inanimate direct object	iets, veel, een ding, ...
OBJ2	indirect object	Karin, Wouter, ...

times within The Netherlands in 2010, taken from the Nederlandse Voornamenbank of the Meertens Institute KNAW, www.meertens.knaw.nl/nvb). The remaining items constitute verbs that may be subcategorized as obligatory verb raisers, obligatory extraposition verbs, or verbs that may act both as a verb raiser as well as in extraposition constructions. Finally we also include transitive and intransitive infinitive verbs. The verbs are manually selected from the electronic version of Algemene Nederlandse Spraakkunst (ans.ruhosting.nl). By combinatorics, we end up with a total of 2 740 concrete sentence instantiations; by the fact that each sentence contains multiple verb-subject annotations, we end up with a total of 30 960 cases for which the probing model will need to predict subjecthood.

Probing Model For the model architecture, we draw from prior work [16], and combine a pretrained neural language model with a *probe*, tasked with extracting the verb-subject dependencies "understood" by the model.

The overall architecture receives a tokenized sentence as input, where each word and subword token is further annotated with the set of verb (or resp. noun phrase) indices it helps form, and produces as its output a probability distribution over candidate subjects for each verb, reflecting the inspected language model's comprehension of the input.

Initially, we let the language model process the input and produce contextualized representations for the entire sequence of word and subword tokens. We then use the span annotations to aggregate (possibly non-singleton) sub-sequences into a single vector for each verb and noun phrase in the sentence, using two distinct attention-based pooling functions. The aggregation results in a collection of vectors in the same space (i.e., a matrix) for verbs, and another such for noun phrases; we proceed to compute their inner product, from which we extract the desired distribution by using the smooth argmax operator over the noun axis. A schematic

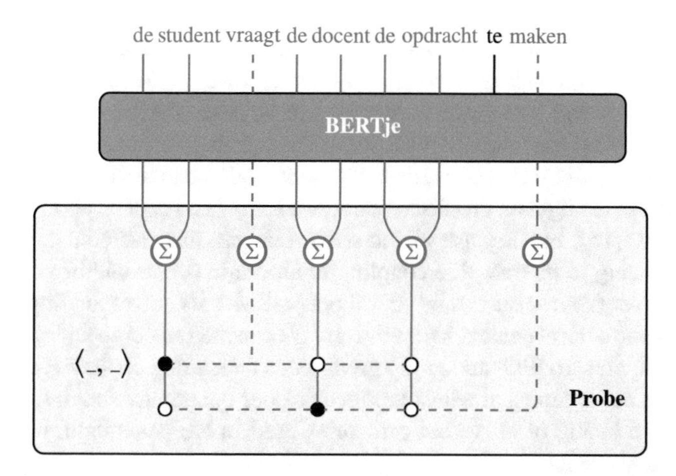

Fig. 12 Illustration of the probing model. Red continuous wires indicate the aggregation of noun phrase tokens (resp. blue dashed wires for verbs). The breadboard-like probe structure computes the inner product between nouns and verbs belonging to the same sentence; one noun is selected per verb by means of the smooth argmax operator

overview is provided in Figs. 1 and 12.

Training and Testing The probe is trained using standard cross-entropy loss against the verb-subject dependencies found in Lassy Small. Lassy Small is a corpus of 65000 Dutch written sentences, manually annotated with parse trees and dependency labelling by experts, making it a sizeable gold-standard resource for Dutch [30]. During training, we cull the back-propagated error signal at the input end of the probe so as to disable weight updates for the inspected neural language model. This ensures that the task-specific optimization only trains the probe to extract the relevant information, without tainting the core language model in the process, allowing us to peek into the latter's semantic capacities unaltered.

After training, we run the model in inference mode on all of the generated test samples to obtain accuracy scores, computed as simply the number of correctly identified subject nouns for each verb in the test set. Random baseline scores are computed as the reciprocal of the number of nouns in a sentence, for a given sample, i.e., the probability of guessing the correct noun given the possible options.

3.2 *Results and Analysis*

In Table 3, we display the total accuracy and baseline on the validation part of Lassy Small and on the test set. We provide a specification of the accuracy by verbal category in Table 4.

Consistent with previous work [16], we observe that the language model's association of verbs with their subjects is significantly lower in the sanitized setup of our artificially generated samples, as compared to its performance in the free ranging text that makes up the validation subset of the training corpus, despite the former being much easier in terms of average competing noun candidates.

In order to shed light on this discrepancy, we follow and expand upon our previous methodology [16], making use of the novel features that our generation pipeline provides. Owing to its traceable coupling to an origin (point within an) AST, each prediction over the test set can be paired not just with its corresponding truth value but rather its structural context in its entirety. We can then select specific higher-order features of interest to filter and group predictions depending on their context, which allows us to better tune and adjust the focus of our quantitative analysis. In Table 4 we provide a listing of all verbal categories used in the generation, together with their accuracy results.

Number of Nouns As a preliminary step, we measure test set performance in relation to the number of subject candidates in the sentence (i.e., number of nouns), which we imagine can confound the model's ability to make a correct semantic judgement, and present our results in Table 5.

As expected, the accuracy does indeed show a correlation to the number of attractors. The correlation is, however, rather weak; accuracy is surprisingly low in comparison to the validation set even in the presence of a single attractor, and only moderately declines as they increase, remaining consistently high above the random

Table 3 Accuracy and baseline results on the validation (Lassy) and test set

	Validation set (Lassy)	Test set (generated)
Accuracy	97.60	79.47
Random Baseline	13.24	39.24

Table 4 Verb categories with their respective accuracy and baseline scores

Category	Description	Accuracy	Baseline
INF0	intransitive infinitive	73.98	45.24
INF1	transitive infinitive with inanimate object	65.17	33.18
INF1A	transitive infinitive, animate object	64.55	38.46
IVR0	obligatory verb raiser	81.32	41.06
IVR1	obligatory verb raiser, subject flipper	87.68	37.09
IVR2	non-obligatory verb raiser	74.90	41.06
INF2	extraposition	92.76	42.58
INF3	extraposition, object control	85.38	35.13
INF4	extraposition, subject control	80.34	35.13

Table 5 Accuracy results by number of nouns

Number of nouns	2	3	4
Accuracy	86.87	75.66	68.76
Random baseline	50.00	33.33	25.00

baseline.

Verbal Type Since the number of nouns is not that telling of a feature in distinguishing correct versus erroneous predictions, the next thing to group results by is the type of verb under inspection. We distinguish nine verb categories, displayed in Table 2 in the Appendix. The verb categories we subdivide in three groups: first, there are infinitives that do not select for a verbal complement, which we refer to as plain infintives (INF0, INF1, INF1A) and that differ in their transitivity, i.e., selecting only a subject or additionally an object, and their selectional preferences (animate vs. inanimate objects). Next, we group three categories of verb raisers (IVR0, IVR1, IVR2), that induce a semantically discontinuous cluster. Finally, we distinguish extraposition verbs (INF2, INF3, INF4) that arrange their arguments in a continuous order, distinguishing between object control (INF3), subject control (INF4) and non-control (INF2) verbs.

Table 6 indicates that the verbal type is a stronger indicator of model performance: accuracy varies between types despite the baselines being comparable. Infinitives governing raising constructions are harder to disentangle compared to their extraposing relatives, and infinitives are remarkably worse off than either. We provide a further specification of accuracy broken down by individual verb categories in Table 4, but note that these results do not show a striking difference between individual verb categories.

Verb Dominance To tell what exactly it is that makes infinitives so difficult for BERTje to understand, we filter predictions of verbs that occur in the context of a *nested* subordinate clause, and group them first by the type of the clause's *governing* verb, and afterwards by the type of the dependent verb.

Table 7 displays the aggregated accuracy scores. As before, we observe a lower overall accuracy in infinitives under an extraposition construction. Despite an extraposing verb being easier to pair to its subject compared to a raising one, verbs under the immediate scope of an extraposition are in fact harder to identify correctly!

Table 6 Accuracy results by verbal type

Verbal type	Raising	Extraposition	Infinitive
Accuracy	81.00	87.03	68.77
Random	39.86	38.27	39.24

Table 7 Accuracy results by dominance, distinguishing verb raisers and extraposition verbs

	Dominated verb, grouped by verbal type			
Dominated by raising	**Overall**	**Raising**	**Extraposition**	**Infinitive**
Accuracy	76.18	76.23	77.76	74.68
Random Baseline	39.86	41.23	38.60	39.76
Dominated by extraposition				
Accuracy	66.70	67.35	85.35	59.62
Random Baseline	38.27	38.60	36.84	38.30

Along the same lines, the disproportionately low performance in the infinitive verbs of Table 6 is now explained by a striking performance drop (ca 20%) in the case of the governing verb being an extraposer. What is making the territory muddier, however, is an inverse trend in the case of an extraposition under the case of an extraposition.

To add to the story, we separately subdivide the overall accuracy for all verbs governed by a raiser or extraposition verb, into the specific categories that make up the raisers and extraposition verbs, in order to offer an orthogonal view on the same problem. The numbers in Table 8 display accuracy results for each subcategory for raisers and extraposition verbs.

The results suggest that it is *not* extraposition per se that is the problem; if anything, performance is higher for clauses in an extraposition setting under the INF2 category rather than a clustering setting under the IVR2 category, despite the two sharing the same lexical vocabulary! The problem, rather, lies in the semantic control properties exhibited by the control verbs in categories INF3 (object control) and INF4 (subject control).

Subject versus Object Control To further deepen the analysis, we perform a variation on the main experiment. In this setup, we gather all pairs of sentences that differ only in the choice of a subject control verb (INF4) or an object control verb (INF3). Such cases are syntactically equivalent, however the control properties of the verb mean that either their subject or their indirect object gets selected as the understood

Table 8 Accuracy results by dominance, for verb raisers and extraposition verbs, where the overall accuracy is broken down by individual categories

	Dominating verb, by subcategory			
Dominated by raising	**Overall**	**IVR0**	**IVR1**	**IVR2**
Accuracy	76.18	78.54	71.41	77.95
Random Baseline	39.86	41.06	37.09	41.05
Dominated by extraposition	**Overall**	**INF2**	**INF3**	**INF4**
Accuracy	66.70	86.74	57.12	47.12
Random Baseline	38.27	42.58	35.13	35.13

Table 9 Results for the masked subject versus object control verb experiment. Each number indicates the percentage of cases in which the model prefers a certain interpretation

	Object (INF3)	Subject (INF4)	Other
Control preference of governed verb	35.08	59.20	5.72

subject in the verbal complement, leading to a different expected prediction of subjecthood for the verb that is under dominance by the control verb. In these instances, we *mask* the control verb, meaning that the language model (and by extension the probing model on top) does not have access to the lexical instantiation that could indicate subject or object control. By asking the prober to make a decision for the verb governed by the masked control verb, we can then quantify whether the model has a preferential bias towards either of the two readings (i.e., measure how many times it selected the subject control interpretation, the object control interpretation, or something different altogether). Table 9 displays these three numbers.

Here we observe a pattern that appears to go against the results in Table 8. Where, in the explicit presence of the control verb, the interpretation of the governed verb was easier to detect in the case of an object control verb, we see that in the masked case, the model typically favours the subject control interpretation.

In other words, the language model shows an inherent bias towards the subject control reading, which gets skewed by the lexical insertion of a concrete verb. Alternatively, if the model is presented a sentence containing a subject control verb, the model would have done a better job identifying subsequent subjecthood predictions if the control verb was masked.

Number of Preceding Flips The above insights guide us towards a potential explanation in the semantic, rather than syntactic, behavior of the governing verb. We note that some verbs *flip* the so-called "understood" subject of the subordinate verbs nested deeper in the clause, with the effect recursing when more than one such verbs take scope (one may imagine the process as a chain of logical negations). In (38), for instance, it is 'hij' that acts as the subject of 'dreigen' and 'laten' but the understood subject of the complement 'vertrekken' is now 'haar'.

hij zal dreigen haar te laten vertrekken
he will threaten her to let leave (38)
'he will threaten to let her leave'

On the data level, we can compute, for each verb-subject dependency, how many subject flips it occurs under, and compute the accuracy accordingly, as presented in Table 10.

The results in the table, as anticipated, show that accuracy drops in the presence of subject flips. This corroborates our previous hypothesis that the semantics of control verbs (and those of the subject flipping verb raisers 'laten' and 'doen') induce linguistically complex cases, that are in turn challenging for the language model to analyze.

Semantic Equivalence, Syntactic Variation Our dataset contains verbs that necessarily induce a cluster or extraposition construction, but also verbs that may induce either. The latter are of particular interest, as they produce drastically differing abstract syntax trees, that in turn get materialized as distinct permutations of the same lexical items, but with identical meanings, i.e., semantic terms; see (39) below.

(a) *hij zal haar*		*proberen*$[IVR2]$	*te willen ontmoeten*			
he will her		try	to want meet			
(b) *hij zal proberen*$[INF2]$		*haar*	*te willen ontmoeten*			(39)
he will try		her	to want meet			
'he will try to want to meet her'						

To examine whether the model exhibits a preference towards either of the two constructions, we identify samples that get assigned identical semantic terms (modulo the IVR2/INF2 distinction), differing only in the word order of their respective surface forms. Across all such pairs, we aggregate accuracy and baseline scores based on the construction type (raising or extraposition), and the position of each inspected verb within the AST (that being the ambiguous verb itself, or a verb occurring above or below it in the tree).

The results in Table 11 solidify our previous evidence that extraposition contexts are overall easier for the model to resolve, regardless of the syntactic position of the verb under scrutiny.

Table 10 Accuracy results by number of preceding flips

Number of preceding flips	0	1	2
Accuracy	86.80	68.47	53.57
Random	40.27	38.46	29.17

Table 11 Accuracy results for raising versus extraposition constructions, with identical semantic terms

	Context in the sentence		
Raising construction	**Above**	**Verb**	**Below**
Accuracy	95.09	86.22	78.15
Random baseline	42.54	41.47	41.44
Extraposition construction			
Accuracy	96.49	93.04	78.50
Random baseline	42.54	41.48	41.44

4 Conclusion

We presented a typelogical modelling of constituency structure and dependency roles by means of a generalization of the non-associative Lambek calculus by adding a multimodal family of unary connectives, each dependency label receiving its own residuated pair of $\diamond, \Box$ operations, plus a unique one for allowing structural movement. In a case study of Dutch verb clusters, we furthermore give an account involving structural rule reasoning, allowing essentially the same abstract meaning to be realized in different surface orders.

To show the practical use of such a modelling, we showed how an ACG-style modelling of verb clusters gives us the kind of patterns that we can then pass to a neural language model to verify how well such a deep learning model can analyze verb-subject dependencies in the presence of patterns of varying complexity. As linguistic theory and our theoretical modelling predicts, in cases of extraposition the verb-subject dependencies are much easier to recognize than in cases of sentence-final verb clustering. This provides empirical support for the kind of modelling proposed here.

In future work we aim to use the kind of modelling we presented as the backbone of a semantic model that may well analyze the kind of task we experimented with, better than the kind of opaque deep learning model used in our current experiment.

References

1. Baillot, P., Mogbil, V.: Soft lambda-calculus: a language for polynomial time computation. In: Walukiewicz, I. (ed.) Foundations of Software Science and Computation Structures. Lecture Notes in Computer Science, vol. 2987, pp. 27–41. Springer (2004). https://doi.org/10.1007/978-3-540-24727-2_4
2. van Benthem, J.: Language in Action: Categories, Lambdas and Dynamic Logic. MIT Press (1995)
3. Blaisdell, E., Kanovich, M.I., Kuznetsov, S.L., Pimentel, E., Scedrov, A.: Non-associative, non-commutative multi-modal linear logic. In: Blanchette, J., Kovács, L., Pattinson, D. (eds.) Automated Reasoning - 11th International Joint Conference, IJCAR 2022, Haifa, Israel, August 8-10, 2022, Proceedings. Lecture Notes in Computer Science, vol. 13385, pp. 449–467. Springer (2022). https://doi.org/10.1007/978-3-031-10769-6_27
4. Boullier, P.: Range concatenation grammars. In: Proceedings of the Sixth International Workshop on Parsing Technologies, IWPT, pp. 53–64 (2000). https://aclanthology.org/2000.iwpt-1.8/
5. Devlin, J., Chang, M.W., Lee, K., Toutanova, K.: BERT: Pre-training of deep bidirectional transformers for language understanding. In: Proceedings of the 2019 Conference of the North American Chapter of the Association for Computational Linguistics: Human Language Technologies. Long and Short Papers, vol. 1, pp. 4171–4186. Association for Computational Linguistics, Minneapolis, Minnesota (2019). https://doi.org/10.18653/v1/N19-1423, https://aclanthology.org/N19-1423
6. Girard, J.: Linear logic. Theor. Comput. Sci. **50**, 1–102 (1987). https://doi.org/10.1016/0304-3975(87)90045-4

7. de Groote, P., Pogodalla, S.: On the expressive power of abstract categorial grammars: representing context-free formalisms. J. Log. Lang. Inf. **13**(4), 421–438 (2004). https://doi.org/10.1007/s10849-004-2114-x
8. Hendriks, H.: The logic of tune – a proof-theoretic analysis of intonation. In: Lecomte, A., Lamarche, F., Perrier, G. (eds.) Logical Aspects of Computational Linguistics, Selected Papers. Lecture Notes in Computer Science, vol. 1582, pp. 132–159. Springer (1997). https://doi.org/10.1007/3-540-48975-4_7
9. Huybregts, R.: The weak inadequacy of context-free phrase structure grammars. In: de Haan, G., Trommelen, M., Zonneveld, W. (eds.) Van Periferie Naar Kern, pp. 81–99. Foris Publications (1984)
10. Kallmeyer, L.: Parsing Beyond Context-Free Grammars. Springer (2010)
11. Kanovich, M.I., Kuznetsov, S.L., Scedrov, A.: The multiplicative-additive Lambek Calculus with subexponential and bracket modalities. J. Log. Lang. Inf. **30**(1), 31–88 (2021). https://doi.org/10.1007/s10849-020-09320-9
12. Kogkalidis, K., Moortgat, M.: Geometry-aware supertagging with heterogeneous dynamic convolutions. CoRR **abs/2203.12235** (2022). https://doi.org/10.48550/arXiv.2203.12235
13. Kogkalidis, K., Moortgat, M., Moot, R.: Æthel: automatically extracted typelogical derivations for Dutch. In: Proceedings of The 12th Language Resources and Evaluation Conference, LREC 2020, Marseille, pp. 5257–5266. European Language Resources Association (2020)
14. Kogkalidis, K., Moortgat, M., Moot, R.: Neural proof nets. In: CoNLL2020, Proceedings of the 24th Conference on Computational Natural Language Learning, pp. 26–40. Association for Computational Linguistics (2020)
15. Kogkalidis, K., Moortgat, M., Moot, R.: Neural proof nets. In: Proceedings of the 24th Conference on Computational Natural Language Learning, pp. 26–40. Association for Computational Linguistics (2020). https://doi.org/10.18653/v1/2020.conll-1.3, https://aclanthology.org/2020.conll-1.3
16. Kogkalidis, K., Wijnholds, G.: Discontinuous constituency and BERT: a case study of Dutch. In: Findings of the Association for Computational Linguistics: ACL 2022, pp. 3776–3785. Association for Computational Linguistics, Dublin, Ireland (2022). https://doi.org/10.18653/v1/2022.findings-acl.298, https://aclanthology.org/2022.findings-acl.298
17. Kurtonina, N., Moortgat, M.: Structural control. In: Blackburn, P., de Rijke, M. (eds.) Specifying Syntactic Structures, pp. 75–113. CSLI, Stanford (1997)
18. Lafont, Y.: Soft linear logic and polynomial time. Theor. Comput. Sci. **318**(1–2), 163–180 (2004). https://doi.org/10.1016/j.tcs.2003.10.018
19. Lambek, J.: The mathematics of sentence structure. Am. Math. Mon. **65**(3), 154–170 (1958)
20. Lambek, J.: On the calculus of syntactic types. Struct. Lang. Math. Asp. **12**, 166–178 (1961)
21. de Marneffe, M.C., Manning, C.D., Nivre, J., Zeman, D.: Universal dependencies. Comput. Linguist. **47**(2), 255–308 (2021). https://doi.org/10.1162/coli_a_00402
22. Montague, R.: Universal grammar. In: Thomason, R.H. (ed.) Formal Philosophy: Selected Papers of Richard Montague, pp. 222–247. Yale University Press, New Haven, London (1974)
23. Moortgat, M.: Constants of grammatical reasoning. In: Bouma, G., Hinrichs, E., Kruijff, G.J.M., Oehrle, R.T. (eds.) Constraints and Resources in Natural Language Syntax and Semantics, pp. 195–219. CSLI Publications (1999)
24. Moortgat, M.: Meaningful patterns. In: Gerbrandy, J., Marx, M., de Rijke, M., Venema, Y. (eds.) JFAK. Assays dedicated to Johan van Benthem on the occasion of his 50th birthday. ILLC, Amsterdam (1999)
25. Moortgat, M., Morrill, G.: Heads and phrases. Type calculus for dependency and constituent structure. Ms Utrecht University (1991)
26. Moortgat, M., Wijnholds, G.: Lexical and derivational meaning in vector-based models of relativisation. CoRR **abs/1711.11513** (2017)
27. Moot, R., Retoré, C.: Natural language semantics and computability. J. Log. Lang. Inf. **28**(2), 287–307 (2019). https://doi.org/10.1007/s10849-019-09290-7
28. Morrill, G.: Type Logical Grammar. Springer (1994). https://doi.org/10.1007/978-94-011-1042-6

29. Morrill, G.: Parsing/theorem-proving for logical grammar CatLog3. J. Log. Lang. Inf. **28**(2), 183–216 (2019)
30. van Noord, G., Bouma, G., Van Eynde, F., de Kok, D., van der Linde, J., Schuurman, I., Sang, E.T.K., Vandeghinste, V.: Large Scale Syntactic Annotation of Written Dutch: Lassy, pp. 147–164. Springer, Berlin, Heidelberg (2013). https://doi.org/10.1007/978-3-642-30910-6_9
31. Pollard, C.: Head grammars, generalized phrase structure grammars, and natural language. Ph.D. thesis, Stanford (1984)
32. Sadrzadeh, M., Muskens, R.: Static and dynamic vector semantics for lambda calculus models of natural language. J. Lang. Model. **6**(2), 319–351 (2018). https://doi.org/10.15398/jlm.v6i2.228
33. Seki, H., Matsumura, T., Fujii, M., Kasami, T.: On multiple context-free grammars. Theor. Comput. Sci. **88**(2), 191–229 (1991). https://doi.org/10.1016/0304-3975(91)90374-B
34. Shieber, S.M.: Evidence against the context-freeness of natural language. Linguist. Philos. **8**, 333–343 (1985). https://doi.org/10.1007/BF00630917
35. de Vries, W., van Cranenburgh, A., Bisazza, A., Caselli, T., van Noord, G., Nissim, M.: BERTje: a Dutch BERT model (2019). arXiv:1912.09582
36. Wansing, H.: Formulas-as-types for a hierarchy of sublogics of intuitionistic propositional logic. In: Pearce, D., Wansing, H. (eds.) Nonclassical Logics and Information Processing, International Workshop, Berlin, Germany, 9–10 November 1990, Proceedings. Lecture Notes in Computer Science, vol. 619, pp. 125–145. Springer (1990). https://doi.org/10.1007/BFb0031928
37. Wijnholds, G., Moortgat, M.: SICK-NL: a dataset for Dutch natural language inference. In: Proceedings of the 16th Conference of the European Chapter of the Association for Computational Linguistics: Main Volume, pp. 1474–1479. Association for Computational Linguistics, Online (2021). https://aclanthology.org/2021.eacl-main.126
38. Wijnholds, G., Sadrzadeh, M.: A type-driven vector semantics for ellipsis with anaphora using Lambek calculus with limited contraction. J. Log. Lang. Inf. **28**(2), 331–358 (2019). https://doi.org/10.1007/s10849-019-09293-4

A Hybrid Approach of Distributional Semantics and Event Semantics for Telicity

Hitomi Yanaka

Abstract The aspectual class of verb phrases plays a key role for tense and aspect semantics like those long studied in formal semantics, determining whether an event is completed and whether its consequences hold at the time. Recently, logic-based inference systems have enabled us to computationally realize the compositional semantics studied in formal semantics. However, since the aspectual class of verb phrases is determined not only by the lexical properties of verbs but also by the context, how to computationally determine the aspectual class is a non-trivial issue. In this paper, we introduce a semantic parsing system to map the aspectual class of English verbs to semantic representations and an inference system that effectively performs inference for aspectual entailment in an effective way. We use Combinatory Categorial Grammar (CCG) as syntax and Neo-Davidsonian event semantics as semantics. We provide an aspectual analyzer to determine the aspectual class of verbs in input sentences and discuss how to handle the ambiguity of telicity.

Keywords Telicity · Event semantics · Distributional semantics · Combinatory Categorial Grammar

1 Introduction

The aspectual class of verb phrases [59] plays a crucial role in tense and aspect semantics, determining whether an event is completed and whether its consequences hold at the time. For example, consider the following two premise–hypothesis pairs.

(1) a. John was drawing
    b. John drew.

H. Yanaka (✉)
The University of Tokyo, 7-3-1 Hongo Bunkyo-ku, Tokyo, Japan
e-mail: hyanaka@is.s.u-tokyo.ac.jp

R. Loukanova et al. (eds.), *Logic and Algorithms in Computational Linguistics 2021 (LACompLing2021)*, Studies in Computational Intelligence 1081,
https://doi.org/10.1007/978-3-031-21780-7_4

(2) a. John was drawing a circle
b. John drew a circle.

The intransitive verb *draw* can be *atelic* (i.e., having no particular endpoint), and (1a) entails (1b). By contrast, the transitive verb *draw* with an objective case can be *telic*. Since it is unclear whether John finished drawing a circle in the situation described in (2a), (2a) does not entail (2b). This problem is known as the *imperfective paradox* [22], in which the telic event does not license entailment from its progressive form to the corresponding non-progressive form.

While tense and aspect have been long studied in formal semantics [22, 31, 33, 48], no implementation of compositional semantics that distinguishes the aspectual class of verbs has been well developed. In fact, in computational semantics, there have been several logic-based inference systems based on formal semantic theories [1, 2, 9, 10, 13, 29, 30, 43, 44] for the Natural Language Inference (NLI) tasks, namely, tasks to judge whether a premise entails a hypothesis. These logic-based approaches use logical formulas for representations of sentence meaning and enable us to computationally realize compositional semantics by providing a semantic lexicon and a pipeline of syntax, semantics, and inference systems. However, these systems do not focus on problems of inferences with the telicity of verb phrases, including the imperfective paradox.

One challenge to handling the imperfective paradox in compositional semantics is that telicity is not lexically determined only by a verb; it also interacts with other words in a sentence. For example, the verb *drank* in (3) is atelic when its object case is uncountable, while the verb in (4) is telic since its object case *two glasses of milk* is countable.

(3) John drank some milk
(4) John drank two glasses of milk
(5) John usually drank two glasses of milk.

Adding the adverb *usually* to (4) makes the verb *drank* in (5) become atelic as a habitual context. As these examples show, the aspectual class of verbs is determined by various syntactic, semantic, and pragmatic factors such as lexical verb type, verb tense, noun phrases of its object case, and adverbials [23, 24]. When we consider lexical entries to represent the aspectual meaning of verbs in logic-based approaches, it is unfeasible to define lexical entries for whole combination patterns.

To classify the aspectual class of verbs, various machine learning approaches based on corpus-based semantics, including distributional semantics [25, 26, 35], have been proposed in computational linguistics. These studies have shown that a classification model using word vector representations based on distributional semantics achieved about from 60 to 85% accuracy in the classification of word-sense level and clause level aspects. Another line of work on distributional semantics has widely developed neural network-based NLI models [17, 20, 49]. These models learn vector representations for sentence meaning from large NLI datasets such as Stanford Natural Language Inference (SNLI) [14] and Multi-Genre Natural Language Inference (MultiNLI) [61], and have achieved high accuracy on these NLI datasets. However, a recent work [36] reported that current neural NLI models fail to consistently perform temporal and aspectual inference. This indicates that while models based on distributional semantics are useful for detecting the aspectual class of verbs, they struggle with obtaining meaning representations and performing inference with aspectuality.

In this paper, we introduce a semantic parsing system to map the aspectual meaning of verbs to semantic representations based on compositional semantics and an inference system that effectively performs inference with aspectuality. Since tense and aspect are complexly related to each other [33], it is necessary to consider the semantics of tense and aspects. With this aim, we map the aspectual class of verbs to semantic representations by integrating Neo-Davidsonian event semantics [47] and distributional semantics. We use Combinatory Categorial Grammar (CCG) [56] syntactic analysis and obtain semantic representations for tense and aspects based on event semantics. To determine the aspectual class of events in input sentences, we propose an aspectual analyzer that extracts a set of logical formulas related to events and classifies their aspectual classes based on distributional semantics. From the results of the CCG syntactic analysis and the aspectual analyzer, we provide semantic representations that account for whether each event includes a culmination point. We demonstrate that our system handles simple inference with the imperfective paradox. The results of the demonstrations described in this paper will be made publicly available to the research community at https://github.com/verypluming/eved.

2 Background and Related Work

2.1 Eventuality

Verb meanings include an eventuality type, namely, the aspectual class of verbs, and there is a long-standing semantic tradition in which properties of eventualities are formally defined and discussed [5, 6, 22, 37, 38, 47, 52, 55, 60, 63]. The aspectual class of verbs is typically categorized into four types [5, 6, 45, 46] according to the time schemata of verbs and verb phrases: activities, achievements, accomplishments, and states. These four verb types are categorized by events and states. For example, a verb like *know* is a stative verb, expressing a state of affairs or being rather than action. By contrast, a verb like *run* is a dynamic, active verb, classifying events such as activities, achievements, and accomplishments.

Telicity (or *actionsart*) refers to the internal temporal property of an eventuality and describes whether it is cumulative (atelic) or not (telic). In other words, telicity is whether an event has a natural endpoint at which the event culminates. The culmination point of the aspectual class of accomplishments comes as the endpoint of a durative and dynamic eventuality. For telic punctual eventualities such as achievements, the culmination point is the moment of transition between two states. Atelic eventualities do not have such a culmination point. States are nondynamic and homogeneous, whereas the internal temporal development of activities is dynamic and homogeneous.

From a quantificational perspective [37, 38, 52, 63], telic events are quantized, while atelic eventualities are non-quantized. There have been other various works for a mathematical formulation designed to represent dynamically changing situations. A small selection of major contributions would include situation calculus for temporal reasoning [51], situation theory [7, 21] with model theory [40, 53], and dependent-type theory [41, 42] for semantics of human languages.

How to lexically distinguish whether an eventuality of verb phrases is telic or atelic, or alternatively, whether a predicate is quantized or cumulative, is a non-trivial issue. For example, both *a cup of milk* in (6) and *a quantity of milk* in (7) are countable NPs, while the former is quantized and the latter is cumulative. Zucchi and White [63] have pointed out that countable NPs are not always quantized. In fact, we can generate many non-quantized complex count nouns (e.g., *a chunk of data* or *a subset of class members*). In addition, the telicity of the event involving the cumulative object *a cup of milk* becomes atelic when the event appears in a habitual context, as in (8).

(6) John drank a quantity of milk
(7) John drank a cup of milk
(8) John drank a cup of milk every day.

These examples show that the telicity often arises from the interplay of lexical, syntactic, semantic, and various contextual factors [23, 24].

2.2 Logic-Based Computational Linguistics

Recently, with further development of formal semantics in linguistic theory and automated parsing and proving in computer science, semantic parsing and inference systems based on compositional semantics, which computationally realize more fine-grained analysis for sentence meaning, have been announced. In what follows, we introduce some previous semantic parsing and inference systems based on compositional semantics.

Bos and Markert [13] use Discourse Representation Structures (DRSs) [33] as semantic representations and provide an inference system based on first-order logic. But while inference systems based on first-order logic are efficient, their expressive power is too limited to cover various temporal and aspectual expressions.

Mineshima et al. [44] and Abzianidze [1, 2] provide inference systems based on CCG syntactic analysis and automated theorem proving in higher-order logic. Although higher-order logic is sufficiently expressive to handle complex expressions involving tense and aspects, inference systems based on higher-order logic tend to rely on hand-coded rules. Bernardy and Chatzikyriakidis [9, 10] provide type-theoretical inference systems using Grammatical Framework and the Coq proof assistant [11] based on higher-order logic, but the theorem-proving components of these type-theoretic systems are not fully automated.

Perhaps the approach that is closest to ours is the typed first-order inference system proposed by Haruta et al. [29, 30], which balances computational efficiency and expressive power to realize a compositional semantics for comparatives and generalized quantifiers in English by combining event semantics and degree semantics. Whereas they focus on analyzing various comparative constructions, we focus on how to computationally treat telicity in a typed first-order inference system.

The compositional semantics approaches aim at assigning a semantic representation to each lexical item in a given phrase and, using the syntax of the language, combining those lexical semantics to create a complete meaning representation. However, these approaches rely on the manual specification of lexical semantics and on precise indications of how to construct meaning representations. To construct meaning representations for the telicity of verb phrases, we should specify lexical items for verb phrases based on all combinations of lexical, syntactic, and semantic factors in an unrealistic way. Our idea is that the specification process could be made simpler by considering the information provided by distributional semantics approaches.

2.3 *Corpus-Based Computational Linguistics*

In distributional semantics, the meaning of a word is based on usage contexts of use in large corpora, which is appropriate for distinguishing the telicity of verb phrases. There have thus been various works in computational linguistics for classifying the lexical aspectual class of verbs based on corpus-based semantics. The early work [34] collected linguistic indicators for lexical aspects (e.g., the presence of in or for-adverbials and the verb tense) from a large corpus. Siegel and McKeown [54] applied different supervised machine learning methods to classify the extracted feature vectors into either states or events, or either telic or atelic events. Friedrich and Palmer [26] also considered a telicity classification model, which uses not only features of linguistic indicators but also features from distributional semantic models. Considering that aspect is part of a verb meaning in Slavic languages, Friedrich and Gateva [25] increased training data by using a cross-lingual annotation projection from parallel English-Czech data and improved the classification of the aspectual class of English verbs.

Along with telicity classification models, there have been datasets of English texts manually annotated with telicity [25–27]. Kober et al. [35] proposed a classification model for categorizing verb phrases as either states or events, or either telic or atelic events by using distributional models. They concluded that distributional models are able to learn representations for word particles and prepositions and that classifiers using composed distributional representations are useful for classifying verb telicity. While these previous studies aimed at modeling a system of classifying the aspectual class of verbs, we apply the classification model to compositional inference systems for aspectual inference.

As another approach to distributional semantics, vector-based sentence representations, including neural network-based NLI models [17, 20, 49], have been widely developed. These models learn vector representations for sentence meaning from large corpora and have achieved high accuracy on NLI tasks [14, 61]. However, a recent work [36] reported that current neural NLI models fail to consistently perform temporal and aspectual inference. This indicates that models based on distributional semantics are useful for classifying the aspectual class of verb phrases, which is determined by diverse contextual factors. However, there remains much room for improvement in learning vector representations for whole sentence meaning and performing inference with aspectuality. We thus take corpus-based approaches for classifying the aspectual class of verb phrases, and take logic-based approaches for representing sentence meaning and performing inference.

3 System Overview

Figure 1 shows an overview of the proposed system for aspectual inference. To compositionally derive semantic representations of sentences, we use ccg2lambda [43, 44], which is a semantic parsing framework based on Combinatorial Categorical Grammar (CCG) [56]. Through syntactic and semantic analyses, we first obtain tentative logical representations of sentence meaning. Note that in the first step, we do not represent whether each event referred to in the sentence includes a culmination point.

Next, we extract subformulas related to events from the logical representations and annotate their aspectual class tags by an aspectual analyzer based on distributional semantics. From the aspectual class tags, we again compositionally derive logical representations again and represent whether the event includes a culmination point. The system outputs *entailment* labels if an automated theorem prover proves entailment relations between a premise and a hypothesis. The system outputs *contradiction* if the negation of the hypothesis can be proved. If both fail, it tries to construct a counter-model and outputs *neutral* if a counter model is found or a timeout occurs. The details of each step are as follows.

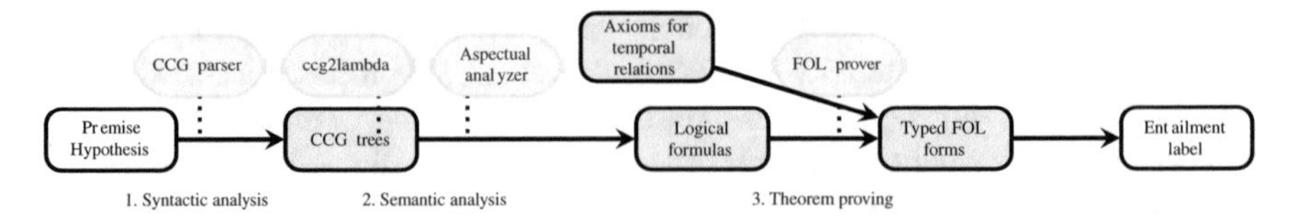

Fig. 1 Overview of the proposed system

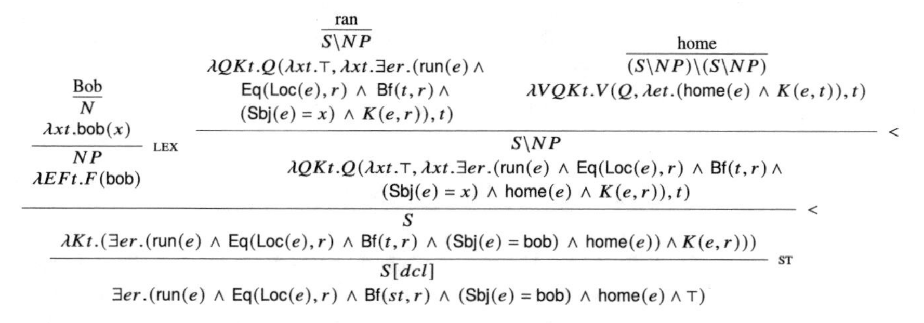

Fig. 2 CCG derivation and semantic composition for *Bob ran home* with lexical entries for tense. $\top$ denotes the tautology, and thus the final formula is equivalent to $\exists er(\mathsf{run}(e) \wedge \mathsf{Equal}(\mathsf{Loc}(e), r) \wedge \mathsf{Before}(st, r) \wedge (\mathsf{Sbj}(e) = \mathsf{bob}) \wedge \mathsf{home}(e))$

3.1 Syntactic Analysis

First, sentences are parsed into syntactic trees based on CCG, which is a syntactic theory suitable for semantic composition from syntactic structures. CCG trees are mapped into semantic representations in a standard way [12], using λ-calculus as an interface between syntax and semantics. We build a lexical template, where a lexical entry for each open word class consists of a syntactic category in CCG and a semantic representation encoded as a λ-term. Another advantage of using CCG as syntax is that there exist robust CCG parsers. To parse sentences, we use the standard CCG syntactic parser and part-of-speech (POS) tagger C&C [16] trained on English CCGbank [32].

In English CCGBank, the basic syntactic categories are N (noun), NP (noun phrase), and S (sentence). Here, a proper noun such as *John* is assigned the category N and shifted to NP by the unary rule LEX (see Fig. 2). Functional categories are described in the form X/Y or $X\backslash Y$, which define functors with an argument Y and a result X, representing meta-variables over syntactic categories. $X\backslash Y$ indicates a function that returns X when combined with Y from its left side, and X/Y is combined with Y from its right side to become X. For example, an intransitive verb has $S\backslash NP$ as its syntactic category, which expects an expression of category NP on its left to become an expression of category S.

3.2 Semantic Analysis

In ccg2lambda [44], a language of standard simple type theory [15] is used as the representation language. The original ccg2lambda uses three basic semantic types: E (Entity), Ev (Event), and Prop (Proposition). To consider tense semantics, we introduce a semantic type for time Time. Thus, the collection of semantic types T of an expression in our system is defined as follows:

Table 1 The mapping from syntactic categories to semantic types. $\rightarrow$ is right-associative

$(NP)^* = (\mathsf{E} \rightarrow \mathsf{Prop}) \rightarrow (\mathsf{E} \rightarrow \mathsf{Prop}) \rightarrow \mathsf{Time} \rightarrow \mathsf{Prop}$
$(S)^* = (\mathsf{Ev} \rightarrow \mathsf{Time} \rightarrow \mathsf{Prop}) \rightarrow \mathsf{Time} \rightarrow \mathsf{Prop}$

$$\mathsf{T} ::= \mathsf{E} \mid \mathsf{Ev} \mid \mathsf{Time} \mid \mathsf{Prop} \mid \mathsf{T} \rightarrow \mathsf{T}$$

where $\mathsf{T} \rightarrow \mathsf{T}$ is a function type. There is a one-to-one mapping between syntactic categories and semantic types: expressions assigned the same category have the same semantic type. The interpretation of every category takes a temporal context as an additional parameter, which serves as a reference for the interpretation of all time-dependent semantics within the phrase. Table 1 defines a mapping $(\cdot)^*$ from syntactic categories to semantic types.

In formal semantics, event semantics [18, 47] is generally used as the framework to account for the semantics of verb phrases. To handle verb tense and aspects, we adopt Neo-Davidsonian event semantics [47], in which every verb is decomposed into a predicate over events and a set of functional expressions relating those events. For example, a sentence containing a transitive verb in (9a), an adverb in (10a), and a quantifier with negation in (11a) would be analyzed as follows:

(9) a. Bob surprises Susan
b. $\exists e(\mathsf{surprise}(e) \wedge (\mathsf{Sbj}(e) = \mathsf{bob}) \wedge (\mathsf{Obj}(e) = \mathsf{susan}))$

(10) a. Bob ran home
b. $\exists e(\mathsf{run}(e) \wedge (\mathsf{Sbj}(e) = \mathsf{bob}) \wedge \mathsf{home}(e)))$

(11) a. No women sang loudly
b. $\neg\exists x(\mathsf{woman}(x) \wedge \exists e(\mathsf{sing}(e) \wedge (\mathsf{Sbj}(e) = x) \wedge \mathsf{loudly}(e)))$.

As examples (10b) and (11b) show, adverbs such as *loudly* and *home* are also represented as predicates over events.[1] Here, the subjective case marker denotes the function Sbj, and the objective case marker denotes the function Obj.

As Parsons discusses, the original Neo-Davidsonian event semantics [47] does not differentiate between temporal instants and intervals, which makes it difficult to consistently analyze the perfect tense. To distinguish between the progressive and perfect in English, we follow Reichenbach's standard theory [50] of interval tense logic, where meanings of verb tenses involve two temporal relations: the relation between reference time r and speech time st and the relation between event time (i.e., location time of an event) $\mathsf{Loc}(e)$ and reference time r. Here, each time constitutes an interval of time. For example, the present progressive is represented as $\mathsf{Equal}(st, r) \wedge \mathsf{During}(\mathsf{Loc}(e), r)$, which indicates that the reference time r is equal to the speech time st during the event time $\mathsf{Loc}(e)$, while the present perfect is represented as $\mathsf{Equal}(st, r) \wedge \mathsf{After}(\mathsf{Loc}(e), r)$, which indicates that the reference time t is after the event time $\mathsf{Loc}(e)$. Here, as described in [19], tense establishes an ordering between the utterance time and the reference time, and aspect orders the reference time and

[1] In Neo-Davidsonian Semantics, the adverb *home* is analyzed as $\mathsf{Target}(e) = \mathsf{home}$. In this paper, we represent $\mathsf{Target}(e) = \mathsf{home}$ as $\mathsf{home}(e)$ for simplification.

Table 2 Basic lexical entries used in semantic analysis for tense. λFxt is an abbreviation for $\lambda F\lambda x\lambda t$. $\top$ denotes tautology. The predicate Eq indicates an abbreviation of the predicate Equal, and the predicate Bf indicates an abbreviation of Before

Example	Category	Logical form
John	N	John
Student	N	$\lambda xt.\mathsf{student}(x)$
Small	N/N	$\lambda Fxt.(\mathsf{small}(x) \wedge F(x))$
Every	NP/N	$\lambda FGHt.\forall x.(Fxt \wedge Gxt \rightarrow Hxt)$
No	NP/N	$\lambda FGHt.\neg\exists x.(Fxt \wedge Gxt \wedge Hxt)$
Run	$S\backslash NP$	$\lambda QKt.Q(\lambda xt.\top, \lambda xt.\exists er.(\mathsf{run}(e) \wedge (\mathsf{Sbj}(e) = x) \wedge K(e, r) \wedge \mathsf{Eq}(\mathsf{Loc}(e), r) \wedge \mathsf{Eq}(t, r)), t)$
Ran	$S\backslash NP$	$\lambda QKt.Q(\lambda xt.\top, \lambda xt.\exists er.(\mathsf{run}(e) \wedge (\mathsf{Sbj}(e) = x) \wedge K(e, r) \wedge \mathsf{Eq}(\mathsf{Loc}(e), r) \wedge \mathsf{Bf}(t, r)), t)$
Drew	$(S\backslash NP)/NP$	$\lambda Q1Q2Kt.Q2(\lambda xt.\top, \lambda xt.Q1(\lambda yt.\top, \lambda yt.\exists er.(\mathsf{draw}(e) \wedge (\mathsf{Sbj}(e) = x) \wedge (\mathsf{Obj}(e) = y) \wedge K(e, r) \wedge \mathsf{Eq}(\mathsf{Loc}(e), r) \wedge \mathsf{Bf}(t, r)), t), t)$

the event time. We treat the event time as a 1-place predicate over events $\mathsf{Loc}(e)$. We represent temporal relations as 2-place predicates over the speech time and the reference time, or 2-place predicates over the event time and the reference time. We introduce six types of predicate types for the temporal relations that are considered in Allen's interval algebra [4]: Before, Equal, Meet, Overlap, Start, Finish, and During.

Table 2 shows the basic lexical entries used in semantic analyses for tense. By using these basic lexical entries for tense, a sentence containing a transitive verb in (12a), an adverb in (13a), and a quantifier with negation in (14a) would be analyzed as follows:

(12) a. Bob surprises Susan
b. $\exists er(\mathsf{surprise}(e) \wedge \mathsf{During}(\mathsf{Loc}(e), r) \wedge \mathsf{Equal}(st, r) \wedge (\mathsf{Sbj}(e) = \mathsf{bob}) \wedge (\mathsf{Obj}(e) = \mathsf{susan}))$

(13) a. Bob ran home
b. $\exists er(\mathsf{run}(e) \wedge \mathsf{Equal}(\mathsf{Loc}(e), r) \wedge \mathsf{Before}(st, r) \wedge (\mathsf{Sbj}(e) = \mathsf{bob}) \wedge \mathsf{home}(e))$

(14) a. No women sang loudly
b. $\neg\exists x(\mathsf{woman}(x) \wedge \exists er(\mathsf{sing}(e) \wedge (\mathsf{Sbj}(e) = x) \wedge \mathsf{loudly}(e) \wedge \mathsf{Equal}(\mathsf{Loc}(e), r) \wedge \mathsf{Before}(st, r)))$.

Figure 2 shows CCG derivation and semantic composition for the sentence (13a). Here, the speech time st is applied as a constant in a root category S of the derivation tree by the unary rule ST from the category S to the category for the declarative sentence $S[dcl]$ [8, 58].

3.3 Aspect Analysis

In aspect analysis, we use an aspectual analyzer to determine the aspectual class of verb phrases in logical representations and construct the meaning of the telicity of verb phrases. The aspectual analyzer has three components: (i) event extraction, (ii) event classification, and (iii) semantic recomposition.

In the first step, we extract a set of predicates related to events from semantic representations. Here, we extract a set of predicates involving event variables (e.g., $\mathsf{draw}(e)$) and a set of predicates that are the accusative case of the events (e.g., $\mathsf{circle}(x)$). Then, we extract surface forms *draw a circle* related to the extracted set of predicates from an original sentence. We verify that event phrases are correctly extracted by checking whether the category of the extracted phrase is $S\backslash NP$. In the second step, we classify the aspectual class of the extracted phrases related to events. Specifically, we classify the aspectual class of the extracted event phrase by a sequence classification model based on distributional semantics. We finetune the pretrained language model BERT [20], which is one of the most standard distributional semantic models, with the English corpora [25–27] annotated with telicity on a two-class classification task (telic or atelic). We use the BERT classification model implemented by using the transformers framework.[2] Using the classification model, we annotate the aspectual class tags telic (TEL) or atelic (ATE) with the event phrase. This is inspired by a previous work [35], which showed that the accuracy when a verb phrase is used as an input is better than the accuracy when a verb or a whole sentence is used.

In the third step, we again compose semantic representations, this time according to the predicted aspectual class. In this step, we represent whether the event includes a culmination point. The original Neo-Davidsonian event semantics account for the distinction of the eventuality by simple 2-place predicates over events, $\mathsf{Cul}(e, t)$ means an event e has a culmination point at time t, and $\mathsf{Hold}(e, t)$ means an event e holds at time t. Zhou [62] showed that we lack an explanatorily satisfactory account of what it means for a temporal phenomenon to be the predicate $\mathsf{Hold}(e, t)$. In this study, we do not use the predicate $\mathsf{Hold}(e, t)$, and the concept of holding an event is just represented as the temporal relation $\mathsf{During}(\mathsf{Loc}(e), r)$ between the event time $\mathsf{Loc}(e)$ and the reference time r. In addition, since the culmination point of an event indicates the end of the event by definition, we simplify the culmination point as being represented as a 1-place predicate over events $\mathsf{Cul}(e)$. Figure 3 shows the CCG derivation and semantic composition for *Bob ran home* via aspect analysis.

Table 3 shows the lexical entries for representing telic and atelic transitive verbs used in the proposed system. CCG categories used in English CCGBank [32] do not distinguish verb tense. Thus, in semantic composition for telicity, we also use POS tagging to supplement the information available from CCG trees and aspectual class tags. The predicate $\mathsf{Cul}(e)$ is introduced only when the POS tag is a verb past tense (VBD) and the aspectual class tag is telic (TEL).

---

[2] https://huggingface.co/transformers.

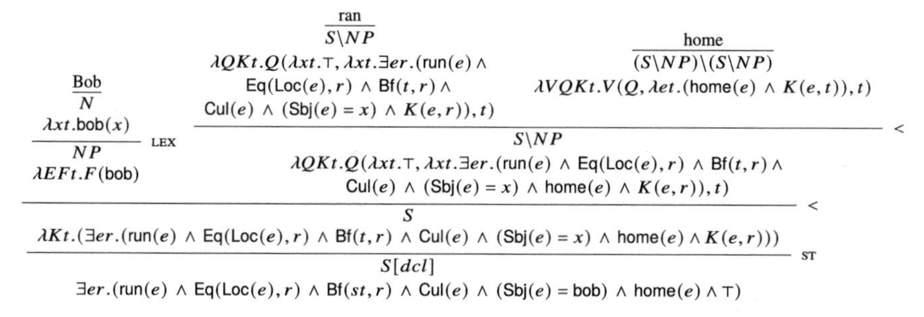

Fig. 3 CCG derivation and semantic composition for *Bob ran home* via aspect analysis

Table 3 Semantic templates for telic and atelic transitive verbs. `cat`, `sem`, `pos`, and `tag` indicate syntactic categories, semantic representations, POS tags, and aspectual class tags, respectively. The predicate Eq indicates an abbreviation of the predicate Equal, and the predicate Bf indicates an abbreviation of Before.

Example	Template

$\text{Drew}_{\texttt{TEL}}$	`cat:` $(S\backslash NP)/NP$
	`sem:`
	$\lambda EQ1Q2Kt.Q2(\lambda xt.\top, \lambda xt.Q1(\lambda yt.\top, \lambda yt.\exists er.(\mathsf{E}(e) \wedge \mathsf{Eq}(\mathsf{Loc}(e), r)$
	$\wedge \mathsf{Bf}(t, r) \wedge \mathsf{Cul}(e) \wedge (\mathsf{Sbj}(e) = x) \wedge (\mathsf{Obj}(e) = y) \wedge \mathsf{K}(e)), t), t)$
	`pos: VBD`
	`tag: TEL`

$\text{Drew}_{\texttt{ATE}}$	`cat:` $(S\backslash NP)/NP$
	`sem:`
	$\lambda EQ1Q2Kt.Q2(\lambda xt.\top, \lambda xt.Q1(\lambda yt.\top, \lambda yt.\exists er.(\mathsf{E}(e) \wedge \mathsf{Eq}(\mathsf{Loc}(e), r)$
	$\wedge \mathsf{Bf}(t, r) \wedge (\mathsf{Sbj}(e) = x) \wedge (\mathsf{Obj}(e) = y) \wedge \mathsf{K}(e)), t), t)$
	`pos: VBD`
	`tag: ATE`

3.4 Theorem Proving

We adopt a typed first-order logic approach [29, 30], which satisfies both computational efficiency and expressive power to perform inference with telicity. We convert semantic representations to typed first-order forms (TFF) in the Thousands of Problems for Theorem Provers (TPTP) format [57]. We try to prove the entailment relation between obtained semantic representations by using Vampire,[3] the most powerful first-order automated theorem prover. Note that although we use λ-calculus for semantic composition, the language of TFF does not allow the use of λ-abstraction. Thus, λ-terms can appear only in the process of semantic composition but not in the resulting logical form. To handle inferences with temporal relations, we use axioms for first-order Theory of Allen's Interval Algebra [4, 28].

[3] https://github.com/vprover/vampire.

4 System Demonstration

In this section, we demonstrate how the proposed system handles the imperfective paradox. As stated in the introduction, consider the two premise–hypothesis pairs (1) and (2).

(1) a. John was drawing
    b. John drew
(2) a. John was drawing a circle
    b. John drew a circle.

First, the semantic representations for (1a) and (1b) via syntactic and semantic analysis are as follows:

(1a') a. John was drawing
    b. $\exists er.(\mathsf{draw}(e) \wedge \mathsf{During}(\mathsf{Loc}(e), r) \wedge \mathsf{Before}(st, r) \wedge (\mathsf{Sbj}(e) = \mathsf{john}))$
(1b') a. John drew
    b. $\exists er.(\mathsf{draw}(e) \wedge \mathsf{Equal}(\mathsf{Loc}(e), r) \wedge \mathsf{Before}(st, r) \wedge (\mathsf{Sbj}(e) = \mathsf{john}))$.

Here, we extract the predicate involving event variables $\mathsf{draw}(e)$ from semantic representations of (1a') and (1b').

Next, we determine whether the extracted event subformula $\mathsf{draw}(e)$ is telic or atelic by using a classification model based on distributional semantics. Here, the subformula can be annotated with an aspectual class tag `ATE`. According to the aspectual class tag and the CCG tree, we derive semantic representations again. In semantic representations for (1a") and (1b") involving the atelic event $\mathsf{draw}(e)$, the predicate $\mathsf{Cul}(e)$ does not appear, so our system correctly predicts entailment relations between (1a) and (1b) with the temporal axiom $\forall er.((\mathsf{During}(\mathsf{Loc}(e), r) \wedge \mathsf{Before}(st, r)) \rightarrow (\mathsf{Equal}(\mathsf{Loc}(e), r) \wedge \mathsf{Before}(st, r)))$.

(1a") a. John was drawing
    b. $\exists er.(\mathsf{draw}(e) \wedge \mathsf{During}(\mathsf{Loc}(e), r) \wedge \mathsf{Before}(st, r) \wedge (\mathsf{Sbj}(e) = \mathsf{john}))$
(1b") a. John drew
    b. $\exists er.(\mathsf{draw}(e) \wedge \mathsf{Equal}(\mathsf{Loc}(e), r) \wedge \mathsf{Before}(st, r) \wedge (\mathsf{Sbj}(e) = \mathsf{john}))$.

On the other hand, the semantic representations for (2a) and (2b) via syntactic and semantic analyses are as follows:

(2a') a. John was drawing a circle
    b. $\exists y.(\mathsf{circle}(y) \wedge \exists er.(\mathsf{draw}(e) \wedge \mathsf{During}(\mathsf{Loc}(e), r) \wedge \mathsf{Before}(st, r) \wedge (\mathsf{Sbj}(e) = \mathsf{john}) \wedge (\mathsf{Obj}(e) = y)))$
(2b') a. John drew a circle
    b. $\exists y.(\mathsf{circle}(y) \wedge \exists er.(\mathsf{draw}(e) \wedge \mathsf{Equal}(\mathsf{Loc}(e), r) \wedge \mathsf{Before}(st, r) \wedge (\mathsf{Sbj}(e) = \mathsf{john}) \wedge (\mathsf{Obj}(e) = y)))$.

Similarly, we extract a set of predicates involving event variables from semantic representations of (2a') and (2b'). In (2a') and (2b'), the set of predicates are $\mathsf{draw}(e)$ and its accusative case $\mathsf{circle}(x)$. We determine the telicity of the set of predicates

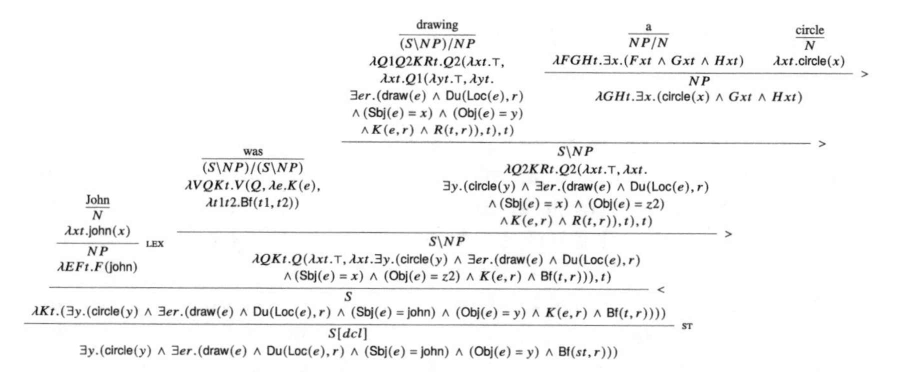

Fig. 4 CCG derivation and semantic composition for *John was drawing a circle*. The predicate Du indicates an abbreviation of the predicate During, and the predicate Bf indicates an abbreviation of Before

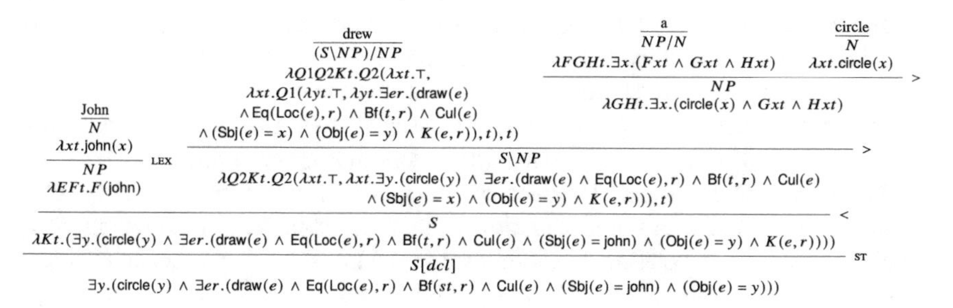

Fig. 5 CCG derivation and semantic composition for *John drew a circle*. The predicate Eq indicates the abbreviation of the predicate Equal

by the classification model and use them to annotate aspectual class tags `TEL` with them. According to the lexical templates with the aspectual class tags and CCG syntactic analysis, we can obtain the semantic representations of (2a") and (2b"). Figures 4 and 5 show semantic compositions of (2a") and (2b"), respectively. Here, the predicate $\mathsf{Cul}(e)$ is introduced when a verb tense is a simple past and the aspectual class tag is telic (`TEL`). The demonstration shows that our proposed system blocks the inference from (2a") and (2b") by the existence of the predicate $\mathsf{Cul}(e)$.

(2a") a. John was drawing a circle
b. $\exists y.(\mathsf{circle}(y) \wedge \exists er.(\mathsf{draw}(e) \wedge \mathsf{During}(\mathsf{Loc}(e), r) \wedge \mathsf{Before}(st, r) \wedge (\mathsf{Sbj}(e) = \mathsf{john}) \wedge (\mathsf{Obj}(e) = y)))$

(2b") a. John drew a circle
b. $\exists y.(\mathsf{circle}(y) \wedge \exists er.(\mathsf{draw}(e) \wedge \mathsf{Equal}(\mathsf{Loc}(e), r) \wedge \mathsf{Before}(st, r) \wedge \mathsf{Cul}(e) \wedge (\mathsf{Sbj}(e) = \mathsf{john}) \wedge (\mathsf{Obj}(e) = y)))$.

5 Discussion and Conclusion

In this paper, we raised the issue of how to computationally handle NLI with telicity, a challenging linguistic phenomenon related to lexical, syntactic, semantic, and various contextual factors, where it is unrealistic to provide lexical entries for whole patterns. To provide computational semantics for representing the aspectual meaning of verbs and an inference system that performs aspectual inference, we introduced a hybrid approach of distributional semantics and event semantics. We represented the telicity of verb phrases by an aspectual analyzer that combines event semantics for logical representations of sentence meaning and distributional semantics for classifying the telicity of event subformulas. A demonstration showed that our proposed system correctly performed inference with the imperfective paradox.

We next address some possible criticisms arising from our proposed system. One is that a classification model based on distributional semantics does not completely predict correct labels for telicity, but simply estimates the posterior probability of the telicity of verb phrases. In addition, the claim of this study is not that theorem proving deterministically proves the entailment relation once the aspectual class of verbs is determined. We argue that telicity is a challenging linguistic phenomenon involving lexical, syntactic, and semantic factors that make it difficult to exhaustively specify lexical items. The idea of this study is to complement the formal semantics approach with the distributional semantics approach to computationally handle semantic composition and inference involving such phenomena. This idea was inspired by universal semantic tags [3, 39], which give information better characterizing lexical semantics for word tokens than do POS tags, for which telicity is beyond the scope. We can interpret a classifier for the aspectual class of verb phrases based on distributional semantics as giving a prior distribution for the results of inference, which might mimic humans performing inference from their own observations and past experience. Such studies are still in their infancy, but we can also subdivide classifiers based on distributional semantics in a more fine-grained way according to whether an object noun phrase is quantized or not from a quantificational perspective, as introduced in Sect. 2.1. This task is left for future work. We believe that distributional semantics is an effective approach to semantic disambiguation based on language use.

In future work, we will extend the proposed approach to cover various temporal and aspectual inferences involving temporal adverbials, habitual aspects, and states where events are negated. We plan to create a temporal and aspectual inference test set and to evaluate the proposed system on that test set. We will also consider an extension of the previous corpora annotated with telicity to improve the performance of the aspectual analyzer.

Acknowledgements We thank the four anonymous reviewers for their helpful comments and suggestions. We are also grateful to Daisuke Bekki, Koji Mineshima, and Roussanka Loukanova for their many comments on this paper. We also thank Takuto Asakura for the technical advice on the LaTeX submission. This work was supported by PRESTO, JST Grant Number JPMJPR21C8, Japan.

References

1. Abzianidze, L.: A tableau prover for natural logic and language. In: Proceedings of the 2015 Conference on Empirical Methods in Natural Language Processing, pp. 2492–2502 (2015). https://doi.org/10.18653/v1/D15-1296
2. Abzianidze, L.: Natural solution to FraCaS entailment problems. In: Proceedings of the Fifth Joint Conference on Lexical and Computational Semantics, pp. 64–74 (2016). https://doi.org/10.18653/v1/S16-2007
3. Abzianidze, L., Bos, J.: Towards universal semantic tagging. In: IWCS 2017—12th International Conference on Computational Semantics—Short papers (2017)
4. Allen, J.F.: Maintaining knowledge about temporal intervals. Commun. ACM (1983). https://doi.org/10.1145/182.358434
5. Bach, E.: On time, tense, and aspect: An essay in English metaphysics. Linguistics Department Faculty Publication Series (1981)
6. Bach, E.: The Algebra of Events, chap. 13, pp. 324–333. John Wiley & Sons, Ltd (2002). https://doi.org/10.1002/9780470758335.ch13
7. Barwise, J., Perry, J.: Situations and Attitudes. MIT Press (1983). https://doi.org/10.2307/2219775
8. Bekki, D.: A Formal Theory of Japanese Grammar: The Conjugation System, Syntactic Structures, and Semantic Composition. Kuroshio. (In Japanese) (2010)
9. Bernardy, J.P., Chatzikyriakidis, S.: A type-theoretical system for the FraCaS test suite: Grammatical framework meets coq. In: IWCS 2017—12th International Conference on Computational Semantics—Long papers (2017)
10. Bernardy, J.P., Chatzikyriakidis, S.: Applied temporal analysis: A complete run of the FraCaS test suite. In: Proceedings of the 14th International Conference on Computational Semantics (IWCS), pp. 11–20 (2021)
11. Bertot, Y., Castéran, P., Huet, G., Paulin-Mohring, C.: Interactive Theorem Proving and Program Development. Coq'Art: The Calculus of Inductive Constructions. Springer (2004). https://doi.org/10.1007/978-3-662-07964-5
12. Bos, J.: Wide-coverage semantic analysis with Boxer. In: Semantics in Text Processing. STEP 2008 Conference Proceedings, pp. 277–286 (2008)
13. Bos, J., Markert, K.: Recognising textual entailment with logical inference. In: Proceedings of Human Language Technology Conference and Conference on Empirical Methods in Natural Language Processing, pp. 628–635 (2005). https://doi.org/10.3115/1220575.1220654
14. Bowman, S.R., Angeli, G., Potts, C., Manning, C.D.: A large annotated corpus for learning natural language inference. In: Proceedings of the 2015 Conference on Empirical Methods in Natural Language Processing, pp. 632–642 (2015). https://doi.org/10.18653/v1/D15-1075
15. Carpenter, B.: Type-Logical Semantics. MIT Press (1997)
16. Clark, S., Curran, J.R.: Wide-coverage efficient statistical parsing with CCG and log-linear models. Comput. Linguist. **33**(4), 493–552 (2007). https://doi.org/10.1162/coli.2007.33.4.493
17. Conneau, A., Kiela, D., Schwenk, H., Barrault, L., Bordes, A.: Supervised learning of universal sentence representations from natural language inference data. In: Proceedings of the 2017 Conference on Empirical Methods in Natural Language Processing, pp. 670–680 (2017). https://doi.org/10.18653/v1/D17-1070
18. Davidson, D.: Truth and meaning. Synthese **17**(1), 304–323 (1967). https://doi.org/10.1007/BF00485035
19. Demirdache, H., Uribe-Etxebarria, M.: Aspect and temporal anaphora. Nat. Lang. Linguist. Theory **32**(3), 855–895 (2014)
20. Devlin, J., Chang, M.W., Lee, K., Toutanova, K.: BERT: Pre-training of deep bidirectional transformers for language understanding. In: Proceedings of the 2019 Conference of the North American Chapter of the Association for Computational Linguistics: Human Language Technologies, Volume 1 (Long and Short Papers), pp. 4171–4186 (2019). https://doi.org/10.18653/v1/N19-1423

21. Devlin, K.: Situation theory and situation semantics. In: Logic and the Modalities in the Twentieth Century, Handbook of the History of Logic, vol. 7, pp. 601–664. North-Holland (2006). https://doi.org/10.1016/S1874-5857(06)80034-8
22. Dowty, D.: Word Meaning and Montague Grammar. Reidel (1979). https://doi.org/10.1007/978-94-009-9473-7
23. Filip, H.: The telicity parameter revisited. In: Proceedings from Semantics and Linguistic Theory (SALT) XIV, pp. 92–109 (2005). https://doi.org/10.3765/salt.v14i0.2909
24. Filip, H.: Events and Maximalization: The Case of Telicity and Perfectivity, chap. 7. John Benjamins Publishing (2008)
25. Friedrich, A., Gateva, D.: Classification of telicity using cross-linguistic annotation projection. In: Proceedings of the 2017 Conference on Empirical Methods in Natural Language Processing, pp. 2559–2565 (2017). https://doi.org/10.18653/v1/D17-1271
26. Friedrich, A., Palmer, A.: Automatic prediction of aspectual class of verbs in context. In: Proceedings of the 52nd Annual Meeting of the Association for Computational Linguistics (Volume 2: Short Papers), pp. 517–523 (2014). https://doi.org/10.3115/v1/P14-2085
27. Friedrich, A., Palmer, A., Pinkal, M.: Situation entity types: automatic classification of clause-level aspect. In: Proceedings of the 54th Annual Meeting of the Association for Computational Linguistics (Volume 1: Long Papers), pp. 1757–1768 (2016). https://doi.org/10.18653/v1/P16-1166
28. Grüninger, M., Li, Z.: The time ontology of allen's interval algebra. In: 24th International Symposium on Temporal Representation and Reasoning (TIME 2017), Leibniz International Proceedings in Informatics (LIPIcs), vol. 90, pp. 16:1–16:16. Dagstuhl, Germany (2017). https://doi.org/10.4230/LIPIcs.TIME.2017.16
29. Haruta, I., Mineshima, K., Bekki, D.: Combining event semantics and degree semantics for natural language inference. In: Proceedings of the 28th International Conference on Computational Linguistics, pp. 1758–1764 (2020). https://doi.org/10.18653/v1/2020.coling-main.156
30. Haruta, I., Mineshima, K., Bekki, D.: Logical inferences with comparatives and generalized quantifiers. In: Proceedings of the 58th Annual Meeting of the Association for Computational Linguistics: Student Research Workshop, pp. 263–270 (2020). https://doi.org/10.18653/v1/2020.acl-srw.35
31. Higginbotham, J.: Tense, Aspect, and Indexicality. Oxford University Press, Oxford (2010). https://doi.org/10.1093/acprof:oso/9780199239313.001.0001
32. Hockenmaier, J., Steedman, M.: CCGbank: a corpus of CCG derivations and dependency structures extracted from the Penn Treebank. Comput. Linguist. **33**(3), 355–396 (2007). https://doi.org/10.1162/coli.2007.33.3.355
33. Kamp, H., Reyle, U.: From Discourse to Logic: Introduction to Model-theoretic Semantics of Natural Language, Formal Logic and Discourse Representation Theory, Studies in Linguistics and Philosophy, vol. 42. Springer (1993). https://doi.org/10.1007/978-94-017-1616-1
34. Klavans, J.L., Chodorow, M.: Degrees of stativity: the lexical representation of verb aspect. In: COLING 1992 Volume 4: The 14th International Conference on Computational Linguistics (1992). https://doi.org/10.3115/992424.992443
35. Kober, T., Alikhani, M., Stone, M., Steedman, M.: Aspectuality across genre: A distributional semantics approach. In: Proceedings of the 28th International Conference on Computational Linguistics, pp. 4546–4562 (2020). https://doi.org/10.18653/v1/2020.coling-main.401
36. Kober, T., Bijl de Vroe, S., Steedman, M.: Temporal and aspectual entailment. In: Proceedings of the 13th International Conference on Computational Semantics - Long Papers, pp. 103–119 (2019). http://dx.doi.org/10.18653/v1/W19-0409
37. Krifka, M.: Nominal reference, temporal constitution and quantification in event semantics. In: R. Bartsch, J.F.A.K. van Benthem, P. van Emde Boas (eds.) Semantics and Contextual Expression, pp. 75–115. Foris Publications (1989). https://doi.org/10.1515/9783110877335-005
38. Krifka, M.: The Origins of Telicity, pp. 197–235. Springer Netherlands (1998). https://doi.org/10.1007/978-94-011-3969-4_9

39. Li, W., Hou, Y., Ye, Y., Liang, L., Sun, W.: Universal semantic tagging for English and Mandarin Chinese. In: Proceedings of the 2021 Conference of the North American Chapter of the Association for Computational Linguistics: Human Language Technologies, pp. 5554–5566. Association for Computational Linguistics (2021). https://doi.org/10.18653/v1/2021.naacl-main.440
40. Loukanova, R.: Situation Theory, Situated Information, and Situated Agents, pp. 145–170. Springer Berlin Heidelberg (2014). https://doi.org/10.1007/978-3-662-44994-3_8
41. Loukanova, R.: Typed theory of situated information and its application to syntax-semantics of human language. In: H. Christiansen, M.D. Jiménez-López, R. Loukanova, L.S. Moss (eds.) Partiality and underspecification in information, languages, and knowledge, pp. 151–188. Cambridge Scholars Publishing (2017). https://www.cambridgescholars.com/product/978-1-4438-7947-7
42. Loukanova, R.: Formalisation of situated dependent-type theory with underspecified assessments. In: E. Bucciarelli, S.H. Chen, J.M. Corchado (eds.) Decision Economics. Designs, Models, and Techniques for Boundedly Rational Decisions, pp. 49–56. Springer International Publishing (2019). https://doi.org/10.1007/978-3-319-99698-1_6
43. Martínez-Gómez, P., Mineshima, K., Miyao, Y., Bekki, D.: ccg2lambda: A compositional semantics system. In: Proceedings of ACL-2016 System Demonstrations, pp. 85–90 (2016). https://doi.org/10.18653/v1/P16-4015
44. Mineshima, K., Martínez-Gómez, P., Miyao, Y., Bekki, D.: Higher-order logical inference with compositional semantics. In: Proceedings of the 2015 Conference on Empirical Methods in Natural Language Processing, pp. 2055–2061 (2015). https://doi.org/10.18653/v1/D15-1244
45. Moens, M., Steedman, M.: Temporal ontology and temporal reference. Comput. Linguist. **14**(2), 15–28 (1988). https://doi.org/10.5555/55056.55058
46. Mourelatos, A.P.D.: Events, processes, and states. Linguist. Philos. **2**, 415–434 (1978). https://doi.org/10.1007/BF00149015
47. Parsons, T.: Events in The Semantics of English: a Study in Subatomic Semantics. MIT Press, Cambridge, USA (1990)
48. Prior, A., Hasle, P., Øhrstrøm, P., Copeland, J., Braüner, T.: Papers on Time and Tense. Oxford University Press (2003)
49. Raffel, C., Shazeer, N., Roberts, A., Lee, K., Narang, S., Matena, M., Zhou, Y., Li, W., Liu, P.J.: Exploring the limits of transfer learning with a unified Text-to-Text Transformer. J. Mach. Learn. Res. **21**(140), 1–67 (2020)
50. Reichenbach, H.: The Direction of Time. Dover Publications (1956). https://doi.org/10.2307/2216858
51. Reiter, R.: Knowledge in Action: Logical Foundations for Specifying and Implementing Dynamical Systems. MIT Press (2001)
52. Rothstein, S.: Structuring Events: A Study in the Semantics of Aspect. Explorations in Semantics. Wiley (2004). https://onlinelibrary.wiley.com/doi/book/10.1002/9780470759127
53. Seligman, J., Moss, L.S.: Situation theory. In: Handbook of logic and language, pp. 239–309. North-Holland (1997)
54. Siegel, E.V., McKeown, K.R.: Learning methods to combine linguistic indicators: improving aspectual classification and revealing linguistic insights. Comput. Linguist. **26**(4), 595–627 (2000). https://doi.org/10.1162/089120100750105957
55. Smith, C.S.: The Parameter of Aspect. Studies in Linguistics and Philosophy. Springer Netherlands (1991). https://doi.org/10.1007/978-94-011-5606-6
56. Steedman, M.: The Syntactic Process. MIT Press (2000). https://doi.org/10.7551/mitpress/6591.001.0001
57. Sutcliffe, G.: The TPTP problem library and associated infrastructure. J. Autom. Reasoning **59**(4), 483–502 (2017). https://doi.org/10.1007/s10817-017-9407-7
58. Utsugi, M.: Analysis of japanese tense in dependent type semantics. Master's thesis, Ochanomizu University (2017)
59. Vendler, Z.: Verbs and times. The Philos. Rev. **66**(2), 143–160 (1957). https://doi.org/10.2307/2182371

60. Verkuyl, H.J.: On the Compositional Nature of the Aspects. Dordrecht, Netherlands: D.Reidel Publishing Company (1972). https://doi.org/10.1007/978-94-017-2478-4
61. Williams, A., Nangia, N., Bowman, S.: A broad-coverage challenge corpus for sentence understanding through inference. In: Proceedings of the 2018 Conference of the North American Chapter of the Association for Computational Linguistics: Human Language Technologies, vol. 1 (Long Papers), pp. 1112–1122 (2018). https://doi.org/10.18653/v1/N18-1101
62. Zhou, Z.: Neo-Davidsonian ontology of events. Linguist. Philos. **44**(1), 1–41 (2019). https://doi.org/10.1007/s10988-019-09292-5
63. Zucchi, S., White, M.: Twigs, sequences and the temporal constitution of predicates. Linguist. Philos. **24**(2), 223–270 (2001). https://doi.org/10.3765/salt.v6i0.2774

Generalized Computable Models and Montague Semantics

Artem Burnistov and Alexey Stukachev

Abstract We consider algorithmic properties of mathematical models which are used in computational linguistics to formalize and represent the semantics of natural language sentences. For example, finite-order functionals play a crucial role in Montague intensional logic and formal semantics for natural languages. We discuss some computable models for the spaces of finite-order functionals based on the Ershov-Scott theory of domains and approximation spaces. As another example, in the analysis of temporal aspects of verbs the scale of time is usually identified with the ordered set of real numbers or just a dense linear order. There are many results in generalized computability about such structures, and some of them can be applied in this analysis.

Keywords Montague semantics · Intensional logic · Functionals of finite types · Generalized computability · Σ-predicates

1 Introduction

This paper continues the work started in [3, 16, 17] on algorithmic issues of formal semantics for natural languages. Montague in [10, 11] proposed a model-theoretic approach to semantics of English known as Montague intensional logic. It is a typed

A. Burnistov · A. Stukachev (✉)
Novosibirsk State University, 1 Pirogova str., Novosibirsk 630090, Russia
e-mail: aistu@math.nsc.ru

A. Burnistov
e-mail: a.burnistov@g.nsu.ru

A. Burnistov
Mines Paris, PSL University, 60 bd Saint-Michel, Paris 75006, France

A. Stukachev
Sobolev Institute of Mathematics, 4 Acad.Koptyug avenue, Novosibirsk 630090, Russia

Novosibirsk State Technical University, 20 Prospekt K. Marksa, Novosibirsk 630073, Russia

R. Loukanova et al. (eds.), *Logic and Algorithms in Computational Linguistics 2021 (LACompLing2021)*, Studies in Computational Intelligence 1081,
https://doi.org/10.1007/978-3-031-21780-7_5

higher-order logic which uses finite types and finite-order functionals to formalize grammar categories of natural languages (in particular, English).

Neither Montague nor other researchers (to our knowledge) studied complexity issues and algorithmic aspects of objects and constructions of this theory. A natural question is to construct computable or effective (in some sense) presentations of rather complicated structures considered in intensional logic. Such presentations allow to regard the meaning of a natural language sentence as an algorithm of checking its relevance (truth value) to a given circumstances (knowledge base).

Our approach is based on the framework of generalized computability formalised via Σ-definability in admissible sets or superstructures developed by Barwise [1], Moschovakis [12], Ershov [8], and also by Montague [9]. To handle finite-order functionals in an effective way we use Ershov-Scott theory of approximation spaces and domains, see [6–8, 14]. As a useful benefit, such approach allows to consider the case when the basic entities can be approached only via approximations. Also, it becomes possible to study spaces of truth values more complicated than just 0 ("no") and 1 ("yes").

Intensionality in models discussed in this paper is limited only to the scale of time as the set of possible worlds, we do not consider modality issues for simplicity reasons. Time in linguistics is usually represented by the ordered set of real numbers (denoted here as $\mathbb{R}$). This model is sufficient to describe formally (and hence analyse effectively) such important features of verbs as tense and aspect, see [2]. There are some properties of dense linear orders (e.g., elimination of quantifiers and decidability), which are well-known for logicians and which could be useful in the analysis of algorithmic properties of interval semantics for verbs in natural languages. The examples of results of this kind can be found in [16, 17].

In this paper, we consider two different kinds of models of entity spaces, namely, rank models and vector models. Vector models are more natural from the approximation-space point of view, and rank models are more natural from the set-theoretical point of view. We prove that type hierarchies based on these classes of models are effectively equivalent. This result is obtained for two different spaces of truth values.

The authors are grateful for anonymous referees for valuable remarks and suggestions. A comparison of our approach with the existing ones will be discussed in a series of forthcoming publications. Here we just present examples of computable models of finite-order functionals and approximation spaces not described before and relevant to linguistics. We believe our approach is closely connected to the problem of understanding the meaning of a natural language sentence via rigorous mathematical formalization. In particular, since we consider effective (in some general sense) models of Montague intensional logic, we plan to describe how exactly our research is connected to the approach by Y. N. Moschovakis named "meaning as an algorithm" [13].

2 Basic Notions of Generalized Computability

Hereditary finite superstructures are the "simplest" examples of models of theory KPU proposed by S.Kripke, R.Platek, J.Barwise, and Yu.L.Ershov for studying generalized computability via Σ-definability in admissible sets (see [1, 8, 15]).

By ω we denote the set of natural numbers. For arbitrary set M, we construct the set $HF(M)$ of hereditarily finite sets over M as follows:

$HF_0(M) = \varnothing$
$HF_{n+1}(M) = \mathcal{P}_\omega(M \cup HF_n(M)), n < \omega$
(here $\mathcal{P}_\omega(X)$ is the set of all finite subsets of X)
$HF(M) = \bigcup_{n<\omega} HF_n(M)$

If $\mathfrak{M}$ is a structure of some relational signature σ then one can define on $M \cup HF(M)$ a structure $\mathbb{HF}(\mathfrak{M})$ of signature $\sigma' = \sigma \cup \{U, \varnothing, \in\}$ (U, $\varnothing$, and $\in$ are some symbols not in σ) with the following interpretation of signature symbols:

$U^{\mathbb{HF}(\mathfrak{M})} = M$
$P^{\mathbb{HF}(\mathfrak{M})} = P^{\mathfrak{M}}, P \in \sigma$
$\varnothing^{\mathbb{HF}(\mathfrak{M})} = \varnothing \in HF_0(M)$
$\in^{\mathbb{HF}(\mathfrak{M})} = \in \cap ((M \cup HF(M)) \times HF(M))$

A class of Δ_0-*formulas* of signature σ' is the least one containing atomic formulas which is closed under $\vee$, $\wedge$, $\rightarrow$, $\neg$, and bounded quantifiers $\forall x \in y$ and $\exists x \in y$ ($\forall x \in y\, \varphi$ and $\exists x \in y\, \varphi$ are abbreviations for $\forall x (x \in y \rightarrow \varphi)$ and $\exists x (x \in y \wedge \varphi)$, respectively).

A class of Σ-*formulas* of signature σ' is the least one containing Δ_0-formulas and closed under $\vee$, $\wedge$, bounded quantifiers $\forall x \in y$, $\exists x \in y$, and $\exists x$. As usual, a set is called Σ-definable if it is definable by some Σ-formula with parameters, and Δ-definable if it and its complement are Σ-definable.

3 Montague Intensional Logic

Let e, t, and s be the some fixed symbols used, correspondingly, as names for basic types of entities and truth values, and for marking an intensional shift, i.e., relativization to a state or situation.

Definition 1 The set $Types_{IL}$ is defined as follows:

- $t \in Types_{IL}$, $e \in Types_{IL}$
- if $a \in Types_{IL}$ and $b \in Types_{IL}$ then $(a \rightarrow b) \in Types_{IL}$
- if $a \in Types_{IL}$ then $(s \rightarrow a) \in Types_{IL}$

The language of intensional logic IL (see [4, 5, 10, 11]) contains countably many constants of any type $a \in Types_{IL}$ and countably many variables of each type $a \in Types_{IL}$.

Table 1 Categories and types of some expressions

Category	Grammar equivalent	Corresponding type	Basic expressions
e	No	e	No
t	Sentences	t	No
IV	Intransitive verbs	(e → t)	Walk, talk
CN	Common nouns	(e → t)	Man, woman
TV	Extensional transitive verbs	(e → (e → t))	Love, find
CN/CN	Extensional adjectives	((e → t) → (e → t))	Tall, young
CN/CN	Extensional adverbs	((e → t) → (e → t))	Rapidly, slowly
T	Noun phrases and proper names	((s → (e → t)) → t)	John, ninety, he
t/t	Sentence determinants	((s → t) → t)	Necessarily, possibly
IV/t	Connective verbs	((s → t) → (e → t))	Believe, assert

A model of intensional logic IL is a quadruple $\langle A, W, T, \leq, F\rangle$ such that A, W, T are nonempty sets, $\leq$ is a linear order on T, F is a function defined on the set of constants of IL as described below. Sets W and T correspond to the sets of possible worlds and time moments correspondingly.

Definition 2 The set D_τ of possible denotations of type $\tau \in Types_{IL}$ is defined by induction on complexity of τ:

- $D_e = A$, $D_t = \{0, 1\}$
- $D_{(a\to b)} = D_b^{D_a}$ (the set of functions from D_a to D_b)
- $D_{(s\to a)} = D_a^{W\times T}$ (the set of functions from $W \times T$ to D_a)

We denote by S_a the set $D_{(s\to a)}$. Function F defines for each constant of type a some element from S_a which is called its *intension*. Elements from D_a are called *extensions* of type a.

Finite types are used to represent grammar categories (parts of speech) of natural languages. Some correspondences between categories and types are listed in Table 1.

For example, proper names correspond to the type $((s \to (e \to t)) \to t)$ – the set of properties true for the individual with this name. Here we do not consider one of the most complex cases, intensional transitive verbs with the type $((s \to ((s \to (e \to t)) \to t)) \to (e \to t)))$.

Extension (the set of denotations) of type a is the set of possible values of the grammar category interpreted by type a in a model of intensional logic. Correspondingly, intension of type a is a function from $W \times T$ to the extension of type a.

4 Ershov-Scott Functional Spaces

To construct an effective model of Montague intensional logic we apply the domain theory proposed by Scott [14] and the theory of functional spaces of finite types proposed by Ershov [6–8]. The definitions below are from [8].

Let $\mathbb{A}$ be a model of KPU (see [8]). If $a \in A$ then $p_l^* a = \{b \mid \exists c(\langle b, c\rangle \in a)\}$, $p_r^* a = \{b \mid \exists c(\langle c, b\rangle \in a)\}$. If $B \subseteq A$ then $B^* = \{b \mid b \subseteq B and b \in A\}$.

The notion of effectively presented functional space is based on the general

Definition 3 Quadruple $\mathcal{B} = \langle B, \leq, Cons, \sqcup\rangle$ is called an *f-base on* $\mathbb{A}$ (see [6, 7]) if the following holds:

(1) B is a Δ-definable subset of $\mathbb{A}$
(2) $\leq$ is a Δ-definable preorder on B;
let $[B]$ be the quotient of set B by the equivalence relation $\equiv$ defined by the preorder $\leq$ $(b_0 \equiv b_1 \Leftrightarrow b_0 \leq b_1 and b_1 \leq b_0)$; as usual, $[b]$ denotes the element of $[B]$ which is the equivalence class of $b \in B$; if $C \subseteq B$ then $[C] = \{[b] \mid b \in C\}$; we also use $\leq$ to denote the order induced on $[B]$ by the original preorder $\leq$
(3) $Cons$ is a Δ-definable subset of $B^* \setminus \{\emptyset\}$, and for any $b_* \in B^*$ holds

$$b_* \in Cons \Leftrightarrow (\exists b \in B)(\forall b' \in b_*)(b' \leq b)$$

(4) $\sqcup : Cons \to B$ is a Σ-definable function such that $[\sqcup b_*]$ for any $b_* \in Cons$ is the least upper bound of $[b_*] \subseteq [B]$ in $\langle [B], \leq\rangle$

Definition 4 Let $\mathcal{B}_0 = \langle B_0, \leq_0, Cons_0, \sqcup_0\rangle$ and $\mathcal{B}_1 = \langle B_1, \leq_1, Cons_1, \sqcup_1\rangle$ be some f-bases on $\mathbb{A}$. A direct product $\mathcal{B}_1 \times \mathcal{B}_2$ of $\mathcal{B}_0$ and $\mathcal{B}_1$ is the f-base $\langle B_0 \times B_1, \leq, Cons, \sqcup\rangle$, where $\leq$, $Cons$ and $\sqcup$ are defined as follows:

(1) $\langle b_0, b_1\rangle \leq \langle b_0', b_1'\rangle$ iff $b_0 \leq_0 b_0' and b_1 \leq_1 b_1'$ for every $b_0, b_0' \in B_0$ and every $b_1, b_1' \in B_1$
(2) $b_* \in Cons$ iff $p_l^*(b_*) \in Cons_0 and p_r^*(b_*) \in Cons_1$ for every $b_* \in (B_0 \times B_1)^*$
(3) $\sqcup b_* \leftrightharpoons \langle \sqcup_0 p_l^*(b_*), \sqcup_1 p_r^*(b_*)\rangle$ for every $b_* \in Cons$

In case the set $Cons$ of mutually consistent fragments (approximations) should be as large as possible, we need

Definition 5 Quadruple $\mathcal{B} = \langle B, b_0, \leq, \sqcup\rangle$ is called an *f^*-base on* $\mathbb{A}$ if $\langle B, \leq, B^* \setminus \{\emptyset\}, \sqcup\rangle$ is an f-base on $\mathbb{A}$, $[b_0]$ is the least element in $\langle [B], \leq\rangle$ and $\sqcup\emptyset = b_0$.

In general, the range (the set of possible values) of a functional can be arbitrary, so the notion of f^*-base is at hand in the following

Definition 6 Let $\mathcal{B}_0 = \langle B_0, \leq_0, Cons_0, \sqcup_0\rangle$ be an f-base, $\mathcal{B}_1 = \langle B_1, b_1, \leq_1, \sqcup_1\rangle$ be an f^*-base. A functional product $F(\mathcal{B}_0, \mathcal{B}_1)$ of f-base $\mathcal{B}_0$ and f^*-base $\mathcal{B}_1$ is the f^*-base $\langle (B_0 \times B_1)^*, \emptyset, \leq, \sqcup\rangle$, where $\leq$ and $\sqcup$ are defined as follows:

(1) $f_0 \leq f_1$ iff $\forall b_0 \in p_l^* f_0(\sqcup_1\{b_1 \mid \exists b_0' \in p_l^* f_0(b_0' \leq_0 b_0\, and \langle b_0', b_1\rangle \in f_0)\} \leq_1$ $\leq_1 \sqcup_1\{b_1 \mid \exists b_0' \in p_l' f_1(b_0' \leq_0 b_0\, and \langle b_0', b_1\rangle \in f_1)\})$ for $f_0, f_1 \in (B_0 \times B_1)^*$
(2) $\sqcup f_* \leftrightharpoons \cup f_*$ for every $f_* \in ((B_0 \times B_1)^*)^*$

Definition 7 For an f-base $\mathcal{B} = \langle B, \leq, Cons, \sqcup\rangle$, the family $I_\Sigma(\mathcal{B})$ of Σ-ideals in $\mathcal{B}$ consists of nonempty Σ-definable subsets $C \subseteq B$ such that

(1) from $c \in C, b \in B, b \leq c$ it follows that $b \in C$
(2) from $c \in C^*$ it follows that $c \in Cons and \sqcup c \in C$

We define a topology on the set $I_\Sigma(\mathcal{B})$ by fixing the basis

$$V_b \leftrightharpoons \{C \mid C \in I_\Sigma(\mathcal{B}), b \in C\}, b \in B$$

The set $I_\Sigma(\mathcal{B})$ together with the topology specified above is called the *space of* Σ*-ideals* of f-base $\mathcal{B}$. The space $I_\Sigma(\mathcal{B})$ is a topological T_0-space.

Let $\mathcal{B}_0$ be an f-base and let $\mathcal{B}_1$ be an f^*-base. For any ideal I of f-base $F(\mathcal{B}_0, \mathcal{B}_1)$ we can define the continuous function $f_I : I_\Sigma(\mathcal{B}_0) \to I_\Sigma(\mathcal{B}_1)$ as follows. Let $I_0 \in I_\Sigma(\mathcal{B}_0)$. We define

$$f_I(I_0) \leftrightharpoons \{b_1 \mid b_1 \in B_1, (\exists c^* \in I)(\exists b_0 \in I_0)\exists b_1'(b_1 \leq_1 b_1' and \langle b_0, b_1'\rangle \in c^*)\}$$

If $\{\langle b_0, b_1\rangle\} \in I, b_0 \in I_0$, then $b_1 \in f_I(I_0)$.

The mapping $I \to f_I$ from $I_\Sigma(F(\mathcal{B}_0, \mathcal{B}_1))$ to $C(I_\Sigma(\mathcal{B}_0), I_\Sigma(\mathcal{B}_1))$ (the set of all continuous functions from the space $I_\Sigma(\mathcal{B}_0)$ to the space $I_\Sigma(\mathcal{B}_1)$) is injective.

To introduce the simplest example of spaces for entities and truth values, let $\mathcal{A} = \langle A, =, P_1(A), \cup\rangle$, where $P_1(A) \leftrightharpoons \{\{a\} \mid a \in A\}$. This quadruple is an f-base with $I_\Sigma(\mathcal{A}) = P_1(A)$. Also, let α be an arbitrary ordinal in $\mathbb{A}$ and let $\mathcal{B}_\alpha = \langle \alpha, \emptyset, \subseteq, \cup\rangle$. This quadruple is an f^*-base with $I_\Sigma(\mathcal{B}_\alpha) = (\alpha + 1) \setminus \emptyset$. Further on we consider the case $\alpha = 2$.

Definition 8 The set of functional types $Types_f$ together with its proper subset $PTypes_f$ are defined as follows:

(1) $\boldsymbol{o} \in Types_f \setminus PTypes_f, B \in PTypes_f \subseteq Types_f$
(2) if $\tau_0, \tau_1 \in Types_f(PTypes_f)$ then $(\tau_0 \times \tau_1) \in Types_f(PTypes_f)$
(3) if $\tau_0 \in Types_f, \tau_1 \in PTypes_f$ then $(\tau_0 \to \tau_1) \in PTypes_f$

Definition 9 For every type $\tau \in Types_f$, the f-base $\mathcal{F}_\tau$ is defined by induction on the complexity of τ:

(1) $\mathcal{F}_{\boldsymbol{o}} \leftrightharpoons \mathcal{A}, \mathcal{F}_B \leftrightharpoons \mathcal{B}_2$
(2) $\mathcal{F}_{(\tau_0 \times \tau_1)} \leftrightharpoons \mathcal{F}_{\tau_0} \times \mathcal{F}_{\tau_1}$
(3) $\mathcal{F}_{(\tau_0 \to \tau_1)} \leftrightharpoons F(\mathcal{F}_{\tau_0}, \mathcal{F}_{\tau_1})$

If $\tau \in PTypes_f$ then $\mathcal{F}_\tau$ is an f^*-base.

Definition 10 By a Σ-predicate of type $\tau \in Types_f$ on A we mean an arbitrary element of $I_\Sigma(\mathcal{F}_\tau)$.

The propositions below easily follow from the definitions. Here $\Sigma(\mathbb{A})$ denotes the set of all Σ-definable subsets of $\mathbb{A}$.

Lemma 1 For any $n > 0$ there is a natural bijective correspondence between Σ-predicates of type $\boldsymbol{o}^n \to B$ and n-ary Σ-predicates on $\mathbb{A}$.

Proposition 1 A mapping $F : \Sigma(\mathbb{A}) \to \Sigma(\mathbb{A})$ is a restriction of a Σ-operator if and only if F is continuous with respect to the strong topology and there is a Σ-function $f : A \to A$ such that $F(Q_{u,a}) = Q_{u,f(a)}$ for all $a \in A$.

Proposition 2 For a family $S \subseteq \Sigma(\mathbb{A})$ the following are equivalent:

1. S is represented by a Σ-predicate of type $((\boldsymbol{o} \to B) \to B)$
2. there is a Σ-formula $\Phi(P^+)$ of signature $\sigma \cup \langle P^1 \rangle$ such that

$$S = \{Q \mid Q \in \Sigma(\mathbb{A}), \langle \mathbb{A}, Q \rangle \models \Phi(P)\}$$

Proposition 3 There is a natural bijective correspondence between Σ-predicates of type $((\boldsymbol{o} \to B) \to (\boldsymbol{o} \to B))$ and unary Σ-operators.

We consider here the most natural case for studying algorithmic issues of Montague intensional logic, namely $\mathbb{A} = \mathbb{HF}(\mathbb{R})$. Indeed, the scale of time in linguislics is usually identified with the ordered set of real numbers $\mathbb{R}$. The correspondences in Table 2 were obtained in [3].

Table 2 Intensional logic types and $\mathbb{HF}(\mathbb{R})$

Category	Grammar equivalent	Type	Object in $\mathbb{HF}(\mathbb{R})$
e	No	e	Sets $\{a\}$ for $a \in \mathbb{HF}(\mathbb{R})$
t	Sentenses	t	No
IV	Intransitive verbs	(e → t)	Unary Σ-predicates
CN	Common nouns	(e → t)	Unary Σ-predicates
TV	Extensional transitive verbs	(e → (e → t))	Binary Σ-operators
CN/CN	Extensional adjectives	((e → t) → (e → t))	Σ-operators
CN/CN	Extensional adverbs	((e → t) → (e → t))	Σ-predicates
T	Noun phrases and proper names	((s → (e → t)) → t)	Σ-definable families Of binary Σ-predicates
t/t	Sentence determiners	((s → t) → t)	Σ-definable families of Σ-predicates on $P_1(\mathbb{R})$
IV/t	Connective verbs	((s → t) → (e → t))	Σ-operators

5 Rank and Vector Models of Intensional Logic

The main result of this paper about isomorphism of rank model and vector model provides a connection between two rather different methods of coding information. Both models are natural, from our point of view, the first because of the set-theoretical simplicity and the second because vectors or finite tuples are the typical approximations for infinite strings which are necessary to represent entities exactly.

Recall that a model of intensional logic is a tuple $\langle A, W, T, \leq, F\rangle$. Here we consider generalized computable models of intensional logic constructed with the help of computable functionals of finite types. In such models, the set A (entities of type e) corresponds to some f-base $\mathcal{A}$, the space of truth values (values of type t) corresponds to some f^*-base $\mathcal{B}$, the sets W and T used to form intensional types (states s) correspond to some f-base $\mathcal{W}$. The valuation function F corresponds to the entire hierarchy of functionals of finite types generated by the triple $\langle \mathcal{A}, \mathcal{B}, \mathcal{W}\rangle$ in accordance with the classical types considered in intensional logic. Thus, we say that the triple $\langle \mathcal{A}, \mathcal{B}, \mathcal{W}\rangle$ defines a (generalized computable) model of intensional logic.

As usual, $P_1(X)$ denotes the set of all one-element subsets of X, i.e., $P_1(X) = \{\{x\} \mid x \in X\}$. In [3] was introduced a model consisting of f-base

$$\mathcal{A}_0 = \langle X, =, P_1(X), \cup\rangle$$

for $X = HF(\mathbb{R}) \cup \mathbb{R}$ corresponding to the space of entities, f^*-base

$$\mathcal{B} = \langle \{0, 1\}, 0, \leq, \max\rangle$$

corresponding to the space of truth values, and f-base

$$\mathcal{W} = \langle W, =, P_1(W), \cup\rangle$$

for $W = P_1(\mathbb{R})$ corresponding to the space of possible worlds. The entities in this model are singleton subsets of X and their structure is not taken into account: trivial equality is considered as a preorder on entities, and the entities themselves are simply "points" or "atoms". The space of truth values of this model can be intuitively interpreted as "0 means that the property does not exist, but may appear in the future" and "1 means that the property is and remains forever".

Also, in [3] was described a model that consists of f-base

$$\mathcal{A}_{vec} = \langle (\mathbb{R} \cup \{\bot\})^{<\omega}, \leq_{vec}, Cons_{vec}, \sqcup_{vec}\rangle$$

where $\alpha_1 \leq_{vec} \alpha_2$ if and only if $lh(\alpha_1) \leq lh(\alpha_2)$ and $\alpha_1(i) \leq \alpha_2(i)$ for all $i \leq lh(\alpha_1)$, while we assume that $\bot \leq a$ for any $a \in \mathbb{R}$ and for $a, b \in \mathbb{R}$ are incomparable for $a \neq b$. Informally, entities are infinite tuples approximated via their finite initial subtuples, and contents of tuples correspond to properties (from categories IV and

CN). In addition, we assume that some encoding is given, which says whether the i-th position of the ordered set is a binary or measurable property. For the space of truth values was used f^*-base

$$\mathcal{C} = \langle \{0, 1, \bot, \top\}, \bot, \leq, \sqcup \rangle$$

where $\bot < 0$, $\bot < 1$, $0 < \top$, $1 < \top$, and 0 and 1 are incomparable (elements of C stand for "no", "yes", "unknown" and "contradiction", correspondingly). Again,

$$\mathcal{W} = \langle W, =, P_1(W), \cup \rangle$$

for $W = P_1(\mathbb{R})$.

The triple $\langle \mathcal{A}_0, \mathcal{B}, \mathcal{W} \rangle$ will be called the *simplest model*, and the triple $\langle \mathcal{A}_{vec}, \mathcal{C}, \mathcal{W} \rangle$ —*vector model* (for brevity, we will omit the space of possible worlds in what follows). In this section, some modification of the simplest model will be considered. It will take into account the structure of entities (that is, the elements that are contained in them as in sets). The entity $\{a\}$, $a \in HF(\mathbb{R})$, will be defined by a set of its properties, which are encoded by the set $\{a_1 \in HF(\mathbb{R}) \mid a_1 \in a\}$. A preorder relation on entities will be introduced, and the space of truth values will also be changed. The preorder relation and the space of truth values will be introduced in accordance with the vector model of intensional logic, so the resulting model (which we will call the *rank model* of intensional logic) will be isomorphic to it.

Consider the simplest model $\langle \mathcal{A}_0, \mathcal{B} \rangle$. Let us indicate a possible way of interpreting the properties of entities in this model. Consider $a \in HF(\mathbb{R})$: let $a = \{a_1, \ldots, a_k\}$. The elements $a_1, \ldots, a_k$ can be considered as properties of the object $\{a\}$ from the basic categories IV and CN. These properties can be decoded based on the ranks of the elements and some of their numerical characteristics.

Let us set $CN = \{cn_1, \ldots, cn_n, \ldots\}$, $IV = \{iv_1 \ldots, iv_n, \ldots\}$. All IV properties are binary (either hold or not), but CN properties can be either binary or take an arbitrary value from real numbers (for example, such properties as height or weight), so we will consider two different categories of CN_{bin} and CN_{cont}. Let us indicate a (possible) encoding of properties by natural numbers. If we associate with each category and element of this category the number ($IV \mapsto 1, iv_n \mapsto n$; $CN_{bin} \mapsto 2, (cn_{bin})_n \mapsto n, (cn_{bin})_n \mapsto n$; $CN_{cont} \mapsto 3, (cn_{cont})_n \mapsto n$), then using the Cantor function $c : \mathbb{N}^2 \to \mathbb{N}$, $c(x, y) = \frac{(x+y)^2+3x+y}{2}$, for the number $n \in \mathbb{N}$ we can restore a category and an element of this category. Let, for example, $IV = \{walk, talk\}$, $CN_{bin} = \{man, woman\}$ (we assume that they are numbered in the order they are listed). Since $c(1, 2) = 7$ and $c(2, 1) = 8$, we get that 7 corresponds to the category IV and the property $talk$, and 8 corresponds to the category CN and the property man. The numerical value of measurable properties can be encoded, for example, using the maximum real number contained in the support of the element a. More generally, we assume that some abstract Σ-function $Val : A \to \mathbb{R}$ is given, which determines the value for measurable properties.

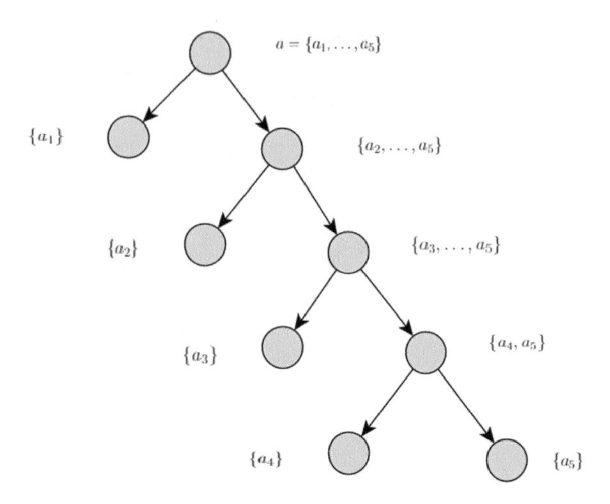

Fig. 1 Rank tree

As an example, consider the following model:

- $IV = \{talk\}$
- $CN_{bin} = \{human, male\}$
- $CN_{cont} = \{height, speed\}$

And the following match:

- $walk \mapsto c(1, 1)$
- $human \mapsto c(2, 1)$, $male \mapsto c(2, 2)$
- $height \mapsto c(3, 1)$, $speed \mapsto c(3, 2)$

Suppose that for some element $a = \{a_1, \ldots, a_5\} \in HF(\mathbb{R})$ the following correspondence is given:

- $rnk(a_1) = c(1, 1)$
- $rnk(a_2) = c(2, 1)$, $rnk(a_3) = c(2, 2)$
- $rnk(a_4) = c(3, 1)$, $Val(a_4) = 180$, $rnk(a_5) = c(3, 2)$, $Val(a_5) = 6$

The element $a = \{a_1, \ldots, a_5\}$ can be visualized using a graph as in Fig. 1.

Then, looking at each vertex of the graph (for example, by breadth- or depth-first searches), we can conclude that

- $a_1 \sim walk$
- $a_2 \sim human$, $a_3 \sim male$
- $a_4 \sim height = 180$, $a_5 \sim speed = 6$

Using this correspondence, one can define Σ-predicates and Σ-operators as follows:

- $(a \in human) \Leftrightarrow (\exists x \in a)(rnk(x) = c(2, 1))$
- $(a \in male) \Leftrightarrow (\exists x \in a)(rnk(x) = c(2, 2))$
- $(a \in man) \Leftrightarrow (a \in male \ \& \ a \in human)$

- $(a \in tall(man)) \Leftrightarrow (a \in man \ \& \ (\exists x \in a)(rnk(x) = c(3,1)\ ; \ \& \ Val(x) \geq 185))$
- $(a \in slowly(walk) \Leftrightarrow (a \in walk \ \& \ (\exists x \in a)(rnk(x) = c(3,2) \ \& \ Val(x) \leq 3))$

5.1 Isomorphism Between Rank and Vector Models

The reasoning above was given within the framework of the simplest model. The vector (or ontological) model of intensional logic uses a different approach to compare entities and different spaces of truth values.

Denote by Nat the set of ordinals in $HF(\mathbb{R})$. Let's assume that some encoding of categories IV and CN is given (for example, as specified in the previous subsection). There exists a partial Σ-function $\nu : Nat \to Nat^2$, which allows one to effectively determine a category and a property from this category by number, and its domain of definition is a Δ-set, as well as some Σ-function $Val : HF(\mathbb{R}) \to Nat$, which determines the numerical characteristics of properties from CN_{cont}. Accordingly, there are Δ-formulas $\varphi_{bin}(x)$ and $\varphi_{cont}(x)$ such that $\varphi_{bin}(x) \Leftrightarrow$ "$xmatchesthebinaryproperty$" and $\varphi_{cont}(x) \Leftrightarrow$ "$xmatchesthecontinuousproperty$". We indicate (following the definition of the vector model) how the presence, absence, or uncertainty of a property can be interpreted in the rank model. If $x \in a$ for $a \in HF(\mathbb{R})$ then:

- if $0 \in x$, then we assume that the property (determined by the rank of the element x) is missing;
- if $0 \notin x$, then we assume that the property is present;
- if an element of rank n is absent in a, then we assume that we do not know about the presence of this property.

In view of what has been said, it is necessary to exclude some of the elements from $HF(\mathbb{R})$ in order to avoid ambiguous interpretation. Namely, to exclude all elements that contain (as sets) different elements of the same rank, as well as elements whose rank does not belong to the domain of the encoding function ν. Thus, we will consider a set S such that

$$a \in S \Leftrightarrow \forall x \in a[\forall y \in a(rnk(x) \neq rnk(y)) \,\&\, rnk(x) \in dom(\nu)]$$

S is a Δ-set. Denote by $rnk^*(a)$ the set of ranks of elements from a, i.e., $rnk^*(a) = \{n \in Nat : \exists x \in a(rnk(x) = n)\}$. On the set S, we introduce the Δ-preorder $\leq_1$ as follows:

$$a_1 \leq_1 a_2 \Leftrightarrow rnk^*(a_1) \subseteq rnk^*(a_2) \,\&\, \forall x \in a_1[(0 \in x \to \exists y \in a_2(rnk(x) = rnk(y))$$

$$\&\, 0 \in y))\&(0 \notin x \,\&\, \varphi_{cont}(x) \to \exists y \in a_2(rnk(x) = rnk(y) \,\&\, Val(x) = Val(y)))]$$

The equivalence relation defined by this preorder makes it possible to consider equal elements with the same properties, but in which these properties are defined by different elements. Therefore, in what follows we will consider the sets of equivalence classes $[S]$ and the (induced on it) order $\leq_1$.

Recall that the elements of the set $Cons$ for the vector model are finite tuples from $(\mathbb{R} \cup \{\perp\})^{<\omega}$. The $\sqcup$ function is clearly defined. Let's denote them by $Cons_{vec}$ and $\sqcup_{vec}$ respectively. As the set $Cons$ for the rank model, we also consider finite tuples (from $[S]$), and the function $\sqcup$ is defined similarly to a vector one. Let's denote them by $Cons_{rnk}$ and $\sqcup_{rnk}$, respectively. The orders on the vector and rank models will also be denoted by $\leq_{vec}$ and $\leq_{rnk}$. Consider f-bases

$$\mathcal{A}_0 = \langle [S], \leq_{rnk}, Cons_{rnk}, \sqcup_{rnk} \rangle$$

$$\mathcal{A}_{vec} = \langle (\mathbb{R} \cup \{\perp\})^{<\omega}, \leq_{vec}, Cons_{vec}, \sqcup_{vec} \rangle$$

and the mapping $\beta : \mathcal{A}_0 \to \mathcal{A}_{vec}$ defined as follows: the element $a \in [S]$ corresponds to an ordered set α of length $rnk(a) - 1$ such that for all $i \leq lh(\alpha)$:

- if $i \notin rnk^*(a)$ then $\alpha(i) = \perp$
- if $\exists x \in a(rnk(x) = i \,\&\, 0 \in x)$ then $\alpha(i) = 0$
- if $\exists x \in a(rnk(x) = i \,\&\, 0 \notin x \,\&\, \varphi_{cont}(x))$ then $\alpha(i) = Val(x)$
- else $\alpha(i) = 1$

Let us show that β is a bijection. Let's show injectivity. Let $a_1, a_2 \in [S]$ and $a_1 \neq a_2$, $\beta(a_1) = \alpha_1, \beta(a_2) = \alpha_2$. Either $rnk^*(a_1) \neq rnk^*(a_2)$ or $rnk^*(a_1) = rnk^*(a_2)$. The first immediately implies $\alpha_1 \neq \alpha_2$, since there is an element x of rank i such that $x \in a_1$ and $x \notin a_2$, so by the definition of the mapping β $\alpha_1(i) \neq \alpha_2(i)$ (or $\alpha_2(i)$ is not defined at all and $lh(\alpha_1) \neq lh(\alpha_2)$). Let $rnk^*(a_1) = rnk^*(a_2)$. Then one of the following is required:

- $\exists x \in a_1 \exists y \in a_2(rnk(x) = rnk(y) \,\&\, 0 \in x \,\&\, 0 \notin y)$
- $\exists x \in a_1 \exists y \in a_2(rnk(x) = rnk(y) \,\&\, \varphi_{cont}(x) \,\&\, Val(x) \neq Val(y))$

Each item immediately follows $\alpha_1 \neq \alpha_2$, so β is an injection. Let's show surjectivity. Let $\alpha \in (\mathbb{R} \cup \{\perp\})^{<\omega}$. Consider an element $a \in [S]$ such that for all $i \leq lh(\alpha)$:

- if $\alpha(i) = \perp$ then $i \notin rnk^*(a)$
- if i corresponds to a continuum property and $\alpha(i) \neq \perp$ then $\exists x \in a(rnk(x) = i \,\&\, Val(x) = \alpha(i))$
- if $\alpha(i) = 0$ then $\exists x \in a(rnk(x) = i \,\&\, 0 \in x)$
- else $\exists x \in a(rnk(x) = i \,\&\, 0 \notin x)$

These conditions uniquely define an element $a \in [S]$ such that $\beta(a) = \alpha$, so β is a surjection.

Let us now show that the mapping β is order-preserving. Let $a_1, a_2 \in S$, $a_1 \leq_{rnk} a_2$, $\beta(a_1) = \alpha_1, \beta(a_2) = \alpha_2$. Since $a_1 \leq_{rnk} a_2$, then $lh(\alpha_1) \leq lh(\alpha_2)$. Let $i \leq lh(\alpha_1)$. If $\alpha_1(i) = \perp$, then $\alpha_1(i) \leq \alpha_2(i)$. If $\alpha_1(i) = 0$, then $\exists x \in a_1(rnk(x) = i \,\&\, 0 \in x)$ is true by the definition of β, and by definition of order $\leq_1$, we get that $\exists y \in a_2(rnk(x) = rnk(y) \,\&\, 0 \in y)$, which gives $\alpha_2(i) = 0$. If now i corresponds to a continuum property, then $\exists x \in a_1(rnk(x) = i \,\&\, \varphi_{cont}(x) \,\&\, Val(x) = \alpha_1(i))$, which gives $\exists y \in a_2(rnk(x) = rnk(y) \,\&\, \varphi_{cont}(x) \,\&\, Val(y) = \alpha_1(i))$, and therefore $\alpha_1(i) = \alpha_2(i)$. So $\alpha_1(i) \leq \alpha_2(i)$ is true for all $i \leq lh(\alpha_1)$, and by definition of the order $\leq_{vec}$ we get that $\alpha_1 \leq_{vec} \alpha_2$. Thus, we have:

Proposition 4 There exists an isomorphism between f-bases:

$$\mathcal{A}_0 = \langle [S], \leq_{rnk}, Cons_{rnk}, \sqcup_{rnk} \rangle$$

and

$$\mathcal{A}_{vec} = \langle (\mathbb{R} \cup \{\bot\})^{<\omega}, \leq_{vec}, Cons_{vec}, \sqcup_{vec} \rangle$$

If we take

$$\mathcal{C} = \langle \{0, 1, \bot, \top\}, \bot, \leq, \sqcup \rangle$$

as the truth space, then we get the following:

Corollary 1 *Hierarchies of computable functionals of finite types for a rank model of the form $\langle \mathcal{A}_0, \mathcal{C} \rangle$ and a vector model of the form $\langle \mathcal{A}_{vec}, \mathcal{C} \rangle$ are equivalent.*

Moreover, from the definition of this isomorphism it is clear that it is a Σ-function.

The ranks of ordered sets of parent elements depend only on the length of the set. Indeed, the following (more general) fact is true

Lemma 2 If $\mathbb{A}$ is an admissible set and $a_1, \ldots, a_n \in \mathbb{A}$, $n \geq 2$ then:

$$rnk(\langle a_1, \ldots, a_n \rangle) = \sup\{ \sup_{i=\overline{1,n-1}} \{rnk(a_i) + 2i\}, rnk(a_n) + 2(n-1)\}$$

Proof 1 Induction on n. For $n = 2$ the assertion is true.
$n \to n + 1$: since $\langle a_1, \ldots, a_{n+1} \rangle = \langle a_1, \langle a_2, \ldots, a_{n+1} \rangle \rangle$, then

$$rnk(\langle a_1, \ldots, a_{n+1} \rangle) = \sup\{rnk(a_1) + 2, \sup_{i=\overline{2,n}} \{rnk(a_i) + 2i\}, rnk(a_{n+1}) + 2n\} =$$

$$= \sup\{ \sup_{i=\overline{1,n}} \{rnk(a_i) + 2i\}, rnk(a_{n+1}) + 2n\}$$

Hence, if $r_1, \ldots, r_n \in \mathbb{R} \cup \{\bot\}$ then $rnk(\langle r_1, \ldots, r_n \rangle) = 2(n-1)$. In the rank model, by definition, the rank of an element corresponding to an ordered set of length n does not exceed n. Thus, even for $n > 2$, elements of lower ranks are obtained. In the general case, the ranks of the elements of the rank model will be much smaller, since if $r_i = \bot$, then the element of rank i is absent in the corresponding object of the rank model. In addition, due to the introduced equivalence relation when defining the rank model, most of the content of the hereditarily finite superstructure becomes insignificant. In particular, since the rank (but not the content) uniquely determines some property, it is sufficient to confine ourselves to considering only one parent element.

6 Interpreting the Semantics of Possible Worlds

In this section, we will show how, for a possible world given by a real number (more precisely, the set $w = \{r\}$, where $r \in \mathbb{R}$), one can reasonably define a Δ-subset in $HF(\mathbb{R})$ and an f-base on this Δ-subset, which will be an interpretation of possible world w. Having such f-bases for all possible worlds, when considering questions of the truth of formulas (depending on the possible world), one can switch from using universal valuation, which, depending on the possible world, assigns one or another value to a variable, to checking the truth of formulas in the constructed f-bases, which are already significantly smaller than the original structure. In addition, this allows us to consider the entire hierarchy of computable functionals of finite types already on the structure corresponding to the possible world.

Various interpretations will be given for f-bases

$$\mathcal{A}_0 = \langle X, =, P_1(X), \cup \rangle$$

where $X = HF(\mathbb{R})$, and

$$\mathcal{A}_{vec} = \langle (\mathbb{R} \cup \{\bot\})^{<\omega}, \leq, Cons_{vec}, \sqcup_{vec} \rangle$$

due to their fundamental differences.

Here we indicate how, given a real number $r \in \mathbb{R}$ in a hereditarily finite superstructure $\mathbb{HF}(\mathbb{R})$, one can define a countable Δ-definable set of real numbers. To do this, we need several auxiliary Σ-functions, defined (by Σ-recursion over ordinals) as follows:

$$f_1 : Nat \to \mathbb{R},\ f_1(0) = 0,\ f_1(n+1) = f(n) + 1$$

$$f_2 : \mathbb{R} \times Nat \to \mathbb{R},\ f_2(r, 0) = r,\ f_2(r, n+1) = \tfrac{f_2(r,n)}{10}$$

$$pr : \mathbb{R} \times Nat \to Nat$$

$$pr(r, 0) = a_0 \Leftrightarrow a_0 \in Nat \,\&\, f_1(a_0) \leq r \leq f_1(a_0) + 1$$

$$pr(r, n+1) = a_{n+1} \Leftrightarrow a_{n+1} \in Nat \,\&\, \tfrac{f_1(a_{n+1})}{10^n} \leq x - \sum_{i=0}^{n} \tfrac{f_1(a_i)}{10^i} \leq \tfrac{f_1(a_{n+1})}{10^n} + 1$$

The function f_1 associates a natural number (ordinal) with a real number (primary element) corresponding to this natural number. The function f_2 divides the real number r by 10 to the power of n, and the function pr determines the natural number corresponding to the n-th digit in the decimal representation of the real number r. From the function pr one can define (by Σ-recursion) the Σ-function $f_3 : \mathbb{R} \times Nat \to Nat$, which enumerates the sequence of the first n numbers in the decimal representation of a real number , i.e., if $r = a_0, a_1 \ldots a_{n-1} \ldots$, then $f_3(r, n)$ is the number of the sequence of natural numbers $\langle a_0, \ldots a_{n-1} \rangle$.

Let some natural number k be given, which indicates how many characters in the decimal representation of real numbers to consider. We will call this number the order of approximation of real numbers. Let $r \in \mathbb{R}$. We define the Δ-set $S_{\{r\}}$ as follows:

$$x \in S_{\{r\}} \Leftrightarrow \exists l(f_3(x,k) = l \,\&\, \forall i \leq k(pr(x,i) = pr(r, p_l^i)) \,\&\, \exists n(n \in Nat \,\&$$

$$\& \, f_1(n) = x \cdot 10^k)$$

$$x \notin S_{\{r\}} \Leftrightarrow \exists l(f_3(x,k) = l \,\&\, \exists i \leq k(pr(x,i) \neq pr(r, p_l^i)) \vee \exists n(n \in Nat \,\&$$

$$\& \, f_1(n) < x \cdot 10^k < f_1(n) + 1)$$

where p_l is the lth prime number. Thus, the set $S_{\{r\}}$ will include all real numbers of the form $a_0, a_1 \dots a_{k-1}$ whose i-th digit in decimal representation is equal to p_l^ith digit in decimal representation of r, where l is the sequence number $\langle a_0, \dots, a_{k-1} \rangle$. In addition, we define the Σ function $\nu_{\{r\}} : S_{\{r\}} \to Nat$ as follows: $\nu(x) = n \Leftrightarrow x \in S_{\{r\}} \,\&\, \forall i \leq k(pr(x,i) = pr(r, p_n^i))$. This function enumerates the set $S_{\{r\}}$.

6.1 Interpretation of Possible Worlds Semantics for Simplest Model

Consider the simplest model

$$\mathcal{A}_0 = \langle X, =, P_1(X), \cup \rangle$$

where $X = HF(\mathbb{R})$. The possible world is given by an element from $P_1(\mathbb{R})$, i.e., set of the form $w = \{r\}$ for $r \in \mathbb{R}$. Based on the possible world w, we define the Δ-set S_w as in the previous section. Based on it, we construct the Δ-set H_w as follows:

$$x \in H_w \Leftrightarrow sp(x) \subseteq S_w$$

where $sp(x)$ is the support of x. It is clear that this is also a Δ-set. Then on the set H_w it is possible to define an f-base

$$\mathcal{H}_w = \langle H_w, =, P_1(H_w), \cup \rangle$$

which we will consider as an f-base, corresponding to the possible world w.

Intuitively, one can consider real numbers as some initial “filling” of $\mathcal{A}$, which contains information about all possible worlds at once. Sets from $HF(\mathbb{R})$ are constructed over this filling, i.e., objects of our structure $\mathcal{A}$. Choosing a real number r and considering the set $S_{\{r\}}$ specified by it, we select a part from the entire content of the structure $\mathcal{A}$ and consider the objects built on this part, which are the objects of the possible world $w = \{r\}$.

6.2 Interpretation of Semantics of Possible Worlds for Vector Model

Let us interpret possible worlds for a vector model of the form

$$\mathcal{A}_{vec} = \langle (\mathbb{R} \cup \{\bot\})^{<\omega}, \leq_{vec}, Cons_{vec}, \sqcup_{vec} \rangle$$

Similarly, we consider the set $w = \{r\}$ as a possible world and construct a Δ-set S_w from it. Given the set S_w, we define the Δ-set H_w as follows (assuming that only ordered sets are considered):

$$x \in H_w \Leftrightarrow \forall i \leq lh(x)[\exists r(r \in S_w \,\&\, \nu_w(r) = i \,\&\, x(i) \in S_{\{r\}})\vee$$

$$\vee(i \notin pr_r^*(\nu_w) \,\&\, \text{“}x(i) \text{ takes any valid value''})]$$

where the formula in quotation marks at the end is written as in Sect. 6.2 using the Σ-formulas φ_{bin} and φ_{cont}. It is also clear that the complement of the set H_w is a Σ-set, so it is indeed a Δ-set. Restrictions of the order $\leq_{vec}$, the set $Cons$, and the function $\sqcup$ to H_w allow us to correctly define the f-base

$$\mathcal{H}_w = \langle H_w, \leq_{vec} \cap (H_w)^2, Cons_{vec} \cap (H_w)^*, \sqcup_{vec} \rangle$$

which will be treat as an interpretation of the possible world w.

The algorithm for constructing a support for a possible world consists in constructing a countable set of real numbers S_w from the possible world $w = \{r\}$. Further, we assume that each number $r_1 \in S_w$ together with its number $i = \nu_w(r_1)$ and a similar countable set of real numbers $S_{\{r_1\}}$ defines the set of admissible values for all i-th coordinates of ordered sets. Next, the set H_w contains all such ordered sets whose coordinate values belong to the sets of their admissible values. The set H_w is considered to be the set of objects in the possible world w.

To analyze time processes, one has to consider both points (moments) and intervals of the form (r_1, r_2), where $r_1 \leq r_2$ are real numbers. Using the above Σ-functions for encoding real numbers and a note about the order of approximation of real numbers, as well as some way of encoding ordered pairs of real numbers (for example, the pair $r_1 = x, x_1x_2\dots, r_2 = y, y_1y_2\dots$ corresponds to the number $r = 0, xyx_1y_1x_2y_2\dots$) we get that there are Σ-predicates $cint(r)$ $int(r, r_1, r_2)$ true if and only if the real number r encodes some interval and when the real number r encodes the interval (r_1, r_2) respectively. Further, following the already defined constructions, we set $S_{\{r\}}^{int} = S_{\{r_1\}} \cup S_{\{r_2\}}$ as a possible world for the interval (r_1, r_2) given by the number r. If it is necessary to use time intervals or a point in time, the possible world $w = \{r\}$ can be treated in one way or another.

6.3 Analysis of Past Simple, Future Simple and Present Continuous in English

Let us give examples of using the interpretation of possible worlds. Consider sentences

(1) Michele barked

(2) Michele will bark

(3) Michele is barking

as well as

(4) Michele barks

It is clear that on possible worlds one can define by a Δ-formula an order relation consistent with the order on real numbers (i.e., $w_1(=\{r_1\}) \leq w_2(=\{r_2\}) \Leftrightarrow r_1 \leq r_2$). In [3] is given an analysis of English sentences using object intensions. Here, we will consider m as a variable with a certain set of its possible values (extensions). Moreover, for each Σ-subset S in $HF(\mathbb{R})$ there is a Σ-subset $S_w \subseteq S$ in H_w (the support of f-base $\mathcal{H}_w$ corresponding to the possible world w). This Σ-subset is obtained in the same way as Σ-subsets were obtained from the Σ-ideals of the functional product $F(\mathcal{A}, \mathcal{B})$, with the replacement by f-base $\mathcal{A}$ to f-base $\mathcal{H}_w$. Then the Σ-subset S_w can be considered an interpretation of the most general Σ-subset S in the possible world w.

Therefore, proposition (1) is true in the model $\mathcal{A}$ and the possible world $w = \{r\}$ if

$$\mathcal{A} \models \exists w_1(=\{r_1\})\exists r_2 r_3(int(r_1, r_2, r_3) \,\&\, r_3 < r \,\&\, m \in S^{int}_{w_1} \,\&\, bark'_{w_1}(m))$$

(when valuing γ), where $bark'(x)$ is the Σ-predicate corresponding to the verb *bark*. In other words, sentence (1) is true at time r if there exists a time interval (r_2, r_3) such that $r_3 \leq r$ and sentence (4) is true in this time interval. Proposition (2) is treated similarly, with a change of order. Proposition (3) is true in the model $\mathcal{A}$ and the possible world $w = \{r\}$ if

$$\mathcal{A} \models \exists w_1(=\{r_1\})\exists r_2 r_3(int(r_1, r_2, r_3) \,\&\, r_2 < r < r_3 \,\&\, m \in S^{int}_{w_1} \,\&\, bark'_{w_1}(m))$$

In the case of proposition (3), the possible world w for which the truth of the proposition is checked can also be an interval if necessary.

Algorithmic issues of interval semantics for Perfect tenses in English and the category of aspect in Russian are discussed in [16, 17].

It is interesting that, in contrast to extensional objects, intensional issues (more exactly, temporal aspects) are arranged very different in English and in Russian. We discuss these differences in one of the forthcoming papers.

Acknowlegements The research was supported by the IM SB RAS state assignment, project number FWNF-2022-0012.

References

1. Barwise, J.: Admissible Sets and Structures. Springer Verlag, Heidelberg (1975)
2. Bennett, M., Partee, B.: Toward the logic of tense and aspect in english. In: Partee, B. (ed.) Compositionality in Formal Semantics: Selected Papers by Barbara H. Partee, pp. 59–109. Blackwell Publishing (2004). https://doi.org/10.1002/9780470751305.ch4
3. Burnistov, A., Stukachev, A.: Computable functional of finite types in montague semantics. Lecture Notes in Computer Science, to appear (2022). http://www.math.nsc.ru/stukachev/CompFunct_MS.pdf
4. Dowty, D.: Word Meaning and Montague Grammar. D. Reidel Publishing Company, Dodrecht (1979)
5. Dowty, D.: Introduction to Montague Semantics. D. Reidel Publishing Company, Dodrecht (1989)
6. Ershov, Y.: The theory of A-spaces. Algebra and Logic **12**(4), 209–232 (1973). https://doi.org/10.1007/BF02218570
7. Ershov, Y.: Theory of domains and nearby. Formal Methods in Programming and Their Applications. Lecture Notes in Computer Science **735**, 1–7 (1993). https://doi.org/10.1007/BFb0039696
8. Ershov, Y.: Definability and Computability. Plenum, New York (1996)
9. Montague, R.: Recursion theory as a branch of model theory. In: Proceedings of the Third International Congress for Logic, Methodology and Philosophy of Science, pp. 63–86. North-Holland, Amsterdam (1967)
10. Montague, R.: The proper treatment of quantification in ordinary english. In: Hintikka, J., Moravcsik, J., Suppes, P. (eds.) Approaches to Natural Language, pp. 221–242. D. Reidel Publishing Company, Dodrecht (1973)
11. Montague, R.: English as a formal language. In: B.V. et al. (ed.) Linguaggi nella Societa a nella Tecnica, pp. 189–224. Edizioni di Comunita, Milan (1974). Reprinted in: Formal Philosophy: selected papers of Richard Montegue, pp. 108–121
12. Moschovakis, Y.: Elementary Induction on Abstract Structures. North-Holland (1974)
13. Moschovakis, Y.N.: A logical calculus of meaning and synonymy. Linguist. Philos. **29**, 27–89 (2006). https://doi.org/10.1007/s10988-005-6920-7
14. Scott, D.: Outline of a mathematical theory of computation. Proceedings of the 4th Annual Princeton Conference on Information Sciences and Systems pp. 169–176 (1970)
15. Stukachev, A.: Effective model theory: an approach via Σ-definability. Lecture Notes in Logic **41**, 164–197 (2013). https://doi.org/10.1017/CBO9781139028592.010
16. Stukachev, A.: Interval extensions of orders and temporal approximation spaces. Siberian Math. J. **62**(4), 730–741 (2021). https://doi.org/10.1134/s0037446621040157
17. Stukachev, A.I.: Approximation spaces of temporal processes and effectivenes of interval semantics. Adv. Intell. Syst. Comput. **1242**, 53–61 (2021). https://doi.org/10.1007/978-3-030-53829-3_5

Multilingual Text Generation for Abstract Wikipedia in Grammatical Framework: Prospects and Challenges

Aarne Ranta

Abstract Abstract Wikipedia is an initiative to produce Wikipedia articles from abstract knowledge representations with multilingual natural language generation (NLG) algorithms. Its goal is to make encyclopaedic content available with equal coverage in the languages of the world. This paper discusses the issues related to the project in terms of an experimental implementation in Grammatical Framework (GF). It shows how multilingual NLG can be organized into different abstraction levels that enable the sharing of code across languages and the division of labour between programmers and authors with different skill requirements. The plan is to start with a simple but functional multilingual NLG system and to proceed towards more and more sophisticated language and wider coverage of topics, also allowing a human in the loop to create content via a Controlled Natural Language (CNL).

Keywords Abstract wikipedia · Controlled natural language · Grammatical framework · Natural language generation · Text robots · Wikidata · Wikipedia

1 Introduction

Abstract Wikipedia is a recent initiative launched by the Wikimedia Foundation [1]. Its purpose is to support the universal availability of Wikipedia in different languages. At the time of writing, Wikipedia has 328 languages, most of which have only very few articles available.[1]

The method chosen in Abstract Wikipedia is Natural Language Generation (NLG) based on Wikidata, which is a database of formalized facts [2]. These facts are stored as **RDF triples** (Resource Description Framework), which are two-place predications

[1] See https://meta.wikimedia.org/wiki/List_of_Wikipedias for an up-to-date list.

A. Ranta (✉)
Department of Computer Science and Engineering, Chalmers University of Technology and University of Gothenburg,Gothenburg, Sweden
e-mail: aarne.ranta@cse.gu.se

R. Loukanova et al. (eds.), *Logic and Algorithms in Computational Linguistics 2021 (LACompLing2021)*, Studies in Computational Intelligence 1081,
https://doi.org/10.1007/978-3-031-21780-7_6

of the form $x R y$, where R is a predicate and x and y are its arguments. As an example, consider the fact

```
wd:Q30 wdt:P1082 331449281
```

whose parts are the unique identifiers `wd:Q30` (for the USA) and the relation `wdt:P1082` (for the predicate "has population") and a numeric constant. Simple NLG can convert this triple to texts such as

The population of the United States is 331,449,281. (English)

Yhdysvaltain asukasluku on 331 449 281. (Finnish)

La population des États-Unis est 331 449 281. (French)

Die Einwohnerzahl der Vereinigten Staaten ist 331.449.281. (German)

The advantages of NLG, as opposed to manual authoring and translation, are several:

- **Consistency**: the content can be guaranteed to be the same in all versions of Wikipedia
- **Speed**: versions in different languages can be produced in a fraction of a second
- **Cost**: no human labour is needed to create new articles, but only for developing the algorithms
- **Updates**: the content of articles can be kept up to date when facts change
- **Customization**: different views of the same content, e.g., summaries and local adaptations, can be produced via suitable parameters in the algorithms.

All of these advantages are relevant for Wikipedia. Inconsistencies are a known problem in the current set-up based on manual work: versions in different languages can vary in size and content, and even contradict each other. Producing content in new languages is slow. Cost is one of the main reasons of this: it is labour-intensive to write and translate articles. Delayed updates cause errors in even well-supported languages. Customization is very limited, as it has to be done manually.

Now, NLG is an old idea with well-known algoritms [3]. It has even been used in Wikipedia, often under the name of **text robots**, which have produced millions of articles in several languages [4]. Hence, given the advantages of NLG, why is it not the standard way to produce and translate articles? Anyone who has tried to use NLG or read articles produced by text robots can probably list a number of problems that explain this:

- **Style**: NLG-produced text is "robotic"—boring, repetitive, unidiomatic
- **Lack of data**: most of the content included in Wikipedia articles is not available in Wikidata or in any other database of formalized facts, and much of it might not even be possible to formalize
- **Cost**: developing NLG algorithms might be a one-time cost, but so high that it is cheaper to write articles by hand
- **Human resources**: writing NLG algorithms is a skill that might not be available for all languages

- **Community resistance**: text robots, machine translation, and other automatic language processing methods are often discouraged or even prohibited in the Wikipedia community [5].

In this paper, we will outline an approach to Abstract Wikipedia that demonstrates the advantages and addresses the challenges. The presentation is based on actually existing code, which is publicly available.[2] However, as the code is a moving target under constant development, we will neither show all details of it here nor guarantee that the examples are completely up to date. The code repository itself contains updated documentation and also a tutorial for readers who want to try it out or develop it further.

2 From Templates to Rendering Functions

The simplest kind of NLG, often used in text robots, is **templates**: texts with slots for variable arguments. Thus the following template could express the population of any country (or other geographical area) in English:

```
The population of {X} is {Y}.
```

Templates work reasonably well in English, where words need not often be inflected. However, a familiar exception is shown when one of the variables is a number attached to a noun:

```
You have {X} new messages.
```

If X=1, the result is grammatically incorrect and reveals the robotic origin of the text. In other languages, the limits of templates are reached much more often. Thus in French, one should produce

La population de la Suède for X=Suède (Sweden)
La population du Danemark for X=Danemark (Denmark)
La population des États-Unis for X=États-Unis (United States)

The variable in the template could of course contain not only the country name but also the preposition with the article (*de la*, *du*, *des*). However, the country name is also used in other contexts, with other prepositions and possibly without articles. For example, to express "in a country", we write

en Suède
au Danemark
aux États-Unis

[2] https://github.com/aarneranta/NLG-examples

Yet different forms are needed when the country name appears as a subject or an object of a sentence.

A purely template-based solution to the country name problem is to insert a "carrier noun", *pays* ("country"), which is inflected in a uniform way:

```
La population du pays {X} est {Y}.
{Y} habite dans le pays {X}.
```

(the latter means "*Y* lives in the country *X*.") This technique is very commonly used in internet services, which do not always need to hide their robotic origin. But it is also used for rendering person profiles in social media. Since such media try to give a friendly impression, robotic language in them can be disturbing. In Wikipedia, it would result in a kind of text that the community, or readers in general, would have difficulties to accept.

A solution to the problem, proposed for Abstract Wikipedia, is to replace templates by proper **rendering functions**. Such a function could for instance wrap template variables by calls of grammatical case:

```
La population {GENITIVE(X)} est {Y}.
{Y} habite {LOCATIVE(X)}.
```

The system then needs, in addition to the templates, definitions of the `GENITIVE` and `LOCATIVE` functions for every possible value of *X*. This *can* be done as long as there is a limited number of such values, but it adds to the cost and the human resources needs of the system. Creating the templates also becomes more demanding, because the author needs to know where to use which of the cases. In fact, the problem is even more complex: think about facts of the form "*X* was born in *Y*". The French template would need to make a difference between male and female values of *X* to get the **agreement** of the word for "born" right:

```
{X} est {IF FEMININE(X) THEN née ELSE né} {LOCATIVE(Y)}
```

In addition to knowledge required about agreement in French grammar, the template notation itself starts to get complicated. And this is just the simplest case, with a choice from two forms: if the changing part is a verb, the template has to choose from six forms (two numbers times three persons).

As one more problem to be solved in rendering functions, even the order of words may need to be varied. Consider the German template for "*X* was born in *Y*", for simplicity without cases (of which German has four):

```
{X} wurde in {Y} geboren.
```

Now, in Wikipedia, it is customary to indicate the sources of facts by links or references. When a fact is disputed, it can be appropriate to describe different opinions by phrases such as *nach* Z ("according to Z"), *Z glaubt, dass* ("Z believes that"). In such contexts, German grammar requires the word order of the "born" template to be changed:

```
Nach {Z} wurde {X} in {Y} geboren.
{Z} glaubt, dass {X} in {Y} geboren wurde.
```

Hence, not only do we need templates for thousands of predicates, but there have to be (at least) three templates for every predicate to get the word order right in all contexts, plus a device in the template notation that enables us to select the correct alternative.

3 Rendering Functions in Grammatical Framework

Agreement and word order are familiar from school grammar and hence by no means advanced concepts. But their precise treatment in formal grammars in computational linguistics is considered specialist knowledge, and in real-world NLG templates, they are usually avoided altogether by using techniques such as carrier nouns.

Grammatical Framework (GF, [6]) is a programming language that aims to make linguistic knowledge accessible to programmers. Abstract Wikipedia has mentioned GF as a possible technology, and the goal here is to investigate how far it can reach. In GF, grammatical features such as inflection, agreement, and word order can be defined by linguistically knowledgeable programmers in the form of software libraries and reused by non-linguist engineers for different purposes such as NLG [7]. In this way, an equivalent of linguistics-aware rendering functions can be written in a format that essentially looks as templates (see Sect. 6 below).

GF is a special-purpose functional programming language with many features, which can be learned from tutorials and manuals on the web[3] and from the GF book [6]. It inherits much of its syntax from Haskell,[4] but adds some constructs relevant for grammars, such as regular expression pattern matching used for morphology. Like Haskell, GF is statically typed, which is a guarantee that grammars do not fail at runtime. A more special feature is **reversibility**: GF grammars can be used for both parsing and generation, as well as their composition, translation.

Viewed as a grammar formalism, a special feature of GF is that it divides grammar specifications into **abstract syntax** and **concrete syntax** parts. An abstract syntax defines a set of **trees**, and a concrete syntax specifies how they are **linearized** in different languages. One and the same abstract syntax can have several concrete syntaxes, which results in **multilingual grammars**. In a multilingual grammar, **translation** is defined as parsing with one concrete syntax and linearization with another one.

[3] http://www.grammaticalframework.org/

[4] https://www.haskell.org/

The largest set of languages covered by a GF grammar known to us has over 90 languages and defines their numeral systems with a shared abstract syntax [8]. A more general grammar, the GF Resource Grammar Library, defines syntactic structure, morphology, and basic lexicon for 55 languages, of which around 40 have complete implementations of a comprehensive shared abstract syntax [9]. This library has played a major role in almost all multilingual text generation projects in GF, including academic projects on topics such as software specifications [10], mathematics [11, 12], cultural heritage [13], law [14], and healthcare [15], as well as various commercial projects.[5]

As a first example of GF, consider templates of the form

```
the {F} of {X} is {Y}
```

which can express many RDF triples of Wikidata. Its implementation in GF consists of an **abstract syntax function** and its **linearization function**, marked by the keywords `fun` and `lin`, respectively:

```
fun AttrFact : Attr -> Obj -> Val -> Fact
lin AttrFact attr obj val =
    "the" ++ attr ++ "of" ++ obj ++ "is" ++ val
```

In words, the function `AttrFact` takes three arguments—an attribute, and object, and a value—and returns a fact. Its linearization combines these arguments by concatenation (operator ++), with some string literals added in between. The types of these arguments must match the argument types specified in the `fun` definition. Like Haskell, GF uses the arrow syntax for function types, and the prefix notation for function application.

The abstract function `AttrFact` can be used for building infinitely many trees of type `Fact`. An example is the tree

```
AttrFact population_Attr (NameObj USA_Name)(IntVal 331449281)
```

in GF's prefix notation. An equivalent graphical representation is

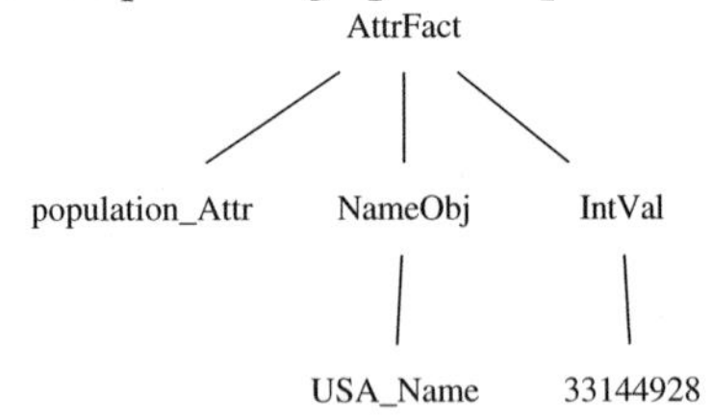

Given obvious linearization functions for `population_Attr`, `NameObj`, `USA_Name` and `IntVal`, this tree is linearized to

the population of the United States is 33144928.

[5] Some of them can be traced from http://grammaticalframework.org.

Stadt, n.f.

	Sg	**Pl**
Nom	*Stadt*	*Städte*
Acc	*Stadt*	*Städte*
Dat	*Stadt*	*Städte*
Gen	*Stadt*	*Städten*

```
{s = table {
     Sg => table {_ => "Stadt"} ;
     Pl => table {Dat => "Städten" ; _ => "Städte"} ;
 g = Fem
}
```

Fig. 1 Inflection table and inherent gender of German *Stadt* and its representation in GF

The above kind of purely concatenative linearization is a special case of GF, corresponding to templates with slots. Similar linearization functions could be defined for other languages as well—but, as we saw in Sect. 1, this would not generalize well over languages. To enable grammatically correct linearizations in all contexts and in all languages, GF generalizes linearization from string concatenation by adding three concepts: **parameters**, **tables**, and **records**.

Grammatical number, case, and gender are examples of parameters, defined as enumerated (and, more generally, finite algebraic) datatypes with the keyword `param`. The following definitions are suitable for German:

```
param Number = Sg | Pl
param Case = Nom | Acc | Dat | Gen
param Gender = Masc | Fem | Neutr
```

These definitions are used in type checking to guarantee the consistency of grammars. But unlike `fun` definitions, they belong to the concrete syntax: different languages can define Number, Case, and Gender in different ways—or not at all, if the language lacks the feature in question.

GF tables are used for expressing inflection tables such as noun declensions. Technically, they are functions over parameter types. As an example, consider the table for the German noun *Stadt* ("city") and its encoding in GF, in Fig. 1. Notice the **wildcard patterns** _ that match cases that are not mentioned explicitly, with a standard notation used in functional programming. Wildcard patterns make it possible to avoid the repetition of similar forms, which typically appear in inflection tables written in the full form. The type checker of GF makes sure that all of the four cases and two numbers are matched.

In addition to the table, Fig. 1 indicates that *Stadt* is a feminine noun (n.f.). The gender of nouns is not a **variable feature** that produces different forms like number and case do (with some exceptions such as *König—Königin* "king—queen"): it is an **inherent feature** of nouns. Inherent features can be collected into GF records, together with all other information about a word. The record shown in Fig. 1 contains both the inflection table and the inherent gender.

Morphological features are used in syntax to implement **agreement**. To give one example in full detail, let us consider a simple one: *one item* vs. *two items*. A possible abstract syntax function for *you have X Ys* is

```
fun YouHaveItems : Numeral -> Item -> Statement
```

A proper linearization requires that every `Numeral` has a grammatical number (singular or plural) attached as an inherent feature. For example,[6]

```
lin one_Numeral = {s = "one" ; n = Sg}
lin two_Numeral = {s = "two" ; n = Pl}
```

In addition, every `Item` (noun) has grammatical number as a variable feature, producing different forms in a table:

```
lin message_Item = table {Sg => "message" ; Pl => "messages"}
```

The linearization function of `YouHaveItems` implements agreement as an interplay between inherent and variable features:

```
lin YouHaveItems num it = "you have" ++ num.s ++ it!num.n
```

Here, `num.s` is the `s`-field projected from the record, and `it!num.n` is the value selected from the table to match the `n` field of the record.

4 Abstraction Levels in GF

The simple mechanism of tables and records has turned out sufficient to model all kinds of agreement and other variation found in the languages that have been implemented in GF so far, including, in addition to several Germanic, Romance, and Slavic languages, also Fenno-Ugric [17], Indo-Iranian [18], Semitic [19, 20], Bantu [21–23], and East-Asian [24, 25] languages.

To give one more example, German word order can be defined by a table that reorders the subject, verb, and complement, as a function of a parameter that stands for main clause, inverted clause, or subordinate clause (which are three values of an Order parameter type):

```
table {
  Main => subj ++ verb ++ compl ;
```

[6] These are special cases of a more general recursive definition of numerals [8]. Different languages may require more distinctions than just singular/plural: Arabic, for instance, has five different agreement patterns, whose GF implementation is explained in [16].

```
Inv => verb ++ subj ++ compl ;
Sub => subj ++ compl ++ verb
}
```

At least as important for the current task as the record and table mechanism itself is the possibility to hide it. This is provided by the **Resource Grammar Library** (RGL), which defines the details of syntax and morphology and exports them via a high-level API (Application Programming Interface) [9]. The complete API is at the time of writing available for 40 languages. It also contains extensive lexical resources for more than half of them. Table 1 relates the available RGL resources to Wikipedias in different languages.

With the RGL API, the linearization of `YouHaveItems` can be defined as follows:

```
lin YouHaveItems num it = mkCl you_NP have_V2 (mkNP num it)
```

The API function `mkCl` builds a clause (`Cl`) from a noun phrase (`NP`), a two-place verb (`V2`), and another noun phrase. Its name follows the convention in the RGL that syntactic functions building values of type *C* have the name `mk`*C*, i.e., "make" *C*. These names are overloaded: all such functions can have the same name, as long as they have different lists of argument types so that the type checker can resolve them. Thus it is often possible to guess a function name without looking it up.

Also the pronoun `you_NP` and the verb `have_V2` are directly available in the API. These functions, as lexical functions in general, take no arguments and cannot thus be resolved by type checking. The RGL convention is to denote them by English words suffixed with part of speech tags.

The object noun phrase is built by the function `mkNP`, which combines a numeral with a noun. Under the hood, `mkNP` takes care of the choice of the singular or the plural, whereas `mkCl` takes care of subject-verb agreement (*have* vs. *has*).

The concept of verb in the RGL is more abstract than in traditional grammars: the abstract two-place verb `have_V2` is in some languages implemented with non-verbal constructions such as prepositional phrases in Arabic, where "I have a dog" is rendered *ladayya kalbun*, literally "with me a dog".

So far, we have seen how GF can express agreement and other kinds of variation and how the RGL can hide the details from the application programmer. But we have not seen the ultimate abstraction yet: the language-independence of the RGL API. The above linearization rule for `YouHaveItems` has in fact *exactly the same code* for every language that implements the RGL API, but compiles into different records and tables under the hood. Here are some examples of what happens:

- `mkNP` selects the gender of the numeral as a function of the noun in Romance and Slavic languages
- `mkNP` selects the number and case of the noun as a function of the number in Slavic languages and Arabic
- `mkNP` adds a classifier to the noun in Chinese and Thai

Table 1 Wikipedias in different languages, according to https://meta.wikimedia.org/wiki/List_of_Wikipedias retrieved 30 August 2022, and their coverage in the current GF RGL, sorted by the number of articles. The third column shows the status of the RGL: +++ means full API coverage with a large lexicon, ++ means full or almost full API coverage with a smaller lexicon, +* partial API coverage with a large lexicon, + means implementation started,—means not started. * and ** mean corresponding coverage in a closely related language: Arabic for Egyptian Arabic, Malay for Indonesian. The languages on the left are the 35 top languages of Wikipedia, including those without RGL. The languages on the right are those after the top 35 that have some RGL coverage. Four of the RGL languages were not found in the list of Wikipedias, but this may be due to different names used for them

Language	Articles	RGL	Language	Articles	RGL
English	6,545,975	+++	Esperanto	323,608	+
Cebuano	6,125,812	–	Hebrew	321,316	+
German	2,719,877	+++	Danish	284,290	++
Swedish	2,552,522	+++	Bulgarian	283,953	+++
French	2,450,741	+++	Slovak	241,847	+
Dutch	2,099,691	+++	Estonian	229,915	+++
Russian	1,849,325	+++	Greek	212,862	++
Spanish	1,798,346	+++	Lithuanian	204,111	+
Italian	1,769,757	+++	Slovenian	177,533	+*
Egyptian Arabic	1,597,544	**	Urdu	176,166	+++
Polish	1,534,113	+++	Norwegian (Nynorsk)	162,695	++
Japanese	1,340,051	++	Hindi	152,475	+++
Chinese	1,300,293	+++	Thai	149,693	+++
Vietnamese	1,275,688	–	Tamil	148,547	+
Waray-Waray	1,265,938	–	Latin	136,958	+*
Ukrainian	1,190,703	–	Latvian	115,349	+++
Arabic	1,184,349	++	Afrikaans	104,596	+++
Portuguese	1,094,514	+++	Swahili	74,639	+*
Persian	925,446	++	Icelandic	54,803	++
Catalan	709,317	+++	Punjabi	38,549	++
Serbian	662,099	–	Nepali	32,241	++
Indonesian	627,502	*	Interlingua	24,231	++
Korean	603,549	+*	Mongolian	21,436	+++
Norwegian (Bokmål)	597,046	++	Sindhi	15,251	++
Finnish	537,889	+++	Amharic	15,051	++
Turkish	514,410	+*	Zulu	10,583	+
Hungarian	511,089	+	Somali	8,467	+*
Czech	509,155	+	Maltese	4,842	+++
Chechen	481,958	–	Xhosa	1,240	+
Serbo-Croatian	456,901	+	Tswana	773	+
Romanian	433,112	+++	Greenlandic	244	+
Min Nan	431,714	–	Greek (Ancient)	–	+
Tatar	417,595	–	Chiga (Rukiga)	–	+
Basque	397,843	++	Kikamba	–	+
Malay	360,146	+	Egekusii	–	+

- `mkCl` selects the word order in German, Dutch, and Scandinavian
- `mkCl` selects complement cases and prepositions of two-place verbs as needed in almost all languages
- `mkCl` implements the ergative agreement in Basque, Hindi, and Urdu
- `mkCl` selects and orders clitic pronouns in Romance languages and Greek.

To sum up, using RGL API implies that

- Those who write rendering functions do not need to worry about low-level linguistic details, but only about the abstract syntax types of their arguments and values
- A rendering function written for one language is ready to be used for all RGL languages.

These two things together have made GF into a productive tool for multilingual NLG. The code sharing for rendering functions is formally implemented by means of **functors**, a.k.a. **parameterized modules**, which are instantiated to different languages by selecting different instances of the RGL API [26]. In this way, the actual code for the rendering functions does not even need to be seen by the programmer that uses the code for a new language.

Scaling up GF and RGL to the Wikipedia task requires "only" that the RGL be ported to all remaining Wikipedia languages. What this means in terms of effort and skill is a topic worth its own discussion, to which we will return in Sect. 10.

5 Smart Paradigms and the Lexicon

The RGL API offers functions such as `mkCl` and `mkNP` shown above to combine phrases into larger phrases. The smallest building blocks of phrases are **lexical units**, i.e., words with their inflection tables and inherent features.

We have seen *one*, *two*, and *message*, as examples of lexical units represented as records and tables. The RGL provides high-level APIs for constructing them. For most languages, it provides **smart paradigms**, which build complete tables and records from one or few characteristic forms [27]. For example, English has two smart paradigms for nouns (`N`):

```
mkN : Str -> N
mkN : Str -> Str -> N
```

The former paradigm takes one string as its argument, the singular form of the noun. It returns a table that also contains the plural form, where the usual stem alternations are carried out, such as *baby-babies*, *bus-buses*. To form the plural, **regular expression pattern matching** is used. A slightly simplified function for this, also usable for the 3rd person singular present indicative of verbs, is

```
add_s w = case w of {
  _ + ("a"|"e"|"o"|"u") + "y" => w + "s" ;        -- boy, boys
  stem + "y"                  => stem + "ies" ; -- fly, flies
  _ + ("ch"|"s"|"sh"|"x"|"z") => w + "es" ;       -- bus, buses
  _                           => w + "s"          -- cat, cats
  }
```

(the notation—marks comments in GF, here showing examples of words matching each pattern). If a noun has an irregular plural (like *man-men*), the two-argument function is used. Thus the programmer who builds a lexicon just needs to use expressions such as

```
mkN "continent"
mkN "country"
mkN "Frenchman" "Frenchmen"
```

with no worries about the internal records and tables or pattern matching. In German, a particularly useful paradigm is

```
mkN : Str -> Str -> Gender -> N
```

which can for instance generate the record shown in Fig. 1 above, by

```
mkN "Stadt" "Städte" Fem
```

For languages other than English, smart paradigms typically have to do more work, such as produce the 51 forms of the French verb or over 200 forms of the Finnish verb. Evaluations have shown that even in highly inflected languages, all forms of most words can be inferred from just one or two characteristic forms [27].[7] This means that a morphologically complete lexicon can be built rapidly and on a low level of skill. What is more, lexicon building can often be automated: a list of words equipped with part of speech information (noun, adjective, verb) can be mechanically converted into a list of smart paradigm applications.

Many languages have existing morphological dictionaries independent of GF, for instance in Wiktionary as well as in Wikidata itself.[8] Such resources can often be converted to GF records and tables, which means that rendering functions can just use abstract syntax names such as `country_N` and not even care about smart paradigms. However, previously unseen words can always be encountered in texts, especially ones containing specialized terminology, and smart paradigms are then needed to add them to the lexicon. Moreover, for many of those 300 languages that

[7] A typical exception are Indo-European verbs that may need three or more forms, but there are usually just a few hundred of them, and they can be collected into a static lexicon.

[8] Wikidata lexicographical resources:
https://www.wikidata.org/wiki/Wikidata:Lexicographical_data

Abstract Wikipedia targets, comprehensive morphological dictionaries do not exist, and defining smart paradigms for them is an essential part of the RGL building effort.

The lexical items standing for atomic concepts are often not expressible by single words: depending on language, they may be **multiword expressions** with internal syntactic structure. As an example, consider the concept "standard data protection clause" from the General Data Protection Regulation (GDPR) of the European Union.[9] It is a typical example of a legal concept that has established translations into different languages. Its syntactic category is common noun (CN), equipped with a plural form and, in many languages, a gender. To form the plural and identify the gender, one needs to know the syntactic structure—in particular, the **head** of the phrase. Thus we have:

- *Standard data protection clause(s)* (English, head last)
- *Clause(s) type(s) de protection des données* (French, head first, first two words inflecting)
- *Clausol(a/e) tipo di protezione dei dati* (Italian, head first, only the first word inflecting)
- *Standarddatenschutzklausel(n)* (German, single word).

For multiword terms, a handful of syntactic functions are needed in addition to the morphological paradigms in order to define inflection and gender. Thus for instance the French linearization is defined by adding layers of modification to the head noun `clause_N`:

```
mkCN (mkCN type_A clause_N)
  (mkAdv de_Prep (mkNP (mkCN protection_N
    (mkAdv (mkAdv de_Prep (mkNP thePl_Det donnÃ©e_N))))))
```

Writing such complex GF expressions by hand can be demanding, but they can fortunately often be obtained by using RGL grammars for **parsing**. In the present case, the parser must convert the string *clause type de protection des données* into an abstract syntax tree of type CN. This technique, known as **example-based grammar writing**, has been used to enable native speakers to provide grammar rules without writing any code [28].

6 More Abstraction Levels

We have gone through three abstraction levels that a GF rendering function (i.e., linearization) can be defined on:

- Records and tables, mostly needed just inside the RGL

[9] Over 3000 GDPR concepts in five languages have been collected to a GF lexicon in a commercial project, https://gdprlexicon.com/

- RGL API functions, used for building new rendering functions
- Wikipedia rendering functions, such as `AttrFact`.

Higher levels can use the lower levels as **libraries** (in the sense of software libraries), which means that they can take earlier work for granted. As explained in more detail in Sect. 10, we do not expect Abstract Wikipedia authors to use GF on the level of records and tables. Even the level of RGL API is too low for most authors: the main level to work on will be by using the high-level rendering functions, built by a smaller group of experts.

On the level that uses Wikipedia rendering functions, only a small fragment of GF notation is needed: function applications and strings. These constructs are so ubiquitous that they do not even need the GF programming language. Instead, they can be used directly in a general purpose language via **bindings** that are available as a part of the GF software. This technique is called **embedded grammars** and enables programmers to use GF grammars without writing any GF code.

To give an example, consider a rendering function for "the *F* of *X* is *Y*" and its RGL linearization for different languages,

```
fun AttrFact  : Attr -> Obj -> Val -> Fact
lin AttrFact attr obj val =
  mkCl (mkNP the_Det (mkCN attr obj)) val
```

A grammar module containing this function can be called from Python by importing it as a Python module, calling it (for example) `G`, and writing

```
G.AttrFact(G.population_Attr, G.NameObj(name), G.IntVal(pop))
```

where the variables `name` and `pop` get their values directly from Wikidata. The only API that the programmer needs to know are the abstract syntax types of the linearization functions.

Embedded GF grammars are available for C, C#, Haskell, Java, and Python.[10] A further abstraction level on top of this is **Wikifunctions**, which is an emerging technology for accessing all kinds of functions via a web API, hiding the underlying programming language [1]. The plans for Abstract Wikipedia include making GF rendering functions available on this level.

Another direction in which the abstraction from GF code can be extended is by using the parser of GF from a general purpose language. The above call of the rendering function can thus be accessed via its linearization, where slots are left for the values. Here is the equivalent code in Python:

```
str2exp("the population of {X} is {Y}".format(X=name, Y=pop))
```

---

[10] http://www.grammaticalframework.org/doc/runtime-api.html

The string argument of `str2exp` looks exactly like a template, but `str2exp` calls the GF parser to convert strings to GF abstract syntax trees that can be linearized in multiple languages, of course obeying all their grammatical rules beyond string concatenation.

With the parser, the programmer who implements NLG rules can thus use GF rendering functions without even knowing the names of those functions. She just needs to know what can be parsed by the grammar. However, since this can be difficult and error-prone, and since parsing can be ambiguous, the more precise use of imported modules may still be needed as a back-up.

7 Improving the Style

In classical rule-based NLG, data is converted to text in several steps [3]:

- **Content determination**: what to say
- **Text planning**: in what order to say it
- **Aggregation**: merging atomic facts into compound sentences
- **Lexical choice**: selecting words for concepts
- **Referring expression generation**: using pronouns and other alternative expressions for entities
- **Surface realization**: selecting the final word order and morphological forms.

We have by now mostly focused on surface realization. When working on highly multilingual tasks, this is the most demanding component, because of the huge differences between the surface grammars of languages.. We have shown how the RGL gives a solution to this problem. Surface realization is also the clearest contribution that GF itself can make to NLG; most of the other steps can be easier to perform in a general purpose language embedding GF in the way shown in Sect. 6. These steps can operate on the abstract syntax of the GF and thereby deal with several languages at the same time.

Lexical choice is also defined by GF functions and their linearizations for all data entities. One improvement above the monotonic phrasing *the F of X is Y* is to define predicate-specific functions, such as

```
fun PopulationFact : Obj -> Int -> Fact
```

linearized *X has Y inhabitants*. Such functions can be language-specific, if a language happens to have a nice idiom for a certain concept. But they can also be cross-lingual and defined by functors, possibly for a subset of languages for which they are natural and which can implement them. Thus a new language added to the system can start with baseline, monotonic renderings early in the process and get incremental stylistic improvements later.

Starting with the beginning of the pipeline, content determination at its simplest is to take all facts in Wikidata about some object, such as a country, and convert them into sentences of the form "the *F* of *X* is *Y*", rendered as "the *F*s of *X* are *Y*" when *Y* contains multiple values. The resulting text is extremely boring, but it "does the job" in the sense of expressing the information in a grammatically correct way in multiple languages.

Predicate-specific rendering functions are perhaps the simplest way to improve style. A more general way is to add functions that implement text planning, aggregation, and referring expression generation:

```
OneSentDoc : Sent -> Doc              -- S.
AddSentDoc : Doc  -> Sent -> Doc      -- D. S.
ConjSent   : Sent -> Sent -> Sent -- S and S
NameObj    : Name -> Obj              -- Argentina
PronObj    : Name -> Obj              -- it
```

These functions can be used to implement document-level templates, such as the following creating small but reasonably fluent articles about countries:

```
str2exp("Doc",
    ("{c} is a {co} country with {p} inhabitants . "
     "its area is {a} . "
     "the capital of {c} is {ca} and its currency is {cu} ."
     ).format(
         c=countr, co=cont, p=pop, a=area, ca=cap, cu=curr))
```

Our experimental implementation[11] has around 40 GF functions, which can be combined to text templates for different purposes; the main test cases have been geographical data and Nobel prize winners.[12]

While the GF functions and templates are language-independent, the NLG system can customize their use for individual languages. As an obvious example, the area of a country may be converted from square kilometres to square miles for some countries. As a more intricate one, the referring expression generation may utilize the gender systems of different languages to enable the most compact expressions: in English, the pronoun *it* is often ambiguous and therefore not adequate, whereas each of German *er*, *sie*, *es* can be unambiguous in the same context.

[11] https://github.com/aarneranta/NLG-examples

[12] Notice that the `co` argument expects an adjective such as *Asian*. Such adjectives, known as demonyms, are in a natural way included in the linearization records of geographical names. Also notice that the string has been manually tokenized to help the GF parser.

8 Selecting Content

The document template for countries in the previous section builds a text from atomic facts: continent, population, area, capital, currency. All these facts are directly available as RDF triples in Wikidata and can therefore be automatically picked into documents. However, texts in general can choose to drop out some facts and also to state facts that are not directly available in the data. For example, *China has the largest population in the world* is a fact verifiable in Wikidata, but requires a more complex query than an individual triple, for instance, a Python expression

```
maxpop = max(cont_data, key=lambda c: c.population).country
```

which is an example of **aggregation** in the database (rather than NLG) sense. Once this query has been performed and the fact established, the value of `maxpop` can be reported in a text, instead of stating the exact populations of all countries.

Here is a Python template for summaries of facts about continents and the whole world:

```
doc = factsys.str2exp("Doc",
     ("there are {n} countries in {co} ."
      "the total population of {co} is {p} ."
      "{mp} has the largest population "
      "and {ma} has the largest area .".format(
        n = ncountries, co = cont, p = totalpop,
        mp = maxpop, ma = maxarea)
```

Here is a text generated in English, Finnish, and German. The last sentence of these texts is not shown in the template. It is included only in regions that actually have countries with over a billion inhabitants.[13]

> *There are 194 countries in the world. The total population of the world is 7552 million. China has the largest population and Russia has the largest area. India and China are the only countries with over a billion inhabitants.*
>
> *Maailmassa on 194 maata. Maailman yhteenlaskettu asukasluku on 7552 miljoonaa. Kiinalla on suurin asukasluku ja Venäjällä on suurin pinta-ala. Intia ja Kiina ovat ainoat maat, joissa on yli miljardi asukasta.*
>
> *Es gibt 194 Länder in der Welt. Die gesamte Einwohnerzahl der Welt ist 7552 Millionen. China hat die größte Einwohnerzahl und Russland hat die größte Fläche. Indien und China sind die einzigen Länder mit über einer Milliarde Einwohnern.*

Notice that all the facts stated in the above texts may change over time. For example, India is predicted soon to have a larger population than China. A new text can then be generated by exactly the same grammar, template, and queries, to reflect this change.

---

[13] The expression *7552 million* is a result of rounding the exact population. Different rounding functions are available in the grammar, and the NLG has the task to select an approriate one.

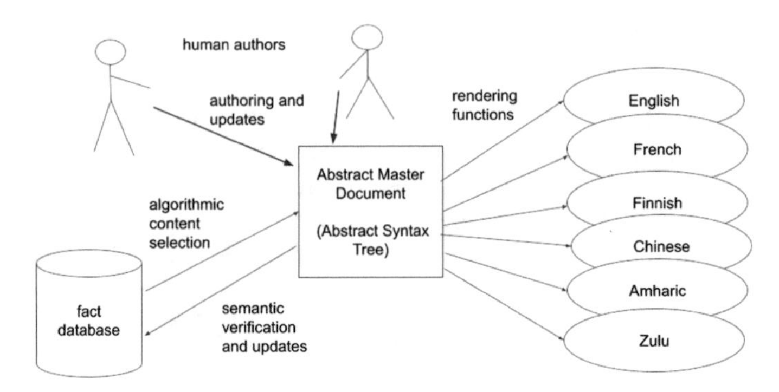

Fig. 2 Text generation from a fact database with a human author in the loop

9 Authoring

In the previous examples, as usual in traditional NLG, the content was selected automatically by an algorithm that decided which facts where interesting. The algorithm also decided the text structure. This is undeniably the fastest and cheapest way to make information available in natural language.

However, fully automatic NLG has two serious shortcomings. First, algorithms cannot always predict which facts are interesting for human readers. Secondly, the content in Wikipedia articles is not always available in Wikidata. To solve these problems, we invoke the parser of GF once again and move from automated NLG to **interactive authoring**. The idea is simply that articles are written by human authors, just like in traditional Wikipedia. But unlike in traditional practices, they are parsed into abstract syntax, verified with respect to Wikidata, and automatically translated to all languages that have a concrete syntax in GF. What is more, the document that is stored as abstract syntax can be edited later by other authors, possibly via input in other languages. Figure 2 shows the architecture of the authoring system.

The idea of interactive authoring was in fact the starting point of whole GF. GF was first created at Xerox in the project entitled Multilingual Document Authoring [29], which in turn was inspited by WYSIWYM ("What You See is What You Mean", [30]). GF was designed to be a general formalism for defining Controlled Natural Languages (CNL). A CNL is a fragment of natural language with (often, even if not always) formal grammar and semantics [31, 32]. In the case of GF, the formal grammar is moreover divided into abstract and concrete parts, which makes it multilingual [33].

One challenge in CNL-based authoring is to make sure that the authors' input can be parsed in the fragment recognized by the grammar. Tools for this have been developed throughout the history of GF [34–38], but there is still room for improvement.

Another challenge, more specific to the Wikipedia case, is that the formal semantics is never going to be complete. As already mentioned, many things that authors want to express might not be present in Wikidata. It can also happen that Wikidata is wrong, or even contradictory. The authoring system can accommodate this by enabling the author to extend or change Wikidata when such situations occur.

10 Roles and Skills

The success of Abstract Wikipedia depends ultimately on getting the community involved. No person or research group could ever recreate all the content in abstract form and define all the required rendering functions in all languages. Community members should be enabled to contribute at different levels of skills:

- **GF implementers**: a specialist group working on the internal algorithms of GF and their integration with Wikipedia software
- **RGL writers**: linguistically proficient programmers implementing new languages
- **NLG programmers**: implementers of rendering functions for new concepts and new textual forms
- **Domain experts**: programmers who adapt rendering functions to new domains and terminologies
- **Content creators**: authors of articles and providers of Wikidata.

As we go down the list, more and more people are needed in each role, while their GF and programming skill requirements decrease.

Starting with GF implementers, this has always been a group of just a few persons. We do not expect that many more are necessarily needed, because GF is already a mature and stable software infrastructure that has been used in numerous NLG projects. Some new functionalities are planned as research projects, in particular dynamic loading and updating of grammars, automated extraction of parts of grammars, and support for Wikifunctions. But even the current implementation is sufficient to get started.

RGL writers are needed for every new language, one or two per language, which means that a few hundred need to be trained to reach the goal of 300 languages. The series of GF Summer Schools organized since 2009[14] has shown that a two-week intensive training is enough for this task. Writing an RGL implementation for a new language itself typically takes two to six person months. In the Abstract Wikipedia project, this task can, however, be started with a subset of the RGL: the experimental implementation described here uses less than 20% of the common abstract syntax functions. Hence a basic rendering system for a new language can be released after a few weeks' work. By using functors (see Sect. 4), one can moreover share most of the concrete syntax code inside language families, so that only one language in the family needs a fully procifient RGL programmer, whereas the others can be added more easily by informants for those languages who provide values to functor parameters. This method has been extensively applied in Scandinavian, Romance [26], Indo-Iranian [18], and Bantu [22] families.

[14] http://school.grammaticalframework.org

NLG programmers implement rendering functions by using the RGL, as well as text templates written in a general purpose host language such as Python. These programmers need to have skills in the host language but can do with a fragment of GF sufficient for using the RGL API.

Domain experts have as their main task to define rendering functions for technical terms, that is, lexical items and multiword expressions. They do not necessarily need GF knowledge, but can just provide strings to be parsed and converted to records and tables via morphological lexica and RGL functions, including smart paradigms. In other words, the domain expert provides the translations of the terms, whereas the RGL generalizes them to rendering rules. The process can be helped by existing sources such as Wikidata lexical resources and WordNets [39, 40].

Content creators, finally, will be the vast majority of Abstract Wikipedia contributors. All Wikipedia authors should be enabled to create abstract content via an easy to use authoring system. Ideally, the overhead of generating abstract content should be small enough to be compensated by the availability of automatic translations to other languages and semantic checking.

11 First Results

Most of the examples above come from an experiment where Wikidata about 194 countries was used for generating articles in four languages: English, Finnish, German, and Swedish. To show how the methods generalize, the experiment was completed by articles about 818 Nobel prize winners:

> *Gerty Cori won the Nobel Prize in Physiology or Medicine in 1947. She was born in Prague in the Czech Republic in 1896. Cori died in 1957. Gerty Cori was the first woman that won the Nobel Prize in Physiology or Medicine.*

The main new feature shown here is the use of tenses. Apart from that, the rendering functions originally written for geographic data were sufficient for the new domain with only a few additions.[15]

The first experiment was carried out by the author of this paper. But how easy is the task for people with no previous GF experience? This was tested in Spring term 2022 by a group of six Bachelor students in Computer Science, who built a system addressing 1235 localities in Sweden [41]. For this group, a 4-hour crash course in GF was given, as well as weekly 1-hour supervision sessions during 12 weeks. The group used the same basic grammar as in the original experiment. In addition to extending the lexicon in the domain expert role, they also extended the texts with new kinds of information, for example, about famous persons and their professions coming from the localities. This experiment can be considered successful in showing how much can be done with a minimal specific training.

[15] One problem should be noticed: in 1896, the Czech Republic did not exist, but Prague was a city in the Austro-Hungarian Empire. Thus more care would have been needed when combining Nobel prize data with geographical data.

Four other student projects in Spring term 2022 were Masters theses addressing research questions related to Wikipedia. One was about formal semantics, interpreting abstract syntax trees in a model based on Wikidata [42]. This was a standard kind of Montague-style logical semantics [43, 44] extended with anaphora resolution and a treatment of sentences that could not be decided in the model. What results from the latter is technically a third truth value. The authoring interface can warn the user about it, or also give her the option to update the semantic model with new facts.

A second Masters thesis built language models, both n-gram and neural, for Wikipedia in 46 languages, with the purpose to assess the fluency of generated texts and help select the most fluent ones of alternative renderings in each language [45]. Unlike purely statistical or neural generation, which has no guarantees to match a non-linguistic reality [46], this use of language models is controlled by the semantics. In the resulting pipeline, rule-based NLG generates several renderings of the same facts, all of them semantically correct, and the language model ranks them in terms of fluency.

The third project was about multilingual treebanks built by parsing a set of Wikipedia articles in 58 languages with Universal Dependencies (UD, [47]). The result was a ranked set of syntactic structures used in Wikipedias, as well as a cross-lingual comparison of how similar the structures are [48]. This makes it possible to estimate how often a functor-based linearization is adequate. The results were moreover used in a fourth project, still in progress at the time of writing, to extract GF rendering functions from UD trees that are actually used in the current Wikipedia.

In addition to the directly NLG-related projects, resource development has started to export GF lexical data into Wikidata, in connection with the GF-WordNet project.[16] At the time of writing, such resources have been exported for 24 languages, around 590k entries per language.

12 Conclusion

The goal of this paper has been to show that the advantages of NLG—consistency, speed, cost, updates, and customization—are relevant for Wikipedia and can be scaled up to a multilingual setting by using GF. We have gone through some results from an experimental implementation and addressed some known problems and how to mitigate them:

- Robotic style: can be improved by adding more NLG functions (Sect. 7) and, in particular, by enabling human authoring via a CNL (Sect. 9).
- Lack of data: can be completed as a by-product of authoring (Sect. 9). But even now, there is enough Wikidata to create useful content, even though the whole Wikipedia cannot be covered.

[16] http://cloud.grammaticalframework.org/wordnet/

- Cost: can be reduced by abstract syntax and functors, which enable code sharing between languages (Sect. 3).
- Human resources: can be helped with tools such as parsing and authoring, so that most of the work needs only minimal training (Sect. 10).
- Community resistance: many of the earlier problems of text robots and automatic translation can be avoided. But it remains to be seen how the wider community responds to this initiative.

A baseline implementation already exists and can be applied to new areas of knowledge and adapted to new languages. Even though perfection might never be achieved, gradual improvements can enable more and more content to be created with better and better style in more and more languages.

Using GF for this task has a number of advantages. First, the Resource Grammar Library is a unique resource for multilingual generation functions. As it contains even less-represented languages (see Table 1), adding more such languages looks like a realistic goal.

Secondly, GF has programming language support for linguistic constructs, and the community has a long experience of using it for many kinds of languages. Using GF, rather than a general purpose programming language, for grammar implementation is analogous to using YACC-like tools for programming language implementation. Even though general purpose languages by definition are theoretically sufficient for implementing grammars, doing this would require many times more effort and low-level repetitive coding.

Thirdly, GF has APIs that allow it to be embedded in general purpose programming languages. Thus programmers can use GF functionalities—in particular, the RGL—without writing any GF code at all. Reproducing the knowledge contained in the RGL, rather than importing it via embedded grammars, would involve dozens of person years of duplicated work.

This said, Abstract Wikipedia is a challenge that exceeds previous applications of GF, or any other NLG project, by at least two orders of magnitude: it involves almost ten times more languages and at least ten times more variation in content than any earlier project.

Acknowledgements I am grateful to Denny Vrandečić for inspiring discussions about the Abstract Wikipedia project and to the anonymous referees for insightful comments on this paper. Special thanks go to Roussanka Loukanova for her extraordinarily supportive editorial help.

References

1. Vrandečić, D.: Building a multilingual wikipedia. Commun. ACM **64**(4), 38–41 (2021). https://cacm.acm.org/magazines/2021/4/251343-building-a-multilingual-wikipedia/fulltext
2. Vrandečić, D., Krötzsch, M.: Wikidata: A free collaborative knowledgebase. Commun. ACM **57**(10), 78–85 (2014). https://doi.org/10.1145/2629489
3. Reiter, E., Dale, R.: Building Natural Language Generation Systems. Cambridge University Press (2000)

4. WikipediaContributors: Lsjbot. https://en.wikipedia.org/wiki/Lsjbot (2022). Accessed 20 April 2022
5. WikipediaContributors: Wikipedia:Content translation tool. https://en.wikipedia.org/wiki/Wikipedia:Content_translation_tool(2022). Accessed 20 April 2022
6. Ranta, A.: Grammatical Framework: Programming with Multilingual Grammars. CSLI Publications, Stanford (2011)
7. Ranta, A.: Grammars as software libraries. In: Bertot, Y., Huet, G., Lévy, J.J., Plotkin, G. (eds.) From Semantics to Computer Science. Essays in Honour of Gilles Kahn, pp. 281–308. Cambridge University Press (2009)
8. Hammarström, H., Ranta, A.: Cardinal numerals revisited in GF. In: Workshop on Numerals in the World's Languages, Department of Linguistics, Max Planck Institute for Evolutionary Anthropology, Leipzig (2004)
9. Ranta, A.: The GF resource grammar library. Linguist. Lang. Technol. **2** (2009)
10. Burke, D.A., Johannisson, K.: Translating formal software specifications to natural language/a grammar-based approach. In: Blache, P, Stabler, E., Busquets, J., Moot, R. (eds.) Logical Aspects of Computational Linguistics (LACL 2005), *LNCS/LNAI*, vol. 3492, pp. 51–66. Springer (2005). http://www.springerlink.com/content/?k=LNCS+3492
11. Ranta, A.: Translating between language and logic: what is easy and what is difficult. In: Bjørner, N., Sofronie-Stokkermans, V. (eds.) Automated Deduction—CADE-23, pp. 5–25. Springer, Berlin Heidelberg, Berlin, Heidelberg (2011)
12. Saludes, J., Xambo, S.: The GF mathematics library. In: THedu'11 (2011)
13. Dannélls, D., Damova, M., Enache, R., Chechev, M.: Multilingual online generation from semantic web ontologies. In: Proceedings of the 21st international conference on World Wide Web, pp. 239–242. ACM (2012)
14. Angelov, K., Camilleri, J., Schneider, G.: A framework for conflict analysis of normative texts written in controlled natural language. The J. Logic Algebraic Program. **82**, 216–240 (2013)
15. Marais, L., Pretorius, L.: Exploiting a multilingual semantic machine translation architecture for knowledge representation of patient information for covid-19. In: Proceedings of the 1st South African Conference for Artificial Intelligence Research, (SACAIR 2020), pp. 264–279 (2021)
16. Dada, A.: Implementation of the Arabic Numerals and Their Syntax in GF. In: Proceedings of the 2007 Workshop on Computational Approaches to Semitic Languages: Common Issues and Resources, Semitic '07, p. 9-16. Association for Computational Linguistics (2007)
17. Listenmaa, I., Kaljurand, K.: Computational Estonian grammar in grammatical framework. In: 9th SaLTMiL Workshop on Free/open-Source Language Resources for the Machine Translation of Less-Resourced Languages, LREC 2014, Reykjavík (2014)
18. Virk, S.: Computational linguistics resources for Indo-Iranian languages. Ph.D. thesis, Dept. of Computer Science and Engineering, Chalmers University of Technology and Gothenburg University (2013)
19. Camilleri, J.J.: A computational grammar and lexicon for Maltese. Master's thesis, Chalmers University of Technology, Gothenburg, Sweden (2013). http://academic.johnjcamilleri.com/papers/msc2013.pdf
20. Dada, A.E., Ranta, A.: Implementing an open source Arabic resource grammar in GF. In: 20th Arabic Linguistics Symposium. Western Michigan University March 3–5 2006 (2006)
21. Bamutura, D., Ljunglöf, P., Nabende, P.: Towards computational resource grammars for runyankore and rukiga. In: Language Resources and Evaluation (LREC) 2020, pp. 2846–2854. Marseille, France (2020)
22. Kituku, B., Nganga, W., Muchemi, L.: Leveraging on cross linguistic similarities to reduce grammar development effort for the under-resourced languages: a case of Kenyan Bantu languages. In: 2021 International Conference on Information and Communication Technology for Development for Africa (ICT4DA), pp. 83–88 (2021). https://doi.org/10.1109/ICT4DA53266.2021.9672222
23. Pretorius, L., Marais, L., Berg, A.: A GF miniature resource grammar for Tswana: modelling the proper verb. Lang. Resour. Eval. **51**(1), 159–189 (2017)

24. Ranta, A., Tian, Y., Qiao, H.: Chinese in the grammatical framework: grammar, translation, and other applications. In: Proceedings of the Eighth SIGHAN Workshop on Chinese Language Processing, ACL, pp. 100–109. Beijing, China (2015). http://www.aclweb.org/anthology/W15-3117
25. Zimina, E.: Fitting a Round Peg in a Square Hole: Japanese Resource Grammar in GF. In: JapTAL, vol. 7164, pp. 156–167. LNCS/LNAI, Kanazawa (2012)
26. Ranta, A.: Modular grammar engineering in GF. Res. Lang. Comput. **5**, 133–158 (2007)
27. Détrez, G., Ranta, A.: Smart paradigms and the predictability and complexity of inflectional morphology. In: EACL 2012 (2012)
28. Ranta, A., Détrez, G., Enache, R.: Controlled Language for Everyday Use: the MOLTO Phrasebook. In: CNL 2012: Controlled Natural Language, *LNCS/LNAI*, vol. 7175 (2010)
29. Dymetman, M., Lux, V., Ranta, A.: XML and multilingual document authoring: Convergent trends. In: Proceedings of the Computational Linguistics COLING, Saarbrücken, Germany, pp. 243–249 (2000)
30. Power, R., Scott, D.: Multilingual authoring using feedback texts. In: COLING-ACL (1998)
31. Fuchs, N.E., Kaljurand, K., Kuhn, T.: Attempto controlled English for Knowledge Representation. In: Baroglio, C., Bonatti, P.A., Małuszyński, J., Marchiori, M., Polleres, A., Schaffert, S. (eds.) Reasoning Web, Fourth International Summer School 2008, no. 5224 in LNCS, pp. 104–124. Springer (2008)
32. Kuhn, T.: A survey and classification of controlled natural languages. Comput. Linguist. **40**(1), 121–170 (2014)
33. Angelov, K., Ranta, A.: Implementing controlled languages in GF. In: CNL-2009, Controlled Natural Language Workshop, Marettimo, Sicily, 2009 (2009)
34. Angelov, K., Bringert, B., Ranta, A.: Speech-enabled hybrid multilingual translation for mobile devices. In: Proceedings of the Demonstrations at the 14th Conference of the European Chapter of the Association for Computational Linguistics, pp. 41–44. Gothenburg, Sweden (2014)
35. Hallgren, T., Ranta, A.: An extensible proof text editor. In: Parigot, M., Voronkov, A. (eds.) LPAR-2000, *LNCS/LNAI*, vol. 1955, pp. 70–84. Springer (2000). http://www.cse.chalmers.se/aarne/articles/lpar2000.pdf
36. Kaljurand, K., Kuhn, T.: A multilingual semantic wiki based on Attempto Controlled English and Grammatical Framework. In: The Semantic Web: Semantics and Big Data, pp. 427–441. Springer (2013)
37. Khegai, J., Nordström, B., Ranta, A.: Multilingual syntax editing in GF. In: Gelbukh, A. (ed.) Intelligent Text Processing and Computational Linguistics (CICLing-2003), Mexico City, February 2003, *LNCS*, vol. 2588, pp. 453–464. Springer (2003)
38. Ranta, A., Angelov, K., Hallgren, T.: Tools for multilingual grammar-based translation on the web. In: Proceedings of the ACL 2010 System Demonstrations, pp. 66–71. Uppsala, Sweden (2010). https://aclanthology.org/P10-4012.pdf
39. Angelov, K., Lobanov, G.: Predicting translation equivalents in linked WordNets. In: The 26th International Conference on Computational Linguistics (COLING 2016), p. 26 (2016)
40. Fellbaum, C.: WordNet: An Electronic Lexical Database. MIT Press (1998)
41. Diriye, O., Folkesson, F., Nilsson, E., Nilsson, F., Nilsson, W., Osolian, D.: Multilingual text robots for abstract wikipedia. Bachelor's thesis, Chalmers University of Technology, Gothenburg, Sweden (2022)
42. Stribrand, D.: Semantic Verification of Multilingual Documents. Master's thesis, Chalmers University of Technology, Gothenburg, Sweden (2022)
43. Montague, R.: Formal Philosophy. Yale University Press, New Haven (1974). Collected papers edited by Richmond Thomason
44. Van Eijck, J., Unger, C.: Computational Semantics with Functional Programming. Cambridge University Press (2010)
45. Le, J., Zhou, R.: Multilingual language models for the evaluation and selection of auto-generated abstract wikipedia articles. Master's thesis, Chalmers University of Technology, Gothenburg, Sweden (2022)

46. Bender, E.M., Koller, A.: Climbing towards NLU: On meaning, form, and understanding in the age of data. In: Proceedings of the 58th Annual Meeting of the Association for Computational Linguistics, pp. 5185–5198. Association for Computational Linguistics, Online (2020). https://aclanthology.org/10.18653/v1/2020.acl-main.463
47. Nivre, J., de Marneffe, M.C., Ginter, F., Goldberg, Y., Hajic, J., Manning, C.D., McDonald, R., Petrov, S., Pyysalo, S.m., Silveira, N., Tsarfaty, R., Zeman, D.: Universal Dependencies v1: A Multilingual Treebank Collection. In: Proceedings of the Tenth International Conference on Language Resources and Evaluation (LREC 2016), pp. 1659–1666. European Language Resources Association (ELRA), Portorož, Slovenia (2016). https://www.aclweb.org/anthology/L16-1262
48. Grau Francitorra, P.: The linguistic structure of wikipedia. Master's thesis, University of Gothenburg, Gothenburg, Sweden (2022)

Decomposing Events into GOLOG

Symon Jory Stevens-Guille

Abstract Linguists working in a variety of semantic theories have converged on a non-atomic view of predicates. According to these theories, the object denoted by a verb can be decomposed into some complex expression which might or might not include one or more primitive predicates expressing acts or results of acts. I develop an account of event-semantic decomposition into GOLOG programs, a variety of the Situation Calculus. The result is an implemented theory of decomposition which can compute well-known inferences concerning telicity.

Keywords Event semantics · Situation calculus · Categorial grammar · GOLOG

1 Introduction

If Bill is running, then it would be reasonable to conclude that Bill has run, even if only a short way. However, if Bill is running a kilometre, then one is likely to conclude he's run *per se* or run some distance less than a kilometre. But one surely cannot conclude that there is any point prior to Bill's passing the one kilometer mark that he's run a kilometre. How does this discrepancy in reasoning arise? That much is up for debate. But what is widely settled is that the source of this discrepancy must be *aspect*, the linguistic expression of 'the internal constituency of actions, events, states, processes or situations' [16].

In the present paper I propose to decompose verb meaning into GOLOG logic programs. In fact, my aim is even more modest: I propose that several notable aspectual properties of predicates, notably their division into classes and the inferences between such classes, can be modelled in GOLOG. GOLOG implements a version of [27]'s Situation Calculus (SC) due to [34] in multi-sorted second order logic. In brief, I define schemes for a typology of verb frames [17, 36, 46], which are distinguished by sets of licit inferences and modifier restrictions discussed in [8]. Among

S. J. Stevens-Guille (✉)
Department of Linguistics, The Ohio State University, Columbus, US
e-mail: stevensguille.1@buckeyemail.osu.edu

R. Loukanova et al. (eds.), *Logic and Algorithms in Computational Linguistics 2021 (LACompLing2021)*, Studies in Computational Intelligence 1081,
https://doi.org/10.1007/978-3-031-21780-7_7

linguists, verb frames are frequently conceived of in terms of events or, in the more fine-grained terminology of [1], eventualities.[1] I suggest that the inferential distinctions between types of eventualities can be made more explicit by modelling them in a tractable inference system. Underlying this suggestion is hope in the possibility of rigorously testing semantic theories by reference to the inferences they provably license.

In Sect. 2, I introduce the linguistic background to the present work, including a summary of the inference patterns which typify different sorts of predicates denoted by verbs and verb phrases.[2] In Sect. 3, I introduce a version of the Situation Calculus due to [33] and show how to read GOLOG programs from decomposed representations with syntax quite close to the extended Motague Grammar of [8]. I describe how the inference patterns I intend to license are indeed licensed by the execution and querying of GOLOG programs. In Sect. 4, I introduce the Lambek Calculus and provide a lexicon, the semantic terms of which reduce to GOLOG programs. Section 5 discusses future work. Section 6 concludes.

2 Data and Background

In this section, I introduce the milleiu in which events came to occupy a priveleged position in the ontology of linguistic semantics and AI. Section 2.2 briefly introduces Davidson's thinking about events. Section 2.3 introduces McCarthy's notion of situation, its subsequent refinement by Reiter, and how actions in the Situation Calculus could be conceived of in terms of events.

2.1 Events in Philosophy

The early 1960s witnessed a convergence of thinking about things that happen in terms of things which might be quantified over. In the early days of formal semantics, Donald Davidson, in a paper on the philosophy of action, argued that 'there is a lot of language we can make systematic sense of if we suppose events exist' [7]. Going on to give an account of the semantics of 'action sentences', he exemplified the benefits of reifying events into objects by showing they could be used to provide antecedents to anaphors, license various intuitive entailment patterns, and provide a simple hook for the 'variable polyadicity' of predicates in the form of the object–the event–which could be the argument of predicate modifiers. The puzzle of 'variable polyadicity' had been given the limelight by Anthony Kenny's work on the philosophy of action in [17]. It describes the phenomenon whereby a predicate can be seemingly indefinitely

[1] See [11] for discussion of linguistic theories of verb decomposition.

[2] By verb phrase I mean nothing more nor less than the syntactic composition of a verb with zero or more of its arguments and/or modifiers.

extended, e.g., [7]'s famous sentence 'He did it slowly, deliberately, in the bathroom, with a knife, at midnight'.

Davidson's conception of events would go on to be nurtured by both philosophers and linguists. In linguistics, events–or the extended notion eventualities [1]–are ubiquitous, so much so there is now a handbook devoted entirely to their use in linguistic theory [43]. But less frequently cited in the linguistics literature is the development of a conception of events and situations in AI that paralleled what now goes under the heading of (neo-)Davidsonian semantics. It is this conception of events which I will argue is ripe for use in semantics, especially for those semanticists who aspire to build theories which could be tested (semi-)automatically.[3]

2.2 *Events in AI*

John McCarthy, in a report just a few years before Davidson's paper, wrote '[h]uman intelligence depends essentially on the fact that we can represent in language facts about our situation, our goals, and the effects of the various actions we can perform' [27]. His report went on to propose the Situation Calculus. The Situation Calculus will be presented formally in the next section, it is worth noting the similarities and dissimilarities to events in formal semantics. While events in formal semantics are sometimes considered primitive, other times composed of more primitive objects [3], the original conception of a situation in the Situation Calculus is a complete state of the universe at a given time [27, 28]. Of course, since the domain involved in physical computing is finite and since the closed-world assumption is often in force, this conception is more philosophical than practical.[4] More recent versions of the Situation Calculus, like the one used here, conceive of situations more modestly in terms of sequences of actions. Moreover, these situations can be extended to incorporate time and participants [34], making them quite close in principle to the conception of event which seems common to (neo-)Davidsonian semantics [10, 32].

In the sequel I will try to bring out the similarites between Situation Calculus actions and (neo-)Davidsonian events by using the expression 'event' where the Situation Calculus literature would use 'action'. Likewise I will subsequently use the term 'situation' to refer to a sequence of events, while in the contemporary Situation Calculus literature these would be sequences of actions. My contention,

[3] The connection between situations in the SC and the research programme of Situation Semantics [2] is worth further study. The differences and similarities between events and situations in Barwise and Perry's sense constitute a rich vein in semantics, mined to some reward by [5, 51]. Due to my focus here on events I defer the discussion of Situation Semantics to future work. For recent novel developments in the situation-theoretic tradition of semantics the reader is referred to [23–25], where the notion of situation is explored in the type-theoretic setting.

[4] The closed-world assumption holds that every sentence which is true is known to be true and moreover every sentence the truth of which is unknown is not true.

Table 1 Vendler-style typology of predicate frames

	Telic	Atelic
Period	Accomplishments *Run five kilometres*	Activities *Run*
Point	Achievements *Die*	States *Be dead*

piggybacking on Davidson, is there is a lot of language we can make systematic sense of if we conceive of events in terms of actions in the Situation Calculus.[5]

2.3 Data

The patterns which concern me, though not entirely novel to them, were brought to the attention of Anglophone philosophers, and subsequently Anglophone linguists, by [17, 36, 46]. They emphasized distinctions in well-formedness conditions and seemingly licit inferences between different sets of predicates. Subsequently [8, 9, 47] developed these ideas in terms borrowed from Generative Semantics and Montague Grammar. One division of verbal predicates due to [46] is common in introducing linguistic theories of aspectual class; I show a version of his typology borrowed and modified from [43] in Table 1.[6]

Telicity refers to whether there is some inherent endpoint to the description of the event. Intuitively there is an obvious end to the process of dying, namely the point at which one becomes dead, beyond which there can be no extension of the process. On a happier note there is a distinct endpoint to a run of five kilometres: the point at which the runner reaches the five kilometre mark. Of course the runner might continue running, but they would no longer be running five kilometres.

---

[5] Previous approaches to aspect modelled in terms close to the Situation Calculus include [16, 40, 45] and [39, 49, 50]. This comparative dearth of work in formal semantics using the Situation Calculus is somewhat surprising.

[6] *Aspectual class* is defined by properties that are often sourced to the verb or verb phrase but can even be properties of sentences once the contribution of the subject is taken into account. But [12] notes it should not be confused with *lexical aspect*, which is *aspectual class* restricted to just verbs, or, for some authors, verb phrases; nor should it be reduced to *aspectual form*, which concerns the syntactic and morphological representation of *grammatical aspect* [12]. *Lexical aspect* is often used to describe intrinsic semantic properties of verbs, while *grammatical aspect* is used to describe those aspects which are explicitly marked by morphology or syntactic constructions, i.e., those aspects which have an *aspectual form* or *forms*. Distinctions among these terms of art are discussed in [12]. For the purposes of this paper, I am restricting myself to aspectual class in the sense of verbal predicate frames, i.e., syntactic and semantic generalizations concerning combinations of verbs, their objects, and modifiers, but not their subjects. Hopefully the restriction of my modelling to English will reduce the risks of using this terminology, which could distort the perception of data if used unreflectively to describe other languages.

The intuition underlying the category of atelic predicates is simply their lack of a definitive endpoint. Thus the runner could set out with no determinate restriction on the run, stopping only when they tired. Irrespective of where and for how long the runner runs, if they stop, they have run. Likewise the state of death persists. At the risk of being morbid, the state of death is unlimited, without expected or intended end, while even the best runner must stop running sometime. Beyond intuitions, linguists have developed distributional tests which provide a guide to distinguishing telic and atelic predicates.

Distinction 1 Atelic but not Telic predicates can be modified by *for*-adverbs.[7]

Consider the following examples which use Distinction 1 to distinguish accomplishments and achievements from activities and states. Here and in the sequel I use the symbol # prefixed to sentences to indicate the sentence is judged infelicitous.

1. *#Bill ran five kilometres for one hour.*
2. *#Bill died for one hour.*
3. *Bill ran for one hour.*
4. *Bill was dead for one hour.*

In fact, Vendler used four tests to distinguish the predicates from one another: *in* modification, *for* modification, *at* modification, and the existence of a felicitous progressive form [30], points out that these tests are not independent; of the possible 16 combinations of positive or negative values for the tests, we get just the four Vendler proposed.

Distinction 2 below is another consequence of telicity.

Distinction 2 Telic but not Atelic predicates can be modified by *in*-adverbs.

This is exemplified by the following judgements:

5. *Bill ran five kilometres in one hour.*
6. *Bill died in one hour.*
7. *#Bill ran in one hour.*
8. *#Bill was dead in one hour.*

I note that there seems to be some level of variation in judgements concerning *dead*–for speakers who find *Bill was dead in one hour* fine, it seems likely they are getting an accomplishment reading rather than a stative reading, e.g., they understand the *in*-adverb to denote the point by which the state of being *dead* holds.

Vendler's third distinction cross cuts the Telic and Atelic distinction.

Distinction 3 Point but not Period predicates can be modified by *at*-adverbs.

My own view is that this test is less robust than the others in that the apparently infelicitious examples seem less infelicitous than those used in the other tests. Nonetheless, the distinction predicts the following distribution of judgements for the sentences below:

---

[7] Note this is not a sufficient condition for distinguishing predicates.

9. *#Bill ran five kilometres at seven.*
10. *Bill died at seven.*
11. *#Bill ran at seven.*
12. *Bill was dead at seven.*

The intuition is that Period predicates must happen over some extended time, while Point predicates are either defined by some climactic moment or else are timeless, which timelessness is conceivable in terms of property persistence under every arbitrary choice of moment.

Rounding out the syntactic and morphological distinctions between these types of predicates in English is whether they have progressive forms. Note that this condition is neither necessary nor sufficient for distinguishing the predicate frame.

Distinction 4 Period but not Point predicates can have progressive forms.

To see the force of this distinction, consider the following distribution of judgements.

13. *Bill is running five kilometres.*
14. *#John is dying.*[8]
15. *Bill is running.*
16. *#John is being dead.*

Statives quite strikingly resist progressive forms, while with some achievements it is possible to coerce a reading of the predicate under which it is the result of some activity. No such restriction is found for accomplishments or activities.

Further distinctions between predicate frames have been proposed. Much evidence, compiled subsequent to Vendler's work, is supposed to merit further subdividing the frames or reducing some of them to others; see [29] for an overview of this evidence. For present purposes I will suppose that the Vendler typology is worth modelling. I defer study of how to modify my account with respect to proposed refinements of Vendler's typology to some future effort.

Once tense is considered, there are inferences within and between the cells of Vendler's table. It is these inferences which I intend to model in the sequel, hoping distinctions one through four will suffice to convince the reader the grammar of aspect might have some semantic source or consequence. Consider the following seemingly licit set of inferences.

Pattern 1

- *Bill ran for five minutes.* → *Bill ran for three minutes.*
- *Bill was dead for one hour.* → *Bill was dead for thirty minutes.*

[8] 'Dying' is well-formed when construed in terms of the preceding injury or illness producing death. Notice neither injury nor illness seem to require the agentivity found in predicates like 'run'. However, one might suggest the activity licensing the progressive form of 'die' is 'kill', which requires a killer but not some living being who kills.

What these examples seem to show is that simple past for atelic durative predicates licenses inferences to simple past for subsets of the length of the event denoted by the predicate, which property is sometimes dubbed 'homogeneity'[18, 29].

Notice, moreover, that when considering inferences within the same cell, atelic predicates, but not telic predicates, license the inference to the simple past from the progressive ([1, 12, 41]):

Pattern 2

- *Bill is dying.* $\nrightarrow$ *Bill (already) died.*[9]
- *Bill is running* $\rightarrow$ *Bill (already) ran.*

Considering inference between cells, notice that simple past and progressive of accomplishments license the simple past and progressive of the corresponding activities.

Pattern 3

- *Bill ran five kilometres.* $\rightarrow$ *Bill ran.*
- *Bill is running five kilometres.* $\rightarrow$ *Bill is running.*

Not only are there inferences to one cell from another, there are some equivalencies too. Thus the simple past of achievements and the present of the corresponding states seem to license inferences between themselves.

Pattern 4

- *John died.* $\leftrightarrow$ *John is dead.*

Of note when considering inferences between cells is the so-called 'imperfective paradox' [8, 9], which relies on the distribution of (non)inferences from present progressive to simple past between and within activities and accomplishments. Thus within both activities and accomplishments we get the following judgements:

Pattern 5

- *Bill is running.* $\rightarrow$ *Bill ran.*
- *Bill is running five kilometres.* $\nrightarrow$ *Bill ran five kilometres.*

But then when we combine these judgements with the judgements concerning inferences from accomplishments to activities, we get the following series of (non)inferences, cf. [42].

17. *Bill is running five kilometres.*

$\rightarrow$ *Bill is running.*
$\rightarrow$ *Bill has run.*
$\nrightarrow$ *Bill has run five kilometres.*

---

[9] Note that this presumes *Bill is dying* is well-formed and still expressing an achievement–otherwise it's not telic, cf. footnote 7.

Evidently activities do not inherit the telicity of the accomplishments they are inferred from. Within the Situation Calculus tradition, the imperfective paradox has been accounted for by both [49, 50] and [16, 40, 45], either in the context of the Situation Calculus or in a novel aspectual calculus [49]. The Situation Calculus account reduces the 'paradox' to the Frame Problem, which enjoys a nice solution in Reiter's version of the Situation Calculus. I expect that the Event Calculus solution can be carried over to the Situation Calculus with some time and effort, though I defer justifying this suggestion to future work.

In the sequel I restrict myself to those inferences (reflexively) to/from simple past and simple present [8], argued that the four Vendler categories could be schematized in the following form:

- States: ϕ
- Achievements: $BECOME(\phi)$
- Activities: $DO(x, \phi)$
- Accomplishments: $CAUSE(DO(x, \phi), BECOME(\pi))$

Without knowing what the predicates mean, these schemes aren't terribly informative but they are suggestive. It will become evident in the sequel how the intuitions underlying Dowty's choice of syntax are fruitfully interpreted in terms of the Situation Calculus. Dowty himself synthesizes [22]'s theory of counterfactuals, [48]'s logic of change, and [35]'s Generative Semantic theory of DO, embedding the result in a version of Montague Grammar. I do not intend to detract from Dowty's ingenuity and I stress that by synthesis I do not mean some kind of hodgepodge. Strictly speaking Dowty's account is couched in an extended Montague Grammar where the extensions are inspired by, among others, the formentioned antecedents. But the semantics of the extensions are defined within the same model-theoretic apparatus Dowty proposes for the rest of his 'natural logic'. Dowty's theory, though not without issue, provides the most worked-out purely model-theoretic account of the sorts of inferences discussed herein. In the next section I show how Dowty's schemes can be given a nice correspondence in the Situation Calculus. Subsequently I show how some of the (non)inferences discussed in the foregoing can be deduced.

3 The Situation Calculus

In this section I introduce the Situation Calculus and my account of some of the foregoing inferences. Section 3.1 introduces the language of the Situation Calculus. I present the syntax in terms of the logic undergirding the SC, but Sect. 3.3 shows that GOLOG code is remarkably close to the logic itself, thereby providing the opportunity to implement some of the inferences discussed in the foregoing. Section 3.2 provides some brief remarks concerning the types of axioms in SC and the roles they perform.

3.1 The Language of the Situation Calculus

The language in which the inferences I intend to model will be written is the Situation Calculus (SC) [27]. To be more precise, it is the version of the SC introduced by [33], which is a theory written in second order logic.[10] SC is further developed in [21, 34] in terms of the logic programming language GOLOG. In the sequel I'll frequently refer loosely to GOLOG for this version of the SC, presuming context will distinguish matters of pure logic from implementation.

L_{sc} is a multi-sorted second order language with equality. Sorts of classical multi-sorted second order logic are extended with *s*ituations, *e*vents (considered 'actions' in the literature), and fluents: situation dependent predicates and functions. The sort *o*bject corresponds to individuals. We define L_{sc} below, with sorts restricting the recursion:

$$
\begin{aligned}
&Var &&::= x^o \mid y^e \mid z^s \mid P^{(e \,\cup\, o \,\cup\, s)^n}\\
&Con &&::= \mathrm{a}^o \mid \mathrm{b}^e \mid \mathrm{c}^s \mid \mathrm{Q}^{(e \,\cup\, o \,\cup\, s)^n} \mid \mathrm{S}_0^s\\
&Fun &&::= f^{(o \,\cup\, e)^n \to o} \mid g^{(o \,\cup\, e)^n \to e} \mid h^{(e \,\cup\, o)^n \times s \to (e \,\cup\, o)^n} \mid \mathrm{do}^{e \times s \to s}\\
&Terms &&::= Var \mid Con \mid Fun(x_0^T, ..., x_n^T)\\
&Preds &&::= R^{(e \,\cup\, o \,\cup\, s)^n} \mid \sqsubset^{(s \times s)} \mid \mathrm{Poss}^{(e \times s)}\\
&Atoms &&::= Preds(x_0^T, ..., x_n^T)\\
&Forms &&::= Atoms \mid \forall x^T.Forms \mid \neg Forms \mid Forms \vee Forms
\end{aligned}
$$

I comment on several features of the foregoing. First, the full set of connectives is derived from the foregoing by the standard methods. Second the superscript T represents the sorts of the elements in *Terms*. Second there are a number of distinguished subsets of expressions:

- The 0-ary function S_0^s is the distinguished starting situation
- The function $\mathrm{do}^{e \times s \to s}$ is the function which brings the event executed in S_n to the situation S_{n+1}
- The predicate $\sqsubset^{s \times s}$ is the predicate which holds of situations S_n and S_m when S_n is a proper subsequence of S_m
- The predicate $\mathrm{Poss}^{e \times s}$ denotes the possibility of executing some event X in the situation S_n
- Predicates of the form $\mathrm{P}^{(e \,\cup\, o)^n}$ are situation independent (likewise for functions with return sort o), while predicates of the form $\mathrm{P}^{(e \,\cup\, o)^n \times s}$ are relational fluents (respectively functional fluents when the return sort is $o \cup e$)
- Functions of the sort $g^{(o \cup e)^n \to e}$ are event functions, which will be restricted by event precondition axioms

I now turn to a brief discussion of the sorts of axioms which underly the Situation Calculus. But before doing so I mention here the fundamental notion of regression in

[10] The higher order requirements are minor enough [21, p. 60] considers it a 'first order language (with, as we shall see, some second order features) specifically designed for representing dynamically changing worlds'.

the SC. Regression corresponds to a certain form of rewriting such that each formula denoting a complex event is reduced to an equivalent formula which mentions just the starting situation S_0. It turns out that a formulae F is entailed by a theory Σ ($\Sigma \models F$) if and only if the regression of F, written $G(F)$, is entailed by the subtheory Σ_0 of Σ which mentions no situation term but S_0 and the so-called unique names axiom for events ($\Sigma_0 \cup UNE \models G(F)$), which is one of the indepedent axioms of SC. Unique Names for Events just requires that every two distinct event names name distinct events and every two identical events have identical arguments. I mention the equivalence concerning regressable and regressed formulae since it is fundamental to computing with SC: evaluation of regressable formulae reduces to theorem proving using only $\Sigma_0 \cup UNE$; this makes implementing SC in Prolog quite transparent, given the execution of Prolog programs is itself reducible to theorem proving. Since an exhaustive discussion and formalization would be lengthy, we refer the interested reader to [34] for an explicit set of axioms, proofs concerning the metamathematical properties of the SC, and thorough exposition of interpreting the SC in Prolog, i.e., implementing GOLOG.

3.2 Domain Dependent and Independent Axioms

The Situation Calculus includes a collection of axioms which are independent of any particular domain. I introduce these axioms below.

18. $\mathrm{do}(e_1, s_1) = \mathrm{do}(e_1, s_1) \supset e_1 = e_2 \wedge s_1 = s_2$

Unique Names for Situations

19. $\forall P.P(\mathrm{S}_0) \wedge (\forall e, s)[P(s) \supset P(\mathrm{do}(e, s)] \supset (\forall s)P(s))$

Second Order Induction

20. $\neg s \sqsubset \mathrm{S}_0$

Subhistory I

21. $s \sqsubset \mathrm{do}(e, s') \equiv s \sqsubset s' \vee s = s'$

Subhistory II

The effects of these axioms are the following. The Second Order Induction axiom restricts situations to the smallest set including the starting situation S_0 closed under the function do on events and situations. In conjunction with The Second Order Induction axiom the Unique Names for Situations axiom ensures two situations are equal iff they're the consequence of the exact same sequence of events executed from S_0. The two axioms Subhistory I and Subhistory II ensure situations are (partially) ordered. The foregoing set of axioms serves to fix the behaviour of the primitive notions of situation and event in every SC theory.

For a given domain, e.g., the theory of aspect being sketched here, domain dependent axioms specifying i) what events are possible, i.e., event preconditions, ii) event effects, i.e., axioms concerning the consequence for the fluents in the domain if event X is executed in situation S, iii) frame conditions, which specify how fluents do not change if event X is executed in situation S.[11] This third set of axioms are best understood in terms of the well-known Frame Problem in AI [28].

To be very brief the problem is that when modelling some world in e.g., first order logic, unless one specifies what doesn't change in addition to what does change, there will be numerous models which satisfy the sentences concerning events yet obviously do not correspond to the world one intends to be modelling. For example, if John paints himself blue, we could give the obvious effect axiom that subsequent to the painting event John will be painted blue. But suppose John then runs five kilometres. Intuitively none of the effects of this event include John turning grue.[12] However, without further axioms nothing prevents John's being grue subsequent to his run.[13] Seemingly for every event, we have to specify the nonconsequences of executing that event for every fluent in every situation, which, besides being inefficient, is profoundly tedious.

There are now numerous solutions to the Frame Problem but the solution pursued in [21, 33, 34] is novel to the Situation Calculus. Space prohibits a proof but Reiter's solution involves the set of domain independent axioms which determine how situations and events *per se* work and the provision of successor state axioms. The reader is referred to Sect. 3 of [34] for discussion of the domain independent axioms. The successor state axioms restrict when fluents are true by reference to the fluents which hold in the preceding situtation and the events which can be executed there to bring about the succeeding situation for which the fluent holds. It turns out there is a nice normal form into which the whole set of effect axioms can be compiled. Once these are embedded into successor state axioms, the frame axioms can be deduced.

Per the foregoing a given domain can be axiomatized by writing the event preconditions, the successor state axioms for fluents, and the properties holding in S_0. The next subsection exemplifies these requirements with an axiomitization written in GOLOG for some of the inference patterns surveyed in the previous section. Some bonus inferences concerning different verb frames are derived for free.

3.3 Aspect in GOLOG

The following code is the domain specific component of the GOLOG program for deriving some of the inferences discussed in the foregoing.[14]

---

[11] See [19] for a compressed but lucid introduction to the axioms defining the Situation Calculus.

[12] On grue see [13].

[13] See [38] for a very similar example on which this one is based.

[14] The domain indepedent component for this GOLOG program is the GOLOG interpreter. My code is tested with the version written in SWIProlog [14], available here: https://www.cs.toronto.edu/cogrobo/main/systems/index.html.

Listing 1 Golog Code
```
1  % Primitive control actions
2  primitive_action(die(X)).
3  primitive_action(run(X)).
4  primitive_action(move(X,N)).
5
6  % Definitions of Complex Control Actions
7  proc(runn(X,N), if(N-1 > 0, runn(X,N-1):move(X,N), move(X,N))).
8
9  % Preconditions for Primitive Actions.
10 poss(move(X,N),S) :- mobile(X,S).
11 poss(run(X),S) :- poss(move(X,N),S).
12 poss(die(X),S) :- \+ dead(X,S).
13
14 % Successor State Axioms for Primitive Fluents.
15 dead(X,do(A,S)) :- A = die(X).
16 ran(X,do(A,S)) :- (A = run(X)) ; (A = runn(X,Y)).
17 rann(X,Y,do(A,S)) :- A = runn(X,Y) ; (A = runn(X,Z), Z > Y).
18 moved(X,N,do(A,S)) :- A = move(X,N).
19 mobile(X,do(A,S)) :- mobile(X,S).
20
21 % Initial Situation.
22 mobile(j,s0).
23 -dead(b,s0).
24
25 % Restore suppressed situation arguments.
26 restoreSitArg(die(X),S,die(X,S)).
27 restoreSitArg(run(X),S,run(X,S)).
28 restoreSitArg(move(X,N),S,move(X,N,S)).
```

This code requires exposition. The primitive control actions correspond to some of the events from Sect. 2.3—*to die*, *to run*—but are extended to include *to move* in the transitive sense. The complex control actions, of which there is only one in this program, are events composed from the composition of further events. The event run exemplifies the treatment of accomplishments. It relies on constructs defined within GOLOG, namely recursion, sequencing (represented by ':'), and conditional event execution (represented by 'if'). In brief running five kilometres corresponds to indefinitely running every number of kilometres less than five but above zero prior to moving to the fifth kilometre position. Since the final subevent of executing the event of running N kilometres is to move to position N, and since the event is recursive, it amounts to first moving one kilometre, then two, etc. In other words we reduce the transitive sense of running to the sequence of moving multiple times, ignoring how the movement is a running movement beyond the uniqueness of the event's name, which is bestowed by the axiom for unique event names. I note `restoreSitArg` ensures events must be executed in the context of situations.

The program here puts quite mild preconditions on the execution of events. To move in some situation requires merely that the mover is mobile in that situation. To run in the intransitive sense requires the possibility of moving in the transitive sense, but only there be some position to which the mover can move. Since the possibility moving *per se* is just that the mover is mobile prior to moving, the possibility of running intransitively is reduced to the requirement of mobility in the state from which the movement is to be executed. The possibility of dying is, unsurprisingly, that the one who dies is not dead prior to the event of their dying. What then is it to be dead?

In fact we don't need to spell out precisely what being dead means, we only need to encode that one can infer death holds of someone in some situation only if they died in that situation. Nothing prohibits death holding in richer situations but the present successor state axioms ensure death in the minimal situation of being the one who dies in the event of dying. This ensures pattern 1 is captured: from the achievement of dying comes the morbid prize of the state of death and possession of that prize implies the achievement required to get it.

From the successor state axiom for ran we immediately get the first inference in pattern 3 that if one ran transitively then one ran intransitively, since one ran in situation S if S is the result of executing some sense of run, either transitive or intransitive–the semi-colon in the axiom corresponds to disjunction while the commas correspond to conjunction. Likewise we get the inference that one ran some specific distance only in the situation resulting from the sequence of transitive runs up to and including the run to the specific distance.

In the initial situation `john` is mobile. Moreover bill isn't dead. While there is surely some connection between mobility and not being dead we won't explicitly model this for present purposes. Consider now the result of querying the first inferences in pattern 3.[15]

Listing 2 John ran five kilometres → John ran (three kilometres)
```
?- rann(j,5,do(runn(j,5),s0)).
true.
?- rann(j,3,do(runn(j,5),s0)).
true.
?- rann(j,6,do(runn(j,5),s0)).
false.
?- ran(j,do(runn(j,5),s0)).
true.
```

It is worth dwelling on what the situation to which rann is predicated looks like. Consider the situation resulting from executing `do(runn(j,5),s0)`:

- `S = do(move(j, 5), do(move(j, 5-1), do(move(j, 5-1-1), do(move(j, 5-1-1-1), do(move(j, ... - ... - 1-1-1), s0)))))`

Evidently `rann(j,6,do(runn(j,5),s0)))` could not be true: notice that `do(runn(j,5),s0)` is a proper subsequence of every situation `S'` which would make `rann(j,6,S')` true. Why `rann(j,3,do(runn(j,5),s0)))` is true is likewise evident: the event of running 3 kilometres (5-1-1) is evidently executed in a subseqence of every sequence of events making `rann(j,5,do(runn(j,5),s0))` true. I postpone producing the inferences concerning states in pattern 1 since it would

[15] Notice the second inference in pattern 3 and both sets of inferences in patterns 2 and 5 concern progressive forms, which requires constructing the semantics of the progressive. Since I'm deliberately omitting the account of the imperfective paradox here, I forego modelling the progressive.

require extending the present axioms to explicitly model adjuncts–while this poses no trouble it wouldn't be instructive to introduce further complexity into the exposition.

Nonetheless the current versions of `dead` and `die` suffice to license the inference in pattern 4.

Listing 3 John died ↔ John is dead

```
1 ?- dead(b,s0).
2 false.
3 ?- dead(b,do(die(b),s0)).
4 true.
5 ?- do(die(b),do(die(b),s0),S).
6 false.
```

Thus the current axioms require Bill living in S_0 and being dead after executing the event of dying. Moreover, once Bill dies, the event of his dying can't be executed again. It should be noted that `die` is an event, while `dead` is a relational fluent. Therefore the inference in pattern 4, when reasoning in the present theory, should not be confused with Boolean equivalence. Note further that the third query is an attempt to find a binding of the situation S to the execution of the event of Bill dying in the situation which results from the execution of the event of his dying. It shows it's impossible to find a situation in which one can die twice.

Returning to Vendler's typology while restricting the focus to simple present and simple past, I exemplify the Vendler categories in the Situation Calculus by the following toy expressions.

- states: Bill is dead.

  $\rightsquigarrow$ `dead(b,s0)`

- achievements: Bill died.

  $\rightsquigarrow$ `dead(b,do(die(b),s0))`

- activities: Bill ran.

  $\rightsquigarrow$ `ran(b,do(run(b),s0))`

- accomplishments: Bill ran five kilometres.

  $\rightsquigarrow$ `rann(b,5,do(runn(b,5),s0))`

Notice some of the differences intuitively encoded in Dowty's version of the typology here get compiled into the axioms which underly inferences concerning events. I turn now to deriving these SC expressions directly from proofs of well-formedness of the sentences used to exemplify the Vendler typology.

4 Grammar to GOLOG

I employ a Lambek Categorial Grammar (LCG) [20] to derive proofs of the well-formedness of sentences. While the proofs correspond to the role of syntactic structure in linguistic theorizing, proofs of well-formedness imply neither the requirement

$$\frac{}{word \vdash A\,;\, M^{\rho}}\text{Lex} \qquad \frac{}{A\,;\, x:\rho \vdash A\,;\, x:\rho}\text{Id}$$

$$\frac{\Gamma\,;\,\Phi \vdash B/A\,;\, M^{\rho\to\tau} \qquad \Delta\,;\,\Psi \vdash A\,;\, N^{\rho}}{\Gamma,\Delta\,;\,\Phi,\Psi \vdash B\,;\, [\![MN]\!]^{\tau}}/E \qquad \frac{\Delta\,;\Psi \vdash A\,;\, N^{\rho} \qquad \Gamma\,;\,\Phi \vdash A\backslash B\,;\, M^{\rho\to\tau}}{\Delta,\Gamma\,;\,\Psi,\Phi \vdash B\,;\, [\![MN]\!]^{\tau}}\backslash E$$

$$\frac{\Gamma, A\,;\,\Phi, x:\rho \vdash B\,;\, M^{\tau}}{\Gamma\,;\,\Phi \vdash B/A\,;\, \lambda x^{\rho}.M}/I \qquad \frac{A,\Gamma\,;\, x:\rho,\Phi \vdash B\,;\, M^{\tau}}{\Gamma\,;\,\Phi \vdash A\backslash B\,;\, \lambda x^{\rho}.M}\backslash I$$

Fig. 1 Rules of LCG with term correspondences

nor the belief that there be some specific ur-structure for every reading of every sentence, the existence of which it is the job of the linguist to uncover. Moreover, it is known through the Curry-Howard correspondence that proofs in the simply-typed lambda calculus can be recovered from those in LCG. Following substitution of semantic terms into these proofs, they can be reduced to expressions in whichever logic one is working with for modelling inference.

My presentation of the Lambek Calculus here follows the exposition of LCG in [4]. Given a set $\mathcal{B}$ of basic Lambek categories which includes NP, S, S', and Pred, Lambek categories are defined by the following grammar:

$$L ::= \mathcal{B} \mid L/L \mid L\backslash L$$

Continuing the exposition from [4], Fig. 1 represents the proof system for LCG in sequent-style natural deduction. Sequents have the form $\Gamma\,;\,\phi \vdash A\,;\, M^{\rho}$, where Γ is a list of lambek categories, ϕ a list of typed variables, A a lambek category and M^{ρ} a simply typed λ-term with type ρ. We use Γ and Δ for lists of lambek categories, Φ and Ψ for lists of typed variables, M and N for λ-terms. Finally ρ and τ denotes arbitrary types.

For LCG to be used to prove well-formedness of sentences in e.g., English, it must include a lexicon. Suppose $\mathcal{W}$ is the set of words over an alphabet. Here the set of words is the set of written English words. A lexicon is a function that associates each $w \in \mathcal{W}$ with a finite pair of sets (C, L) where C is a set of lambek categories and L is a set of lambda-terms. I provide a toy lexicon for deriving the sentences used to exemplify Vendler's typology in Fig. 2. Figure 3 exemplifies LCG by proving the well-formedness of 'John ran five kilometres'.[16] I follow the conventions of not writing the terms on the left side of the turnstile, substituting categories on the left side of the turnstile by words in the lexicon, and using superscripts for types.

[16] Reduction is performed implicitly in the / (respectively \) E rules.

word	Lambek Category	Lambda Term
Bill	NP	$\mathtt{b}^o$
five_km	NP	5°
is	(NP\S)/(NP\Pred)	$\lambda P^{o\to s\to t}x^o.P(x)$
dead	NP\Pred	$\lambda x^o y^s.\mathtt{dead}(x,y)$
died	NP\S	$\lambda x^o y^s.\exists z^e.\mathtt{dead}(x,\mathtt{do}(\mathtt{die}(x),y))$
ran	NP\S	$\lambda x^o y^s.\mathtt{ran}(x,\mathtt{do}(\mathtt{run}(x),y))$
ran	(NP\S)/NP	$\lambda x^o y^o z^s.\mathtt{rann}(y,x,\mathtt{do}(\mathtt{runn}(y,x),z))$
Σ	S'/S	$\lambda P^{s\to t}.P(\mathtt{s0})$

Fig. 2 Toy LCG Lexicon

$$\dfrac{\dfrac{}{\Sigma \vdash S'/S\ ;\ \lambda P^{s\to t}.P(\mathtt{s0})}\text{Lex} \quad \dfrac{\dfrac{}{Bill \vdash NP\ ;\ \mathtt{b}^o}\text{Lex} \quad \dfrac{\dfrac{}{ran \vdash (NP\backslash S)/NP\ ;\ \lambda x^o y^o z^s.\mathtt{rann}(y,x,\mathtt{do}(\mathtt{runn}(y,x),z))}\text{Lex} \quad \dfrac{}{five\_km \vdash NP\ ;\ 5^o}\text{Lex}}{ran, five\_km \vdash S\backslash NP\ ;\ \lambda y^o z^s.\mathtt{rann}(y,5,\mathtt{do}(\mathtt{runn}(y,5),z))}/E}{Bill, ran, five\_km \vdash S\ ;\ \lambda z^s.\mathtt{rann}(\mathtt{b},5,\mathtt{do}(\mathtt{runn}(\mathtt{b},5),z))}\backslash E}{\Sigma, Bill, ran, five\_km \vdash S'\ ;\ \mathtt{rann}(\mathtt{b},5,\mathtt{do}(\mathtt{runn}(\mathtt{b},5),\mathtt{s0}))}/E$$

Fig. 3 LCG proof of 'John ran five kilometres'

5 Future Work

The present work focusses on exemplifying Vendler's typology in GOLOG. But further constructions introduce intriguing questions for the model of events pursued here. Vendler considers accomplishments telic, yet there exist non-culminating accomplishments. This is exemplified by the French data below.

22. Pierre lui a expliqué le problème (mais elle n'a rien
Pierre her explain-PFV- 3SG the problem but she NEG has nothing
compris du tout).
understood at all.
Pierre explained the problem to her (but she didn't understand it at all). [26]

[26] considers the defining property of non-culminating accomplishments the defeasability of the inference that the event culminated. Thus in sentence (22) the bracketed text could follow the unbracketed text without eliciting the judgement of contradiction. However if the bracketed text is omitted the inference that the event did indeed culminate is predicted.[17]

Non-culminating accomplishments are intriguing from the SC perspective in that they seem to require a form of non-monotonicity. One possible solution for handling this non-monotonicity would be the use of state constraints–properties that hold in every situation. Appendix B in [34] discusses such constraints, the puzzles they've

[17] The French example is one of what Martin considers 'sublexical modal verbs' which imply rather than entail the existence of the result state corresponding to the verb, e.g., it is implied rather than entailed that Pierre's explanation is successful in producing in Mary the state of understanding. The bracketed text then explicitly blocks the inference to this result state.

posed for basic action theories, and how their effects can be reduced to constraints on S_0 plus successor state axioms.[18] The SC approach to such non-monotonicity effectively compiles it away into the underlying monotonic logic.

In the context of nonculminating achievements I conjecture we could employ state constraints that ensure if some accomplishment produces some result then the result holds in the situation subsequent to executing the accomplishment event if and only if there is no explicit negation of the result state in that situation.

Deferring the question of how these inferences are to be modelled to future work, the contention that they should be modelled is supported by the plethora of languages which include non-culminating accomplishments in their stock of predicate frames–[26] reports non-culminating readings of accomplishments have been documented in 13 specific languages and 4 language families. To add to this list of languages we note that Lithuanian seems to exemplify what [26] dubs incompletive atelic predicates: predicates which introduce defeasible inferences of non-culmination [37, p. 284]. Moreover, Lithuanian includes both periphrastic and morphological causatives [31]. The periphrastic causative is used in object control to express e.g., 'force' c.f., sentence (23). Future work will use Lithuanian to explore how rich morphology, including case, affects the semantics corresponding to different predicate frames.

Further constructions put pressure on the simple picture of events presented here. In English analytic causatives, e.g., (23), and object control constructions, e.g., (24), we find verbs which denote events of (in)directly bringing about other events[19]:

23. John made Bill buy cheese.
24. John forced Bill to buy cheese.

It follows from 'forced' and 'made' that Bill bought cheese. To see this must be a semantic effect even in object control simply substitute 'encouraged' for 'forced'—it doesn't follow from John encouraging Bill to buy cheese that Bill did buy cheese. Evidently for it to be possible for John to force Bill to buy cheese, it must be possible for Bill to buy cheese. But is the event of making Bill buy cheese coextensive with the event of his buying cheese? If the events are distinct the SC account of events pursued here would need to be extended to include underspecified events, e.g., the content of 'make'. If such events were not underspecified it would be mysterious how one could sensibly utter the following:

25. John made Bill sell his stocks on Wednesday by threatening him the previous night.

This sentence suggests the manner of making Bill sell his stocks can be described independently of the event of Bill selling his stocks. But is the manner in which John made Bill sell his stocks itself an event independent of the 'make' event or is it just a description of the 'make' event? This question would need to be addressed before 'make' could be considered an underspecified event. Moreover it could be the best

[18] See references cited therein for work on state constraints up to 2001.

[19] On analytic causatives, see [15]. On control, see [6]. For crosslinguistic discussion of control, see [44], who accounts for object control in Lithuanian.

treatment of these constructions is in terms of complex single events, e.g., 'make buy cheese', which would make the question of the independence of the manner of John's making Bill sell his stocks moot. The work [15] identifies a host of reasons for considering analytic causatives in Romance to form one complex syntactic unit with their infinitive complement. Might this unit then correspond to one complex event? In future work these questions could serve an instructive role in deciding how to model possible dependencies between events.

6 Conclusion

In this paper I review Vendler's typology of predicate frames, commonly understood in linguistics in terms of aspect. I noted some inferences which seem ripe for modelling in the SC. I then showed how to produce some of these inferences in GOLOG, thereby implementing the proposed SC theory of these inferences. Subsequently I showed how the SC meanings I proposed for Vendler's typology could be compositionally derived from proofs in LCG. While there remain a number of outstanding inferences still to model, I hope the present work shows the promise of SC for rigorous semantic theory.

Acknowledgements I thank four anonymous reviewers who significantly improved this paper. I thank Elena Vaiksnoraite for careful reading and suggestions concerning content. I further thank the organizers of the conference for inviting me to submit to the proceedings and this post-symposium book "Logic and Algorithms in Computational Linguistics 2021 (LACompLing2021)". The questions I received during the conference were quite helpful when writing up the content in paper form.

References

1. Bach, E.: On time, tense, and aspect: An essay in English metaphysics. Linguistics Department Faculty Publication Series (1981)
2. Barwise, J., Perry, J.: Situations and attitudes. The J. Philos. **78**(11), 668–691 (1981). https://doi.org/10.2307/2026578
3. Casati, R., Varzi, A.: Events. In: Zalta, E.N. (ed.) The Stanford Encyclopedia of Philosophy, Summer, 2020th edn. Stanford University, Metaphysics Research Lab (2020)
4. Catta, D., Stevens-Guille, S.J.: Lorenzen won the game, Lorenz did too: Dialogical logic for ellipsis and anaphora resolution. In: International Workshop on Logic, Language, Information, and Computation, pp. 269–286. Springer (2021). https://doi.org/10.1007/978-3-030-88853-4_17
5. Cooper, R.: Austinian propositions, Davidsonian events and perception complements. In: The Tbilisi Symposium on Logic, Language, and Computation: Selected Papers, pp. 19–34 (1997)
6. Culicover, P.W., Jackendoff, R.: Turn over control to the semantics! Syntax **9**(2), 131–152 (2006). https://doi.org/10.1111/j.1467-9612.2006.00085.x
7. Davidson, D.: The logical form of action sentences. In: Rescher, N. (ed.) The Logic of Decision and Action, pp. 81–95. University of Pittsburgh Press (1967)

8. Dowty, D.: Word Meaning and Montague Grammar: The Semantics of Verbs and Times in Generative Semantics and in Montague's PTQ. Springer (1979)
9. Dowty, D.R.: Montague grammar and the lexical decomposition of causative verbs. In: Montague grammar, pp. 201–245. Elsevier (1976)
10. Dowty, D.R.: On the semantic content of the notion of 'thematic role'. In: Properties, types and meaning, pp. 69–129. Springer (1989). https://doi.org/10.1007/978-94-009-2723-0_3
11. Engelberg, S.: Frameworks of lexical decomposition of verbs. In: Semantics—Lexical Structures and Adjectives, pp. 47–98. De Gruyter Mouton (2019). http://dx.doi.org/10.1515/9783110626391-002
12. Filip, H.: Lexical aspect. The Oxford handbook of tense and aspect pp. 721–751 (2012). https://doi.org/10.1093/oxfordhb/9780195381979.013.0025
13. Goodman, N.: Fact, fiction, and forecast. Harvard University Press (1983)
14. Group, C.R.: A golog interpreter in swi-prolog (2000). https://www.cs.toronto.edu/cogrobo/main/systems/index.html
15. Guasti, M.T.: Analytical Causatives, pp. 1–36. Wiley (2017). https://doi.org/10.1002/9781118358733.wbsyncom038
16. Hamm, F., Bott, O.: Tense and Aspect. In: Zalta, E.N. (ed.) The Stanford Encyclopedia of Philosophy, Fall, 2021st edn. Stanford University, Metaphysics Research Lab (2021)
17. Kenny, A.: Action, emotion and will. Routledge (2003)
18. Krifka, M.: The Mereological Approach to Aspectual Composition. Uil-OTS, University of Utrecht, Perspectives on Aspect (2001)
19. Lakemeyer, G.: The Situation Calculus and Golog—a tutorial. https://www.hybrid-reasoning.org/media/filer/2013/05/24/hybris-2013-05-sitcalc-slides.pdf
20. Lambek, J.: The mathematics of sentence structure. The Am. Math. Monthly **65**(3), 154–170 (1958)
21. Levesque, H.J., Reiter, R., Lespérance, Y., Lin, F., Scherl, R.B.: Golog: a logic programming language for dynamic domains. The J. Logic Program. **31**(1–3), 59–83 (1997). https://doi.org/10.1016/S0743-1066(96)00121-5
22. Lewis, D.: Counterfactuals and comparative possibility. In: Ifs, pp. 57–85. Springer (1973)
23. Loukanova, R.: Situation Theory, Situated Information, and Situated Agents. In: Nguyen, N.T., Kowalczyk, R., Fred, A., Joaquim, F. (eds.) Transactions on Computational Collective Intelligence XVII, Lecture Notes in Computer Science, vol. 8790, pp. 145–170. Springer, Berlin, Heidelberg (2014). https://doi.org/10.1007/978-3-662-44994-3_8
24. Loukanova, R.: Typed theory of situated information and its application to syntax-semantics of human language. In: Christiansen, H., Jiménez-López, M.D., Loukanova, R., Moss, L.S. (eds.) Partiality and Underspecification in Information, Languages, and Knowledge, pp. 151–188. Cambridge Scholars Publishing (2017). https://www.cambridgescholars.com/product/978-1-4438-7947-7
25. Loukanova, R.: Formalisation of situated dependent-type theory with underspecified assessments. In: Bucciarelli, E., Chen, S.H., Corchado, J.M. (eds.) Decision Economics. Designs, Models, and Techniques for Boundedly Rational Decisions. DCAI 2018, Advances in Intelligent Systems and Computing, vol. 805, pp. 49–56. Springer International Publishing, Cham (2019). https://doi.org/10.1007/978-3-319-99698-1_6
26. Martin, F.: Non-culminating accomplishments. Lang. Linguist. Compass **13**(8) (2019). https://doi.org/10.1111/lnc3.12346
27. McCarthy, J.: Situations, Actions, and Causal Laws. Department of Computer Science, Stanford University, Tech. rep. (1963)
28. McCarthy, J., Hayes, P.J.: Some philosophical problems from the standpoint of artificial intelligence. In: Readings in artificial intelligence, pp. 431–450. Elsevier (1981). https://doi.org/10.1016/B978-0-934613-03-3.50033-7
29. Mittwoch, A.: Aspectual classes. In: The Oxford Handbook of Event Structure, pp. 30–49. OUP (2019). https://doi.org/10.1093/oxfordhb/9780199685318.013.4
30. Naumann, R.: A dynamic logic of events and states for the interaction between plural quantification and verb aspect in natural language. Logic J. IGPL **7**(5), 591–627 (1999). https://doi.org/10.1093/jigpal/7.5.591

31. Pakerys, J.: On periphrastic causative constructions in Lithuanian and Latvian. Argument realization in Baltic pp. 427–458 (2016). https://doi.org/10.1075/vargreb.3
32. Parsons, T.: Events in the Semantics of English: A Study in Subatomic Semantics. MIT Press (1990)
33. Reiter, R.: The frame problem in the situation calculus: a simple solution (sometimes) and a completeness result for goal regression. In: Artificial and Mathematical Theory of Computation, pp. 359–380 (1991)
34. Reiter, R.: Knowledge in Action: Logical Foundations for Specifying and Implementing Dynamical Systems. MIT Press (2001)
35. Ross, J.R.: Act. In: Semantics of natural language, pp. 70–126. Springer (1972)
36. Ryle, G.: The Concept of Mind. Hutchinson's University Library (1949)
37. Seržant, I.A.: The independent partitive genitive in Lithuanian, pp. 257–299. John Benjamins Amsterdam & Philadelphia (2014). https://doi.org/10.1075/vargreb.1.07ser
38. Shanahan, M.: The Frame Problem. In: Zalta, E.N. (ed.) The Stanford Encyclopedia of Philosophy, Spring, 2016th edn. Stanford University, Metaphysics Research Lab (2016)
39. Steedman, M.: The productions of time. Draft. http://www.cogsci.ed.ac.uk/steedman/papers.html (2000)
40. Stenning, K., van Lambalgen, M.: Semantic interpretation as computation in nonmonotonic logic: the real meaning of the suppression task. Cognit. Sci. **29**(6), 919–960 (2005). https://doi.org/10.1207/s15516709cog0000_36
41. Taylor, B.: Tense and continuity. Linguist. Philos. **1**(2), 199–220 (1977). https://doi.org/10.1007/BF00351103
42. Truswell, R.: Introduction. In: The Oxford Handbook of Event Structure. OUP (2019). https://doi.org/10.1093/oxfordhb/9780199685318.001.0001
43. Truswell, R.: The Oxford Handbook of Event Structure. OUP (2019). https://doi.org/10.1093/oxfordhb/9780199685318.001.0001
44. Vaikšnoraitė, E.: Case and non-verbal predication: the syntax of Lithuanian control clauses. Master's thesis, Leiden University (2015)
45. Van Lambalgen, M., Hamm, F.: The Proper Treatment of Events. Wiley (2008). https://doi.org/10.1002/9780470759257
46. Vendler, Z.: Linguistics in Philosophy. Cornell University Press (1967). https://doi.org/10.7591/9781501743726
47. Verkuyl, H.J.: On the Compositional Nature of the Aspects. Dordrecht, Netherlands: D.Reidel Publishing Company (1972). http://dx.doi.org/10.1007/978-94-017-2478-4
48. Von Wright, G.H.: And next. In: Studia logico-mathematica et philosophica, in honorem Rolf Nevanlinna die natali eius septuagesimo 22.X., 18, pp. 293–304. Acta philosophica Fennica (1965)
49. White, M.: A calculus of eventualities. In: Proceedings of the AAAI-94 Workshop on Spatial and Temporal Reasoning (1994)
50. White, M.W.: A computational approach to aspectual composition. Ph.D. thesis, University of Pennsylvania (1994)
51. Zucchi, S.: Events and situations. Ann. Rev. Linguist. **1**(1), 85–106 (2015). https://doi.org/10.1146/annurev-linguist-030514-125211

Generating Pragmatically Appropriate Sentences from Logic: The Case of the Conditional and Biconditional

Renhao Pei and Kees van Deemter

Abstract It is widely assumed that there exist mismatches between the connectives of Propositional Logic and their counterparts in Natural Language. One mismatch that has been extensively discussed is Conditional Perfection, the phenomenon in which a conditional sentence is interpreted as a biconditional under some circumstances. The Pragmatics literature has provided valuable insights into the question of whether Conditional Perfection will happen in a given context. In order to make these insights more explicit and testable, we designed an algorithm to generate pragmatically more appropriate sentences from propositional logical formulas involving material implication and biconditional implication. This algorithm was tested in an evaluation by human participants, in which generated sentences are compared against those generated by a simple baseline algorithm. The evaluation results suggest that the designed algorithm generates better sentences, which capture the semantics of the logical formulas more faithfully.

Keywords Conditional perfection · Propositional logic · Natural language generation

1 Introduction

Mathematical Logic has been extensively used for representing meaning of natural language in Formal Semantics. This is not unproblematic, because a formula in logic is rarely completely equivalent with a sentence of, for example, English. Even in the simplest component of first-order logic (FOL), namely propositional logic, mismatches between the logical connectives and their natural language counterpart are known to exist. These mismatches have long been noticed and studied by logicians

---

R. Pei (✉) · K. van Deemter
Utrecht University, Utrecht, The Netherlands
e-mail: renhaopei@gmail.com

K. van Deemter
e-mail: c.j.vandeemter@uu.nl

R. Loukanova et al. (eds.), *Logic and Algorithms in Computational Linguistics 2021 (LACompLing2021)*, Studies in Computational Intelligence 1081,
https://doi.org/10.1007/978-3-031-21780-7_8

and linguists, and from their studies, valuable insights about these mismatches have been gained. However, these insights have often lacked the kind of formal detail that makes them testable; consequently, they are difficult to make use of in practical applications.

One relevant area of applications, which we will be looking at in this paper, is Natural Language Generation (NLG) [8, 16]. Of particular relevance is a line of work in which NLG programs take a logical formula as their input, and produce a clear and intelligible English sentence as output. Applications of this kind are of substantial present importance in connection with "explainable Artificial Intelligence", whose aim is to make the workings of Artificial Intelligence programs understandable to human users and other stakeholders (see, e.g., [15]). Mismatches between natural language and logic naturally also play a role in this field.

These two lines of study can complement each other well, as NLG from logic can provide a practical framework for testing theoretical insights while these insights can in turn improve the clarity of natural language sentences generated out of logical formulas.

In the following Sects. 1.1 and 1.2, Condition Perfection and NLG from logic will be explained in more details respectively, which serve as the theoretical background for the rest of the paper.

1.1 Conditional Perfection

The term Conditional Perfection (henceforth referred to as CP) was coined by Geis and Zwicky [9],[1] it refers to the phenomenon that the conditional in the form of *if p, then q* often invites an inference of *if not p, then not q*. Through the mediation of this invited inference, the original conditional will express ('be perfected into') its corresponding biconditional *if and only if p, then q*, as illustrated by the example in [9]:

(1) a. If you mow the lawn, I'll give you five dollars
  b. If you don't mow the lawn, I won't give you five dollars
  c. If and only if you mow the lawn, I'll give you five dollars

Here, the sentence (1a) would also suggest (1b), thus conveying the meaning of (1c). The phenomenon of (1a) inviting the inference of (1b) is called CP.[2]

With propositional logic, Table 1 clearly illustrates how CP is derived through the mediation of an invited inference: Here, the original proposition $p \rightarrow q$ invites an inference of $\neg p \rightarrow \neg q$, and the conjunction of $p \rightarrow q$ and $\neg p \rightarrow \neg q$ will have the

[1] Geis and Zwicky [9] coined the term CP, but they were not the first to address this phenomenon in linguistics. See [2] for a historic overview.

[2] There have been two different usages of the term CP, the first one is that the invited inference of (1b) constitutes CP, the other one is that the biconditional in (1c) constitutes CP. In this paper, the term CP is used to refer to the invited inference itself.

Table 1 Truth table illustrating how CP is derived from an invited inference

p	q	$p \rightarrow q$	$\neg p \rightarrow \neg q$	$(p \rightarrow q) \wedge (\neg p \rightarrow \neg q)$	$p \leftrightarrow q$
T	T	T	T	T	T
T	F	F	T	F	F
F	T	T	F	F	F
F	F	T	T	T	T

same truth table as $p \leftrightarrow q$. As a result, the original material implication $p \rightarrow q$ is perfected into a biconditional implication $p \leftrightarrow q$.

Various explanations have been proposed concerning the nature of CP. Van der Auwera [1] regarded CP as a scalar implicature. In that approach, CP invokes the scale ⟨(*if p, q* and *if r, q* and *if s, q*), (*if p, q* and *if r, q*), (*if p, q*)⟩. The assertion of *if p, q* then implicates there is no other antecedent (*r*, *s*, etc.) with *q* as a consequent, thus expressing *if and only p, q*.

A second approach proposed by Horn [11] considers CP as an implicature motivated by the R-based pragmatic strengthening. In this approach, CP is the result of the second Maxim of Quantity (*do not make your contribution more informative than is required*) and the Maxim of Relation (*be relevant*). As the stipulated condition *p* in *if p, q* must be taken to be informative and relevant, *p* would be interpreted as necessary, hence the original sufficient condition is strengthened into a necessary and sufficient condition.

Thirdly, Herburger [10] proposed the 'whole truth theory' to account for CP. In this proposal, a sentence *S* can be silently conjoined with *only S*, resulting in the conjunction *S and only S*, which is then taken to express 'the truth and the whole truth'. Therefore, a conditional sentence with *if* will be taken to express *if and only if*.

In the present study, no position regarding the nature of CP is taken. Whatever the explanation of CP might be, for our purpose of building an NLG system, what matters most is the distribution of CP, i.e., when CP is expected to occur, and when CP is expected to not occur.

While the above-mentioned studies sought to propose a theory about the nature of CP, Van Canegem-Ardijns and Van Belle [18] sought to provide a descriptive account for CP. Van Canegem-Ardijns and Van Belle [18] identified three sub-types of CP: two specific ones (*only if p, q and only if not p, not q*), and a more general one (*if not p, then not q*), and showed the correlation between each of the three sub-types of CP with various speech act or utterance types, as well as some utterance types in which CP will not occur. Among the many utterance types that [18] discussed, three of them are particularly relevant for the purpose of the present study, namely: conditional promise, conditional threat and conditional warning.

According to [18], in a conditional promise, the speaker attempts to get the hearer to do something by offering a potential reward. It involves examples as 'If you get me some coffee, I'll give you a cookie.' Conditional promises have the following semantic characteristics: [*A is desirable for S; H has control over A; S wants H to do A; S considers C desirable for H because S assumes H wants C; S has control over C*] (Here the abbreviations follow the usage in [18]: A = action/event described in the antecedent, C = action/event described in the consequent, S = speaker, H = hearer).

In a conditional threat, the speaker attempts to refrain the hearer from doing something by threatening with a potential consequent, such as 'If you lie to me, I'll kill you.' Conditional threats have the following semantic characteristics: [*A is undesirable for S; H has control over A; S wants H to do not-A; S considers C undesirable for H because S assumes H does not want C; S has control over C*]. It is easy to see that the conditional threat and the conditional promise are parallel to each other, the only difference is that the antecedent and the consequent in a conditional promise are both desirable while they are both undesirable in a conditional threat.

The conditional warning resembles the conditional threat in many ways, as they both share the semantic characteristics of [*H has control over A; S wants H to do not A; S considers C undesirable for H because S assumes H does not want C*]. One crucial difference that sets them apart is that the conditional warning lacks the semantic characteristic of [*S has control over C*]. For instance, in the conditional warning of 'If you park your car on private land, you will get your car wheel clamped,' the speaker does not have actual control over whether the car wheel will be clamped or not.

According to [18], conditional promises readily invite an inference of *only if p, q* (*q* in no event other than that *p*), which entails *if not p, then not q*. Conditional threats invite an inference of *only if not p, not q*, which also entails *if not p, then not q*. On the other hand, conditional warning does not invite the inference of *only if not p, not q*, since the speaker does not have control over the consequent in a conditional warning, she or he cannot justifiably guarantee whether the consequent will or will not take place. Van Canegem-Ardijns and Van Belle [18] did not explicitly mention whether conditional warning invites the more general inference of *if not p, then not q*, while several experimental studies involving human judgements [5, 13, 14] have shown that, conditional warning is indeed less likely to invite the inference of *if not p, then not q*, compared to conditional threats and conditional promises.

In these experimental studies [5, 13, 14], apart from the already mentioned conditional promise, conditional threat and conditional warning, a fourth type of conditionals (or utterance type as in Van Canegem-Ardijns and Van Belle[18]'s terminology) is involved, which is conditional tip. An example of conditional tip could be 'If you show up early for work, you will impress your boss,' in which the consequent is desirable for the hearer, while the speaker does not have control over the consequent. The conditional tip resembles conditional promise, in the same way that conditional warning resembles conditional threat, thus forming a complete tetrachoric table as in Table 2.

Conditional promise and conditional threat can be grouped together under the umbrella term *inducement*, as their utterances offer inducements to alter people's behavior and only differ in that conditional promise encourages actions by potential

Table 2 Classification of conditionals into four sub-types using two features

	Speaker having control over the consequent	Speaker **not** having control over the consequent
The consequent being desirable for the hearer	Conditional promise	Conditional tip
The consequent being **un**desirable for the hearer	Conditional threat	Conditional warning

desirable reward while the conditional threat deters actions by prospective undesirable punishment. On the other hand, conditional tip and conditional warning can be grouped as *advice*, as their utterances offer advice and again only differ with regard to whether the outcome is desirable or undesirable.

Note that in Table 2, both features used for the classification are about consequent, while for the antecedent, whether the antecedent is desirable and whether the hearer has control over the antecedent are not mentioned. Similarly, in [18] the desirability of the antecedent is always in accordance with the desirability of the consequent and the hearer always has control over the antecedent. In the present study, we will follow Table 2 and omit the information whether the hearer has control over the antecedent and whether the antecedent is desirable, since this kind of information can be easily derived and is therefore redundant. The fact that only the consequent plays a role in classification of the conditionals will be reflected in the design of the algorithm discussed in Sect. 2.

Despite different experimental settings, all of the three studies [5, 13, 14] have shown that inducements tend to invite the inference of *if not p, then not q*, while advice is significantly less likely to invite such an inference. In other words, CP is more likely to occur in inducements and less likely to occur in the advice.

1.2 Generating Natural Language from Logical Formulas

Natural Language Generation (NLG) refers to the domain of generating natural language text from some non-linguistic representations of information. For example, a weather forecast NLG system can generate weather forecast in text from raw meteorological data. The present study, however, utilizes a different source of information, namely the logical formula, as the input for an NLG system.

Logic is a useful tool in many different domains. It has been extensively used for representing meaning in the Formal Semantics, as well as for representing argument in formal argumentation. While logical formula has the advantage of being formal and clearly defined, it is certainly not the most understandable format for representing information, since not everyone knows logic.

For example, as has been put forward in [15], when an automatic system is asked to give explanations to a human operator for why it takes a certain action or why it provides a certain plan for the user, then the best that many systems can do is to respond in the form of logical formula, which is however opaque for most non-technical human users. Mayn and Van Deemter [12] have also remarked that in Explainable AI, there exists a focus on making algorithms in Artificial Intelligence transparent. For that purpose, tools that can produce automatic English 'translations' from logical formulas can offer a good way to convey information to users, especially if these users are not familiar with Mathematical Logic. In this case, an NLG system that can translate logical formula into natural language is needed to make the autonomous system scrutable to non-technical users.

The task of NLG from logical forms dates from 1980s [20] and various kinds of logical forms and different approaches have been used, targeting at different practical applications.

Among them, one that is particularly relevant to the present study is the NLG system of [7], which provides feedback to students of logic. The logic form used as input in [7] is the first-order logic (FOL), which is one of the most basic forms of logic and is widely used in domains such as Artificial Intelligence, Linguistics, and Philosophy. FOL is also the logic most often taught [4, 6]. When students learn the FOL, one crucial exercise is to translate between natural language sentences and their corresponding FOL representations. For this kind of translation exercises, an automated NLG system could greatly facilitate a student's learning process by generating natural language sentences from the student's incorrect solution in FOL. This generated natural language sentence could show the student what his or her proposed solution actually says, as a way to prompt repair.

However, to build such an NLG system is not a trivial task. One important obstacle for translating between natural language and FOL faithfully is that there are many mismatches between the two. Even in the simplest component of FOL, namely the propositional logic, mismatches between the logical connectives and their natural language counterpart exist, such as the phenomena *Conditional Perfection* and *exclusive/inclusive disjunction*. These issues have long been noticed and studied by logicians and linguists, and from their studies, valuable insights have been gained, which could then be used to help building a successful NLG system.

The present study focuses on the Conditional Perfection. As explained in Sect. 1.1, the ideas drawn from theoretical and experimental studies about Conditional Perfection will then be used to improve an NLG system that deals with the two logical connectives $\rightarrow$ and $\leftrightarrow$.

1.3 Focus of the Paper

Under the general topic of testing the insights gained from Pragmatics literature in the framework of NLG from logic, the present paper will revolve around how to generate more appropriate sentences to deal with CP.

For the input of propositional logical formulas in the present study, only two logical connectives are included, namely $\rightarrow$ for material implication and $\leftrightarrow$ for biconditional implication, as CP is only relevant to these two connectives. Notably, [3] reported that students of logic were found to have difficulty with distinguishing these two connectives, and the errors caused by confusing them account for 29.77% of total error in the corpus of student translations of natural language into logic. Since this type of error is so common, it would be very helpful to build an automatic program that could show the students what their errored logical formulas containing $\rightarrow$ and $\leftrightarrow$ really say in natural language. Therefore, a simple NLG system that only focuses on these two connectives would already have great merits for practical purpose.

What's more, the present study only uses binary propositions, i.e., there is only one connective for each propositional formula, and the formula would be in the form of [literal[3] + connective + literal]. This is only for convenience's purpose and to avoid potential interactions caused by multiple connectives within one proposition. Among various kinds of conditionals, the present study only focuses on 4 types of conditionals namely promise, threat, warning, and tip, as described in Sect. 1.1. And lastly, the natural language used in this study is restricted to English.

2 A Pragmatic Algorithm for Expressing Propositional Logic Formulas in English

2.1 *Key Ideas Underlying the Algorithm*

As explained in Sect. 1.1, the 4 types of conditionals under discussion (promise, threat, warning, and tip) are distinguished by two features: 1. Whether the speaker has control over the consequent, and 2. Whether the consequent is desirable or undesirable for the hearer. These two features, each having two values (positive or negative), can be abbreviated as [$\pm$control] and [$\pm$desirability].

At the first stage of the algorithm, a set of binary logical propositions is randomly generated by firstly selecting one connective from either $\rightarrow$ or $\leftrightarrow$, and then selecting one antecedent and one consequent from two pre-defined atomic proposition banks (an antecedent bank and a consequent bank) respectively. Each atomic proposition in the consequent bank is associated with its features of [$\pm$control] and [$\pm$desirability], while the atomic propositions in the antecedent bank are not associated with these two features. As explained in Sect. 1.1, this is because the antecedents do not play any role in distinguishing the 4 types of the conditionals. In all the 4 types of conditionals under discussion, the hearer would always have control over the antecedent, i.e., the antecedent would always be [+control], thus making this information redundant. In terms of desirability, the atomic propositions for antecedents are intentionally constructed to be inherently neutral, so that within a binary proposition, the inher-

[3] A literal is an atomic formula p or its negation $\neg p$.

Table 3 Truth table showing the effect of cancelling CP

p	q	p → q	If p, q (inducement)	If p, q, but if not p, might still q
T	T	T	T	T
T	F	F	F	F
F	T	T	F	T
F	F	T	T	T

ently neutral antecedent could be interpreted to be either desirable or undesirable in accordance with the consequent.

By utilizing the information from features of the consequent, each proposition can then be classified as either a promise, a threat, a warning, or a tip. CP is expected to occur in inducements (promises and threats) and not occur in the advice (warnings and tips). After having classified the conditionals into the 4 types, we can have a prediction of whether CP will occur or not.

For the input binary proposition of material implication p → q, the most straightforward way is to translate the connective as 'if p, then q.' But if the sentence is an inducement, then CP will occur, which is not preferable since it would be a distortion of the original meaning of the logical formula, and we want the NLG system to faithfully express the logical meaning in natural language. Fortunately, when CP is present, it can still be cancelled by adding an additional condition as noted by Herburger [10]. As in the following example:

(2) a. If you buy him a drink, I will help you
    b. If you buy him a drink, I will help you, and if you don't, I might still do

Here the sentence (2a) is a conditional promise and would invite an inference of *if you don't buy him a drink, I won't help you*, thus expressing *if and only if you buy him a drink, I will help you*. However, this inference can be cancelled by adding and if you don't, I might still do as in (2b). The meaning of p → q (with p being 'you buy him a drink' and q being 'I will help you') is faithfully preserved in (2b), but not in (2a), in which CP occurs. As shown in the Table 3.

Therefore, for the binary proposition in the form of p → q, if the proposition is classified as an inducement, then CP is expected to occur and would be translated as 'if p, q, but if not p, might still q.' On the other hand, if p → q is advice, then CP is not expected to occur and there is no need to add 'but if not p, might still q.'

As for the biconditional implication p ↔ q, the most straightforward way is to translate it as 'if and only if p, q.' But if the proposition is classified as an inducement, then CP is expected to happen, which means 'if p, q' would express the same meaning as 'if and only if p, q.' In this case, we can simply use 'if p, then q' for expressing p ↔ q, which would be a better choice of words, since 'if and only if' is not a very natural expression and people typically do not use this phrase outside the domain of mathematics or logic. On the other hand, if p ↔ q is advice, then CP will not happen and we have no other choice but to use 'if and only if,' if we want to faithfully preserve the meaning of the original logical formula.

2.2 *The Pragmatic Algorithm and the Baseline*

In this section, to demonstrate what the designed algorithm really does, a comparison will be made between two algorithms: one so called 'pragmatic algorithm' that tries to generate pragmatically more appropriate sentences following the ideas explained in Sect. 2.1, and another baseline algorithm that does nothing pragmatic and only do the most straightforward realization.

Below are some pseudo-codes illustrating the baseline algorithm, in which l is a logical formula that is in the form of either $p \rightarrow q$ or $p \leftrightarrow q$, and r is the realized sentence for the logical formula.

Algorithm 1 The baseline algorithm

if l is in the form of $p \rightarrow q$ **then**
  $r \leftarrow$ 'If p, q.'
else if l is in the form of $p \leftrightarrow q$ **then**
  $r \leftarrow$ 'If and only if p, q.'
end if

The input for the baseline algorithm is a list of randomly generated binary logical formulas that are in the form of either p $\rightarrow$ q or p $\leftrightarrow$ q. If the logical formula l is a material implication (i.e., in the form of p $\rightarrow$ q), it will be realized into 'If p, q'. If the logical formula l is a biconditional implication (in the form of p$\leftrightarrow$ q), it will be realized into 'If and only if p, q'.

What the baseline algorithm essentially does here, is simply translate the logical connectives $\rightarrow$ and $\leftrightarrow$ into their commonly used natural language counterparts 'if...then' and 'if and only if...then', while the pragmatic algorithm does a more sophisticated realization.

The pragmatic algorithm also takes the list of logical formulas as an input. Besides, it also takes in a label dictionary, which contains the information about the corresponding label for each logical formula. These labels include *promise*, *threat*, *tip*, and *warning*, and furthermore, *promise* and *threat* are under the umbrella label *inducement*, while *tip* and *warning* are under the label *advice*. These labels are automatically obtained by a label detector that can label a binary proposition according to the predefined features ([±control] and [±desirability]) of the consequent, with the rationale explained in Sect. 2.1 and Table 2.

Below are some pseudo-codes illustrating the pragmatic algorithm, with a basic setting similar to Algorithm 1.

The pragmatic algorithm realizes the logical formulas differently depending on their labels. Compared to the baseline algorithm, the improvement lies in the situation

Algorithm 2 The pragmatic algorithm

```
if l is in the form of p → q then
    if l has the label 'inducement' then
        r ← 'If p, q, but if not p, might still q.'
    else if l has the label 'advice' then
        r ← 'If p, q.'
    end if
else if l is in the form of p ↔ q then
    if l has the label 'inducement' then
        r ← 'If p, q.'
    else if l has the label 'advice' then
        r ← 'If and only if p, q.'
    end if
end if
```

where the logical formula is labeled as 'inducement'. If a logical formula of material implication p $\rightarrow$ q is labeled as 'inducement', some extra words 'but if not p, might still q' will be appended after the basic sentence 'If p, q', in order to cancel the expected CP. If a logical formula of biconditional implication p $\leftrightarrow$ q is labeled as 'inducement', the connective will be translated as 'If…then' rather than 'If and only if…then', in order to make the realized sentence shorter and more natural. On the other hand, when the logical formula is labeled as 'advice', the pragmatic algorithm will do exactly the same realization as the baseline algorithm does.

2.3 Example of Input and Output

Before showing the input and output example, it is important to note that although the algorithm can inherently work for any proposition of promises, threats, tips, and warnings, it also has some other general requirements in practice. Since the binary logical propositions are all generated by randomly selecting one antecedent and one consequent, it is important to ensure that all the random combinations of an antecedent and a consequent make sense. Not any two atomic propositions can form a good proposition in which the connection between the antecedent and the consequent can be easily inferred through common sense. A bad example (in natural language) could be 'If you drink this bottle of milk, I will book a ticket to Iceland.' The connection between drinking milk and booking a ticket is hard to establish, thus making this sentence sound weird. To ensure that all the generated binary propositions make sense, a good practice is to limit the atomic propositions within some specific domain.

This domain of propositions in our example, is the communication between players in a strategic video game, in which promises, threats, tips, and warnings could often be made. Suppose we have three atomic propositions: 1. *You destroy the bridge*, 2. *I will attack you* and 3. *Player C will attack you*, in which the first one is antecedent and the latter two are consequents. Here, the first proposition is inherently neutral regarding desirability. The second and the third propositions are both undesirable for the hearer, and they differ in that the speaker has control in *I will attack you*, but lacks control in *Player C will attack you*. Together with two connectives $\rightarrow$ and $\leftrightarrow$, four binary logical formulas can be generated:

(3) a. You destroy the bridge $\rightarrow$ I will attack you
b. You destroy the bridge $\rightarrow$ Player C will attack you
c. You destroy the bridge $\leftrightarrow$ I will attack you
d. You destroy the bridge $\leftrightarrow$ Player C will attack you

Based on the information of control and desirability of the consequent, they can be easily labeled as:

(4) a. You destroy the bridge $\rightarrow$ I will attack you:'inducement(threat)'
b. You destroy the bridge $\rightarrow$ Player C will attack you:'advice(warning)'
c. You destroy the bridge $\leftrightarrow$ I will attack you:'inducement(threat)'
d. You destroy the bridge $\leftrightarrow$ Player C will attack you:'advice(warning)'

Taking a list containing the four formulas in (3) as the input, the baseline algorithm will yield:

(5) a. If you destroy the bridge, I will attack you
b. If you destroy the bridge, player C will attack you
c. If and only if you destroy the bridge, I will attack you
d. If and only if you destroy the bridge, player C will attack you

With a list containing the four formulas in (4), alongside with the information about the corresponding labels as input, the pragmatic algorithm will yield output as:

(6) a. If you destroy the bridge, I will attack you, but if you don't, I might still do
b. If you destroy the bridge, player C will attack you
c. If you destroy the bridge, I will attack you
d. If and only if you destroy the bridge, player C will attack you

Comparing (6) with (5), the advantage of the pragmatic algorithm can be clearly seen. While both algorithms take the list of logical propositions as input, the pragmatic algorithm yield more appropriate natural language sentences as output by also taking the extra information from the labels. (6a) would express the original semantics of the logical formula more faithfully than (5a), and (6c) would sound more natural compared to (5c), while (6c) and (5c) convey the same meaning.

3 Evaluating the Algorithm

3.1 Design and Materials

As shown in Sects. 2.2 and 2.3, the pragmatic algorithm is designed to be a better algorithm than the baseline algorithm. For material implication, the pragmatic algorithm should express the logical meaning more faithfully by adding some extra phrases to cancel CP. For biconditional implication, the pragmatic algorithm should generate more natural sentences by using 'if...then' instead of 'if and only if...then', while also preserving the logical meaning faithfully. On the other hand, we also hypothesis that the extra phrases added for of material implication would make the sentences longer and therefore slightly more unnatural, as the cost of making the sentences more faithful.

But so far, the merits of the pragmatic algorithm have been purely theoretical. For validation, an evaluation is made by asking human participants to evaluate the natural language output from both algorithms. The evaluation has two metrics: naturalness and faithfulness.

To evaluate the naturalness of the output, the participants are asked to mark a number on a linear scale from 1 to 5 (from 'very unnatural' to 'very natural') for how natural they think the generated natural language sentence sounds.

To evaluate whether generated natural language sentence faithfully conveys the original semantic meaning of the logical formula, an adapted version of *truth table task* is used. As in [17], the *truth table task* refers to the task in which 'participants are given a *rule* and asked to indicate which cases are consistent with that *rule*.' In our adapted version, this task is placed within the setting of the communication between players in a hypothetic strategy game,[4] as described in Sect. 2.3. The *rule* comes in the form of a message sent by one player to another player, and the participant are asked to select all the cases where both players' actions are true and consistent with that message.

To prepare the cases in the checkbox for each logical formula, the antecedent and the consequent in the original formula can be either true (unchanged) or false (negated), and then they are combined through conjunction. Thus, 4 cases (TT, TF, FT, FF) will be automatically created for each logical formula. For example, from a logical formula like *You destroy the bridge* $\rightarrow$ *I will attack you*, the four cases are: *You destroy the bridge* $\wedge$ *I will attack you* (TT), *You destroy the bridge* $\wedge$ $\neg$*I will attack you* (TF), $\neg$*You destroy the bridge* $\wedge$ *I will attack you* (FT), $\neg$ *You destroy the bridge* $\wedge$ $\neg$*I will attack you* (FF). These four cases will be realized into natural language sentences, and then be presented in a random order within each question.

Below is an example question demonstrating how the truth table task is used in the evaluation:

---

[4] The settings of this strategy game are very intuitive and should be easily understood through common sense (See Appendix for the game settings). The settings are presented to the participants before the evaluation begins, to ensure that all the participants will have a unified world knowledge about how the game works.

In a game, one player Sophie sent a message to another player Hans: 'If you destroy the bridge, I will attack you, and if you don't, I might still do.'
Having received the message,
(Tick all that apply)

- ☐ *Hans didn't destroy the bridge, and Sophie attacked him*
- ☐ *Hans destroyed the bridge, and Sophie attacked him*
- ☐ *Hans destroyed the bridge, and Sophie didn't attack him*
- ☐ *Hans didn't destroy the bridge, and Sophie didn't attack him*

If the participant ticks the checkbox of a particular case, it means that the participant judges this case to be true for the message, and if the participant leaves the box unchecked, it means that the participant judges this case to be false for the message. Similar to the naturalness score, a faithfulness score can be derived, through comparing the participant's answers with the truth table value of the original logical formula.

For instance, according to the truth table value of the material implication *You destroy the bridge* $\rightarrow$ *I will attack you*, if the participant's understanding of the generated natural language message 'If you destroy the bridge, I will attack you, and if you don't, I might still do.' is the same as the meaning of the logical formula, the checkbox would be ticked as:

- ☑ *Hans didn't destroy the bridge, and Sophie attacked him*
- ☐ *Hans destroyed the bridge, and Sophie attacked him*
- ☑ *Hans destroyed the bridge, and Sophie didn't attack him*
- ☑ *Hans didn't destroy the bridge, and Sophie didn't attack him*

If the participant leaves any checkbox unticked where it should be ticked, or ticks any checkbox which should not be ticked, it would mean that the participant has a different understanding of the generated natural language sentence, as compared to the semantics of the logical formula, thus making the natural language generation unfaithful. If the judgment of the participant is completely in accordance with the truth table value of the original logical formula, then the participant can receive 4 points. For each checkbox containing a different value than the truth table of the original logical formula, 1 point would be deducted, thus making the faithfulness score in a range of 0–4. The greater the difference, the lower the faithfulness score will be.

Since the pragmatic algorithm and the baseline algorithm yield the same output for advice (tips and warnings), these identical sentences are naturally not included in the evaluation, as the purpose of the evaluation is to see whether the outputs of these two algorithms are evaluated differently in terms of naturalness and faithfulness. For each algorithm, the evaluation material includes both promise and threat, and for each type, one conditional and one biconditional are included. Therefore, the evaluation form has 2 (baseline and pragmatic) $\times$ 2 (promise and threat) $\times$ 2 (conditional and biconditional) = 8 target sentences, plus 8 filler sentences, hence 16 sentences in total. The filler sentences are some conditionals containing disjunction and conjunction, and a concessive conditional with 'Even if', which are not the topic of this present

Table 4 Mean naturalness score comparison between baseline algorithm and pragmatic algorithm

	Baseline algorithm	Pragmatic algorithm
Material implication	4.54	3.35
Biconditional implication	3.95	4.6

study. Each sentence in the evaluation form has 2 evaluation tasks: (a) a truth table task for evaluating faithfulness and (b) a linear scale for evaluating naturalness.

3.2 Participants and Procedure

40 participants[5] took part in the evaluation. Unlike grammatical judgment, this evaluation relies more on conscious thinking and less on linguistic intuition, therefore being a native speaker of English was not required. Nonetheless, all participants were proficient in English (having degree courses taught in English and/or being native English speaker).

The evaluation was done online by asking participants to firstly fill in their personal information of gender, age, and English proficiency, and to read the instructions, and then to fill in the actual evaluation form. No time limit was posed for the evaluation.

3.3 Results and Discussion

The **naturalness score** results confirmed the theoretically motived hypothesis stated in Sect. 3.1, as shown in Table 4: For biconditional implication, the pragmatic algorithm achieved a higher mean naturalness score 4.6 compared to the 3.95 of the baseline algorithm. To test the statistical significance of this difference, a paired t-test is performed, showing a statistical significance between the naturalness scores for biconditional implication of baseline ($M = 3.95$, $SD = 1.03$) and that of the pragmatic algorithm ($M = 4.6$, $SD = 0.61$), $t(39) = 6.00$, $p < 0.001$. On the other hand, for the material implication, the mean naturalness score of the pragmatic algorithm is only 3.35, less than the 4.54 of the baseline algorithm. This difference is also tested using a paired t-test, also showing a statistical significance between the naturalness score for material implication of baseline ($M = 4.54$, $SD = 0.65$) and that of the pragmatic algorithm ($M = 3.35$, $SD = 1.13$), $t(39) = 8.76$, $p < 0.001$. Therefore, the evaluation results show that the pragmatic algorithm generates more natural sentences for biconditional implication. It generates less natural sentences

[5] Of the 40 participants, 20 were male, 18 were female, and 2 indicated as ‘other’ in the form. The average age of the participants was 37.6 years.

Table 5 Mean faithfulness score comparison between baseline algorithm and pragmatic algorithm

	Baseline algorithm	Pragmatic algorithm
Material implication	2.9	3.44
Biconditional implication	3.56	3.78

Table 6 Truth table of material implication and biconditional implication

p	q	$p \rightarrow q$	$p \leftrightarrow q$
T	T	T	T
T	F	F	F
F	T	T	F
F	F	T	T

for material implication, but it can be argued that their lack of naturalness is offset by their substantially enhanced faithfulness, as we shall see.

For the **faithfulness score**, the means are shown in Table 5. The results show that the pragmatic algorithm has achieved a higher mean faithfulness score for both material implication and biconditional implication. For both material implication and biconditional implication, the paired t-test is performed to test whether the difference has statistical significance. The result indicates that there is a statistical significance between the faithful scores for material implication of baseline ($M = 2.9, SD = 0.44$) and that of the pragmatic algorithm ($M = 3.44, SD = 0.74$), $t(39) = 6.60$, $p < 0.001$. And there is also a statistical significance between the faithful scores for biconditional implication of baseline ($M = 3.56, SD = 0.74$) and that of the pragmatic algorithm ($M = 3.78, SD = 0.48$), $t(39) = 2.76$, $p = 0.007$.

These results show that the pragmatic algorithm can express the logical meaning more faithfully for both material implication and biconditional implication. This confirmed our expectation about material implication, while exceeding our expectation about biconditional implication. As prior to the evaluation, the pragmatic algorithm is designed to only improve the naturalness for biconditional, rather than being designed to improve the faithfulness. But the evaluation results suggest that 'if…then' is not only more natural, but also expresses the biconditional connective more faithfully compared to 'if and only if…then'.

Apart from the faithfulness scores and naturalness scores calculated for each question, we can take a closer look at the results of the all the four cases (TT, FF, FT, FF) in the truth table task separately. For each of the four rows in the truth table, if the participant's judgement is the same as in the truth table of the logical connectives, it will be interpreted as the semantic meaning is faithfully convey to the participant, otherwise, it means the meaning is conveyed unfaithfully (Table 6).

Table 7 show the results of material implication for both algorithms. The numbers in the columns of TT, TF, FT, FF stand for the accuracy score for each case

Table 7 Results of accuracies of TT, TF, FT, and FF in truth table task for material implication

	p → q				
	TT	TF	FT	FF	Average accuracy
Baseline algorithm	0.99	0.95	0.08	0.86	0.72
Pragmatic algorithm	0.95	0.83	0.88	0.79	0.86

Table 8 Results of accuracies of TT, TF, FT, and FF in truth table task for biconditional implication

	p ↔ q				
	TT	TF	FT	FF	average accuracy
Baseline algorithm	0.94	0.89	0.9	0.84	0.89
Pragmatic algorithm	0.99	0.94	0.94	0.91	0.94

separately, and the *average accuracy* is the average score of all four cases. All the numbers presented are averaged across all participants.

The results show that the participant generally had high accuracies for both algorithms in the cases of TT, TF and FF, showing a good understanding of the semantics of $p \rightarrow q$ in these three cases. However, the sentences generated by the baseline algorithm seems to pose a great difficulty for participant in the case of FT, scoring an accuracy of only 0.08, which means for 92% of the time the participants believe $p \rightarrow q$ should be false in the case of FT. This is exactly what we predict, because of the effect of CP.

On the hand, the pragmatic algorithm can greatly improve the accuracy from 0.08 to 0.88, result in a higher average accuracy of 0.86, compared to the 0.72 of the baseline algorithm. This shows that the pragmatic algorithm has a great advantage for faithfully expressing the semantics of $p \rightarrow q$, especially for the FT case.

Similar to the Tables 7, 8 shows the results for biconditional implication, that both algorithms have achieved high accuracies for biconditional implication across the four cases of TT, TF, FT, and FF. The accuracies of the pragmatic algorithm are slightly higher, which is in line with the results of faithfulness scores.

To summarize, the evaluation results have shown that for material implication, the pragmatic algorithm can indeed generate sentences that are more faithful at the cost of being less natural. And for biconditional implication, the sentences generated by the pragmatic algorithm are both more faithful and more natural compared to the baseline algorithm. All these differences between the pragmatic and baseline algorithm are statistically significant.

4 Conclusion and Future Work

In this study of Conditional Perfection, we started from insights drawn from the pragmatic literature [5, 13, 14, 18]. Although these insights are interesting and plausible, they are also less than fully explicit, and have never been formally evaluated or implemented in an algorithm that expresses logical information in natural language. In the present work, we have provided an algorithmic formalisation of some pragmatic insights and subjected these to experimental testing with human subjects.

Inspired by the insight that Conditional Perfection (CP) is more likely to happen in *inducements* and less likely to happen in *advice*, an algorithm was designed to generate pragmatically more appropriate sentences from propositional logical formulas involving the two connectives $\rightarrow$ and $\leftrightarrow$. The merits of our algorithm have been tested in an evaluation by human participants, in which the sentences generated by the pragmatic algorithm are compared with the sentences generated by a simple baseline algorithm.

The theoretically motivated expectations behind the pragmatic algorithm have been confirmed in the evaluation. The results suggested that, by adding as 'if p, q, but if not p, might still q,' the effect of CP can be canceled, resulting in a great improvement for the FT case in the truth table (the row where the antecedent is false and consequent is true), thus improving the faithfulness of the sentence. The evaluation results also confirmed that by using 'If…then' instead of 'If and only if…then', the pragmatic algorithm improves the naturalness of the generated sentences.

Furthermore, the evaluation results indicated that the pragmatic algorithm using 'If…then' instead of 'If and only if…then' is not only more natural, but also more faithful. This suggests that, although 'If and only if…then' is conventionally used as the natural language counterpart for $\leftrightarrow$, it actually leads to more misunderstanding compared to simply using 'If…then' for $\leftrightarrow$. A possible explanation is the fact that 'if and only if' is not really a common expression outside logic and mathematics, and it may cause confusion to people who are not very familiar with logic or mathematics.

As expected, the pragmatic algorithm also has a disadvantage when it comes to the naturalness of the sentences involving material implication. Saying 'if p, q, but if not p, might still q' boosts the faithfulness, but it also makes the sentence longer and a bit stilted.

As summarized in Table 9 (the symbol '>' represents 'surpass', 'being better than'), we can see that the pragmatic algorithm is superior to the baseline algorithm in three ways and is inferior in only one way. In practical NLG applications such as providing automatic feedback to the translation between FOL and natural language (see Sect. 1.2), faithfulness tends to be more important than naturalness. Therefore, it seems fair to conclude that our pragmatic NLG algorithm would be a worthwhile alternative for use in practical NLG applications, for example when NLG is used to clarify logical expressions in Artificial Intelligence applications [15].

Our study has addressed two problems: (1) under what precise circumstances does Conditional Perfection arise, and (2) how an NLG algorithm might optimize the wording of the natural language expression to take Conditional Perfection into

Table 9 Summarized comparison between pragmatic and baseline algorithm

	Biconditional implication	Material implication
Faithfulness	pragmatic > baseline	Pragmatic > baseline
Naturalness	pragmatic > baseline	Baseline > pragmatic

account. We believe that this work paves the way for a number of further investigations. First, the present study has only considered four types of conditionals namely promises, threats, tips, and warnings, as the relations between them are systematic and they can form a complete tetrachoric table through two features. In the future, the question of how to generate pragmatically appropriate sentences for other types of conditionals such as background conditionals and concessive conditional (which have also been discussed in [18]) should be further investigated to see whether they can be handled along the lines of the present paper.

Second, the present study has only dealt with the two logic connectives $\rightarrow$ and $\leftrightarrow$. Future work could be done to investigate other connectives such as $\vee$ and $\wedge$, thus covering all the connectives used in propositional logic. An important question for further research is to what extent our findings are sensitive to the choice of connectives. For example, $\neg p \rightarrow q$ and $p \vee q$ are logically equivalent, but these two logically equivalent formulations may not be pragmatically equivalent. Other constraints governing the natural language form will need to be taken into consideration. For example, a conditional of the form $\neg p \rightarrow q$ can make both promises and threats, whereas the logically equivalent disjunction $p \vee q$ can only make threats, but not promises [19]. If more connectives are to be added into the NLG system, then these issues should be taken into account.

Thirdly, the present study has only considered the binary proposition in which two atomic propositions are connected through a single connective. In future work, more complex propositions involving multiple connectives can be tested, to see if the conclusion based on binary propositions still hold when the proposition becomes more complex. This direction could also lead to further investigating of the interaction between different kinds of connectives when they are present within the same complex proposition.

Finally, it would be worth investigating how our work may be extended to enhance existing Natural Language Processing work in paraphrasing and text simplification [21]. The idea would be to take an input text T that contains conditional statements, and to convert it into a clearer and more explicit output text T' in which these conditionals are rendered in a more explicit way, for instance by replacing conditionals by biconditionals if and when this is appropriate, analogous to our pragmatic NLG algorithm. To make this work, an NLP algorithm would first have to detect whether the relevant texts in T express promises, threats, and so on because, as we have seen, these pragmatic factors affect the way in which conditionals are interpreted.

Appendix: Instructions for Participants in our Experiment

This study is about testing people's understanding about some sentences. The target sentences will be embedded in the setting of the communication between players in a strategy game.

In case you haven't played any strategy game, this is a multiplayer game, in which buildings like castles and bridges can be built to gain advantage. Players can attack other players and destroy other players' buildings to beat them in the game. Players can also form alliance with each other against a common enemy. Resources (including food, wood, and stone) and gold coins are valuable in the game and can be exchanged among players.

In one game, one player Sophie has sent some messages to another player Hans. In the following sections, each message will be accompanied with 4 scenarios.

After you have read the sentence, please identify all the possible scenarios in which the actions of both Sophie and Hans are consistent to the message. In other words, if you think the scenario follows the message, tick the box; if you think the scenario is impossible given the message of Sophie, don't tick the box.

Afterwards, please rate the wording of the message in terms of how natural the message sounds to you.

References

1. Van der Auwera, J.: Conditional perfection. Amst. Stud. Theory Hist. Linguist. Sci. Ser. **4**, 169–190 (1997)
2. Van der Auwera, J.: Pragmatics in the last quarter century: the case of conditional perfection. J. Pragmat. **27**(3), 261–274 (1997)
3. Barker-Plummer, D., Cox, R., Dale, R., Etchemendy, J.: An empirical study of errors in translating natural language into logic. In: Proceedings of the Annual Meeting of the Cognitive Science Society, vol. 30 (2008)
4. Enderton, H.B.: A Mathematical Introduction to Logic. Elsevier (2001)
5. Evans, J.S.B., Twyman-Musgrove, J.: Conditional reasoning with inducements and advice. Cognition **69**(1), B11–B16 (1998)
6. Fitting, M.: First-Order Logic and Automated Theorem Proving. Springer Science & Business Media (2012)
7. Flickinger, D.: Generating English paraphrases from logic. In: From Semantics to Dialectometry (2016)
8. Gatt, A., Krahmer, E.: Survey of the state of the art in natural language generation: core tasks, applications and evaluation. J. Artif. Intell. Res. **61**, 65–170 (2018)
9. Geis, M.L., Zwicky, A.M.: On invited inferences. Linguist. Inq. **2**(4), 561–566 (1971)
10. Herburger, E.: Conditional perfection: the truth and the whole truth. In: Semantics and Linguistic Theory, vol. 25, pp. 615–635 (2016)
11. Horn, L.R.: From if to iff: conditional perfection as pragmatic strengthening. J. Pragmat. **32**(3), 289–326 (2000)
12. Mayn, A., van Deemter, K.: Evaluating automatic difficulty estimation of logic formalization exercises (2022). arXiv:2204.12197
13. Newstead, S.E.: Conditional reasoning with realistic material. Think. & Reason. **3**(1), 49–76 (1997)

14. Ohm, E., Thompson, V.A.: Everyday reasoning with inducements and advice. Think. & Reason. **10**(3), 241–272 (2004)
15. Oren, N., van Deemter, K., Vasconcelos, W.W.: Argument-based plan explanation. In: Knowledge Engineering Tools and Techniques for AI Planning, pp. 173–188. Springer (2020)
16. Reiter, E., Dale, R.: Building Natural Language Generation Systems. Studies in Natural Language Processing. Cambridge University Press (2000). https://doi.org/10.1017/CBO9780511519857
17. Sevenants, A.: Conditionals and truth table tasks the relevance of irrelevant (2008)
18. Van Canegem-Ardijns, I., Van Belle, W.: Conditionals and types of conditional perfection. J. Pragmat. **40**(2), 349–376 (2008)
19. Van Rooij, R., Franke, M.: Promises and threats with conditionals and disjunctions. Discourse and Grammar: From Sentence Types to Lexical Categories, pp. 69–88 (2012)
20. Wang, J.-t.: On computational sentence generation from logical form. In: COLING 1980 Volume 1: The 8th International Conference on Computational Linguistics (1980)
21. Xu, W., Callison-Burch, C., Napoles, C.: Problems in current text simplification research: new data can help. Trans. Assoc. Comput. Linguist. **3**, 283–297 (2015)

White Roses, Red Backgrounds: Bringing Structured Representations to Search

Tracy Holloway King

Abstract Search has become a key component of many on-line user experiences. Search queries are usually textual and hence should benefit from improvements in natural language processing. However, many of the NLP algorithms used in production systems fail for queries that require structured understanding of the query and document or that require reasoning. These issues arise because of the way information is stored in the search index and the need to return results quickly. The issues are exacerbated when searching over non-textual documents, including images and structured data. The use of embedding-based techniques has helped with some types of searches, especially when the query vocabulary does not match that of the documents and when searching over images. However, these techniques still fail for many searches, especially ones requiring reasoning. Simply combining classic word-level search and embedding-based search does not solve these issues. Instead, in this position paper, I argue that we need to create hybrid systems from traditional search techniques, embedding-based search, and the addition of structured data and reasoning. Enabling such hybrid systems will require a deep understanding of linguistic representations of meaning, of information retrieval optimization, and of the types of information encoded in the queries and documents. It is my hope that this paper inspires further collaboration across disciplines to improve these complex search problems.

Keywords Search · Information retrieval · Semantic search · Query understanding

1 Introduction

Search has become a key component of many on-line user experiences. These include large web search engines (e.g., Google, Bing, Baidu, Yandex), eCommerce (e.g., Amazon, eBay, Rakuten, Taobao), and search within websites and within documents.

T. H. King (✉)
Adobe Inc., San Jose, CA, USA
e-mail: tking@adobe.com

R. Loukanova et al. (eds.), *Logic and Algorithms in Computational Linguistics 2021 (LACompLing2021)*, Studies in Computational Intelligence 1081,
https://doi.org/10.1007/978-3-031-21780-7_9

Search queries are usually textual and hence should benefit from natural language processing (NLP) and especially from the rapid improvements in NLP over the past decades. This is especially true in situations where there is not enough behavioral data from past users to "memorize" the top search results and so a deeper understanding of the query and documents is required.

However, many of the NLP algorithms currently used in production systems fail for queries that require a structured, whether syntactic or logical, understanding of the query and document, or that require reasoning. Examples of such queries include *dresses between $50 and $100* (eCommerce search), *cake recipes without wheat flour* (web search and recipe site-internal search), and *photo white rose red background* (image search). Even simple queries that only require an understanding of word-order based relations such as *chocolate milk* vs. *milk chocolate* often include many irrelevant search results.

These issues arise because of the way information is stored in the search index, often as single words for text-based documents, and the need for extremely fast computation in order to return search results quickly to the user. The issues are exacerbated when searching over non-textual documents, including images and structured data (e.g., prices for eCommerce products). The use of embedding-based techniques has helped with some types of searches, including when the query vocabulary does not match that of the documents (e.g., misspellings, synonyms) and when searching over images. However, these techniques still fail for many searches, especially ones requiring reasoning. Simply combining classic word-level search and embedding-based search does not solve these issues. There remain classes of queries which require information beyond the representation of words and multi-word expressions stored in an inverted index. These are the structured representations referred to in the title of this paper. They involve associating typed information with the content (e.g., prices) and storing relationships among the entities (e.g., in an image the rose is white, the background is red). These structured representations in turn must support reasoning (see the *snacks without nuts* example query below). Although some of this reasoning can be computed off-line and stored for commonly queried information, to support the broad range of search queries, fast and accurate reasoning at query time is required.

Three Example Queries To understand the scope of the issue, consider three real-world example queries that require information beyond simple keyword matching.

First consider a search for images: *white rose red background.*[1] This query is looking for images of white roses shown on a red background. However, there are many more images of red roses than white roses because red is such a popular color for roses and there are a huge number of images with white backgrounds because these are used when compositing images. This means that there are many more images of red roses on white backgrounds than white roses on red backgrounds and that those images are much more popular (e.g., viewed, downloaded, or purchased

[1] There is no preposition in this query (cf. *white rose on red background*), but the search results are similar even with the preposition. The preposition helps to delineate the two noun phrases, but the techniques to improve the results with the preposition also help when the preposition is absent.

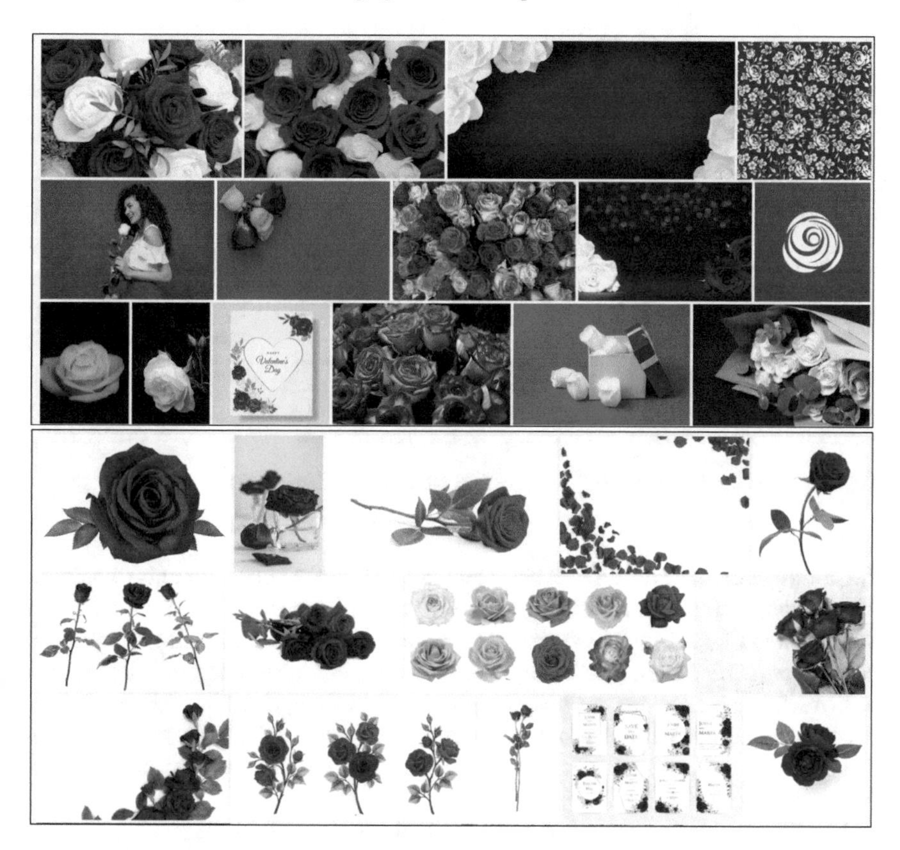

Fig. 1 Results for the query *white rose red background* with (upper) and without (lower) sufficient query and document understanding for search

more). If the words in the query are treated as a bag of words without reference to word order or syntactic structure, then any image associated with roses, backgrounds, and the colors red and white will be returned and the more popular images of red roses on white backgrounds will rank higher. Example search results for this query are shown in Fig. 1. One way to handle this query using structured representations will be discussed in detail in Sect. 4.3.

Next consider an eCommerce search with negation: *snacks without nuts*. The preposition *without* encodes a negation of containing nuts. This is a common query by users who do not like nuts or are allergic to them. However, when treated as a bag of words, the preposition *without* is either dropped entirely because it is so frequent as to be considered useless[2] or will match many items where the *without* applies to some other text in the product description. To make matters worse, the search engine will try to match the word *nuts* and so will return snacks which specifically contain nuts. As a result, the search results both miss relevant items (snacks without nuts that

[2] Such words are referred to as stopwords in search. See Sect. 2.2 on text processing.

Fig. 2 Results for the query *snacks without nuts* where most of the results clearly contain nuts

did not mention the exclusion of nuts) and include many items with nuts (where the word *without* referred to something other than nuts). An example set of search results are shown in Fig. 2. The search results are much better when rephrasing the query as *nut free snacks* but even then many relevant results are missed because only snacks that overtly state they are nut free are returned. In search, checking for the absence of something (nuts in this example) is complicated because there are so many things that could be absent in a document that they cannot be listed: their absence has to be determined at search time.

Finally consider an eCommerce query that requires simple reasoning: *dresses over $100*. Here the price phrase requires the search engine to constrain the price of the returned items (the dresses) to be greater than $100. If the query is treated as a simple textual query, the dresses returned will either be exactly $100 if the dollar sign is searched for or, more likely, any price but containing the word *100* somewhere in the product description.

Given the frequency of price-related queries on eCommerce sites, many of sites have special query processing to detect the price constraint and map it to structured data on each item. In this case, the phrase *over $100* would be mapped to the price information and checked that the value is greater than $100. Example search results where this reasoning has been applied are shown in Fig. 3.

The remainder of this paper examines why search works as it does and how this is being improved. Section 2 provides the basics on how search works, focusing first on inverted indices, then introducing text processing, the use of user behavioral data, and structured data, and finally describing search ranking. Section 3 discusses why pure inverted indices are not sufficient. Section 4 introduces techniques to enhance traditional search to provide more accurate results and provides a detailed example. Finally, Sect. 5 concludes.

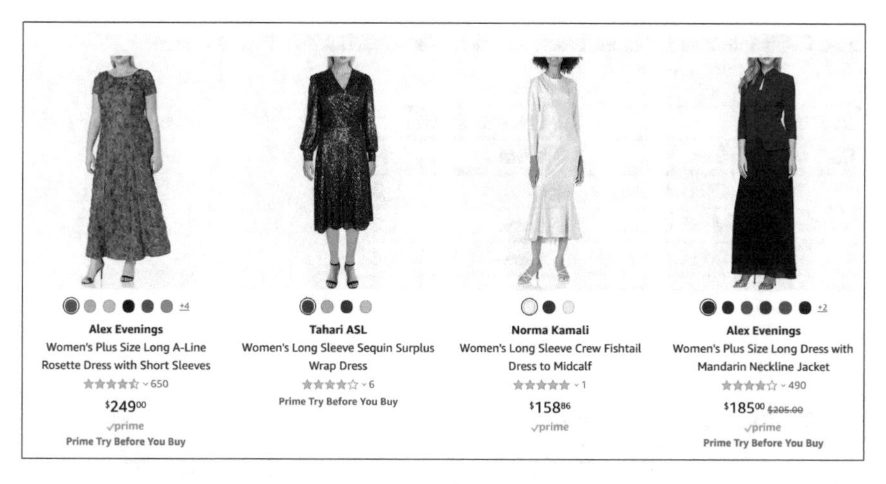

Fig. 3 Results for the query *dresses over $100* where the price constraint has been properly applied. The word *100* does not appear in the titles or prices

2 How Search Works

This section provides a high level overview of how search works, focusing on traditional inverted indices but also introducing other aspects of search used in production search engines. Those familiar with search can skip this section.

2.1 Inverted Indices

Most search engines are based on an inverted index. An inverted index is similar to the index at the end of a book. It allows you to look up a word or phrase and see what pages are associated with it. In search, these pages are documents (e.g., web pages in web search, product listings in eCommerce, documents in document collections like arXiv).

To create a search index for a set of documents, first the system identifies all the words associated with each document. Then it builds an index of the words and associates them with an identifier for relevant documents. Table 1 shows four simple one-sentence documents and the inverted index that is created from them.

When the user issues a query, the query is broken into words. All of the documents associated with each word are found by looking up the word in the index. The document lists for each word are compared. The documents that appear on all the lists are returned. This set of documents is referred to as the recall set or match set. These documents are then ranked so that the most relevant one is first, then the next more relevant, etc. Four sample queries and the documents returned for them are shown in Table 2.

Table 1 Sample inverted index based on 4 documents, each with only one sentence

Documents	
Doc. Id	Text
1	Cats are furry
2	Mice are furry
3	Dogs chase cats
4	Cats chase mice

Inverted index	
Word	Doc. Ids
cats	1, 3, 4
furry	1, 2
mice	2, 4
dogs	3
chase	3, 4

Table 2 Sample queries and the documents they return for the documents and inverted index in Table 1

Query	Initial documents	Final documents
cats	1, 3, 4	1, 4, 3
cats chase	cats: 1, 3, 4; chase: 3, 4	3, 4
cats birds	cats: 1, 3, 4; birds:—	—
mice dogs	mice: 2, 4; dogs: 3	—

The one-word query *cats* returns all the documents with the word *cats* in them. For the ranking, here we ranked the two documents with the word *cats* as the subject higher than the one with it as an object. For the query *cats chase*, there are two documents which contains both words and so those two are returned. We ranked the document which matches the word order in the query higher. In contrast, the query *cats birds* does not return any results because there is no document that contains the word *birds*. Similarly, the query *mice dogs* also returns no results. In this case, there are two documents which contain the word *mice* and one which contains the word *dogs*, but no documents that contain both. Many search engines return "partial matches" where not all the query words are matched if there are no documents that match all the words. This is not shown in Table 2.

2.2 Beyond Inverted Indices

Text Processing In the inverted index example above, the words were indexed in their inflected forms (e.g., plural *cats*). In search, the processing of the text in documents and queries strongly affects search result quality. In general, text is uniformly lower cased, even for proper nouns where it can be distinguishing. This is because queries do not contain canonical capitalization (e.g., English proper nouns are usually lower cased; users may have the caps-lock key on). Accent marks may be removed, mapping accented letters to their unaccented counterpart. Whether to de-accent depends on the language and in particular on whether users reliably type queries with accent marks. Punctuation is stripped except when crucial for meaning (e.g., decimal points within numerals are not removed but sentence- and abbreviation-final periods are).

Stemming[3] or lemmatization is usually applied. Queries often include plural nouns even when the user is looking for a singular instance. For example, in eCommerce, the query *hammers* has more purchases for single hammers than for hammer sets or multiple hammers. The same is true for the query *dresses*: even though many people own multiple dresses, they purchase them one at a time.

Finally, words that are extremely common in the document collection are not indexed. These are referred to as stop words. The linguistically closed class words (e.g., prepositions, determiners) are usually stop words. These are not indexed because they are so frequent as to not discriminate between documents. However, as discussed in Sect. 3, they can be crucial for certain types of reasoning (e.g., *snacks without nuts*) and for media queries (e.g., bands like The Who and The The), which will not be discussed further here.

See [17] for a detailed overview of text processing for search.

Behavioral Data Most search engines have some queries that are much more popular than others. These popular queries are referred to as head queries, and the rare ones as tail queries. In web search and larger eCommerce sites [24], the search engine can effectively memorize the best results for head queries. Although this can be done by editorially providing results, it is usually learned from aggregated user behavior. That is, the search result that users click on the most is put in first position, the next most popular result in second position, etc. In web search, this can be seen with navigational queries where the query is looking for a particular popular site (e.g., the query *bbc*, where most users are looking for the BBC on-line news site). Even when a query is not frequent enough to memorize the top results, user behavioral data can provide signal as to which documents are most popular and which queries are associated with these popular documents.

Meta-data and Structured Data Most search engines involve documents which contain more than just text, or images in the case of image search. This data can provide valuable information for search, both for identifying relevant documents and for ranking them. Web pages have titles and urls (site addresses). In addition, web documents often include data about what queries led to clicks on them and what text was used to link to them from other documents (referred to as anchor text). Products in eCommerce search contain a large amount of structured data (e.g., price, brand, size) [15]. Similarly, publications contain author, year, and publisher information.

Our example query *dresses over $100* demonstrates the importance of structured data. Knowing that a query is restricting the price of the search results is not actionable unless the price is available in the documents in a way that the search engine can

[3] Stemmers remove the endings of words, leaving a stem that can be indexed. This stem may or may not be a word in the language [20]. Lemmatizers map inflected forms to a linguistic or dictionary base form. The difference between these is most obvious with highly irregular inflected forms where stemming cannot normalize a word to a citation form.

apply the restriction. In the case of prices, this means having them in a format that allows basic arithmetic operations to be performed. The prices are also used when ranking the products by price instead of relevance, an option that is available on most eCommerce sites.

2.3 Ranking Features

The set of documents retrieved has to be ranked for display to the user. The order of results is important for ensuring users have easy access to the most relevant results. Users rarely look beyond the first page of results and have a tendency to click on the top result. Here we mention some key factors in search ranking because even if we solve the core relevancy issues described in the introduction around queries like *white roses red background*, *snacks without nuts*, and *dresses over $100*, there is still significant work to be done to order those results to provide an optimal search experience.

Some ranking features focus only on the document, in particular on document popularity and reliability. For example, in web search, results from Wikipedia are often ranked highly for entity queries like proper names (e.g., *tbilisi*, *ruth bader ginsburg*). Other ranking features involve query-document affiliation. For example, where the query word matches within the document matters: documents which match the query words in the title are ranked more highly than ones which match lower in the document. Relatedly, if the query contains multiple words, then documents where the two words are near to each other are ranked more highly. A special subcase of this is multi-word expressions (MWE) where the words should be a MWE both in the query and the document. There is a strong preference in the ranking for MWE to be adjacent in the document text. High-frequency MWE may be indexed as single words (e.g., *san francisco*, *hot dog*), i.e., they are treated as words with spaces for purposes of the index.

In this paper, we focus on the match set since the integration of structured representations in search is fundamental for retrieving all and only documents that match the intent of the user's query. For more on ranking see [7, 17, 24].

2.4 An Example: **Milk Chocolate** *Versus* **Chocolate Milk**

This section walks through an example of how search works, moving beyond the simplistic furry cat example in Sect. 2.1. Consider the queries *milk chocolate* vs. *chocolate milk*. Assume that neither *milk chocolate* nor *chocolate milk* are MWE in the search index.

For the document processing, lower-casing allows us to match titles and section headers which contain initial or all capital letters (e.g., *Berkeley Farms Chocolate Milk*). Lemmatization allows us to match documents with plural words like *assortment of milk chocolates*. After the text normalization, documents that mention the normalized form of *chocolate* or *milk* will have entries in the inverted index for those words.

When the query is issued, it is lower-cased, lemmatized, and stop words are removed in exactly the same way that the documents were. Stop words are not an issue here since neither query contains one. Lemmatization maps *milk chocolate* and *chocolate milk* to themselves since none of the words are inflected. Lower-casing maps any capitalization in the query to the lower-case forms used in the index.

Using a standard inverted index and no MWE for these words, the two queries will include precisely the same documents in the match set. This is because although the word order differs, they contain the same words.

However, ranking will treat the queries differently and rank different documents highly. In both web and eCommerce search, these queries are likely to have been seen before and have user click data associated with them. So, documents that were clicked for those queries by other users will be ranked more highly. In addition, word proximity, including word order, is key to distinguishing the queries: documents with the exact query phrase in the document are much more likely to be relevant and should be ranked higher. A Google search for the queries demonstrates the strength of the phrase matching and user behavioral data. At least in the US (google.com) shopping results are shown at the very top of the page; the Wikipedia article with the exact phrase title is shown as the first result; there is an entity box with a definition of chocolate milk (for *chocolate milk*) and milk chocolate (for *milk chocolate*) as well as additional entity-related information (e.g., wine pairings for *milk chocolate* and serving size for *chocolate milk*).

3 Why Inverted Indices Are Not Enough

In Sects. 1 and 2, we alluded to the fact that inverted indices are not enough to ensure relevant search results for all queries. Even with basic text processing, many relevant documents can be missed and many irrelevant documents can be included. This section describes how these issues arise and some ways to address them. As we will see, even these capabilities are not enough to provide relevant results for many types of queries. Those will be addressed in Sect. 4.

3.1 Vocabulary Mismatches

User queries may contain synonyms and paraphrases of the words used in the documents. These documents should be returned in the search results since they are relevant to the query: they just use different words with the same semantic meaning. For example, there is a piece of furniture that can be referred to in English as a *sofa*, *couch*, or *chesterfield*. In an eCommerce scenario, if a query contains any of these words, all the relevant products should be returned, regardless of which word appears in the product title and what the name of the category that contains the products is.[4]

There are two ways to handle synonyms in inverted indices. The first is to normalize to a specific form (e.g., mapping *couch* and *chesterfield* to *sofa*). This is similar to using lemmatization to normalize singular and plural forms of nouns. Normalization applies to both the index and the query so that they can match. Unlike lemmatization, synonym normalization may add additional forms instead of replacing the form. This allows the ranker to use information about exact form matching in addition to retrieving all the documents via the normalization. When using normalization while maintaining the original form, the query treats the two forms as alternatives. For example, the query *leather couch* would match documents with either the words *leather* and *couch* or with the words *leather* and *sofa*. Table 3 shows the two treatments of normalization. Misspellings are always handled as normalization.

The other way to handle synonyms in inverted indices is via expansion. In expansion, a word is expanded to all of its synonyms, as opposed to normalization where one of the synonyms is the form that all the words map to. Expansion can occur either in the index or the query. If done in the index (Table 4), whenever one of the words is found, all the synonyms are indexed. Expanding in the index has two advantages. First, more context is available. This means the system is more certain that the word should be expanded (e.g., it is not some other meaning of the word which would not have the synonym expansion). Second, the search is faster because the number of words being searched from the query is smaller. If done in the query (Table 5), whenever one of the words is found, all of the synonyms are looked for as alternatives. Query expansion has the advantage that if there is a problem with a synonym it can be quickly fixed because there is no need to reindex the documents.

3.2 Mapping to Structured Data

Many documents contain structured data. Sometimes the structured data is part of the original document (e.g., prices on products in eCommerce, date stamps on papers in document collections). In other cases, the structured data is created as part of the document processing. For example, a named entity detector can extract all the

---

[4] There are two interesting variants of the synonym problem. The first is misspellings, which are a different, albeit technically incorrect, way to refer to a word. Search engines usually employ a separate speller module to correct misspellings in queries [10]. Given how short queries are, spell correction can be difficult, especially in the broad domain of web search. In addition, the autocomplete (also referred to as type-ahead or query suggestion) [8] feature provided in many search engines helps to guide users to properly spelled queries. In domains where there are likely to be misspellings in the documents, the speller may also be applied during document indexing. The second variant of the synonym problem is providing cross-lingual search, i.e., having a search query in one language find documents in another language [19]. In this case the synonyms are the words in the different languages with the same meaning. Cross-lingual search occurs for certain document collections, especially domain-specific archives, and when searching for images, where the images may be tagged in one language, usually English, but searched for in many languages.

Table 3 Normalization approach to synonyms: Documents with any of *sofa*, *couch*, or *chesterfield* will index *sofa* (indexing of the words *brown* and *leather* is not shown). Queries with any of *sofa*, *couch*, or *chesterfield* will search for *sofa*. The pipe (|) indicates an alternative (logical OR). All three documents will be returned for all three queries. The original words can be used in ranking

Document synonym normalization			
Document text	Original word	Indexed words	
		Normalized only	Normalized+original
Brown Leather Sofa	Sofa	sofa	sofa
Brown Leather Couch	Couch	sofa	sofa, couch
Brown Leather Chesterfield	Chesterfield	sofa	sofa, chesterfield

Query synonym normalization		
Original query	Normalized query	
	Normalized-only query	Normalized+original query
sofa	sofa	sofa
couch	sofa	sofa|couch
chesterfield	sofa	sofa|chesterfield

Table 4 Document expansion approach to synonyms: Documents with any of *sofa*, *couch*, or *chesterfield* will index all three forms (indexing of the words *brown* and *leather* are not shown). All three documents will be returned for queries with any of those words in them

Document synonym expansion		
Document text	Original word	Indexed words
Brown Leather Sofa	Sofa	sofa, couch, chesterfield
Brown Leather Couch	Couch	sofa, couch, chesterfield
Brown Leather Chesterfield	Chesterfield	sofa, couch, chesterfield

Table 5 Query expansion approach to synonyms: Queries with any of *sofa*, *couch*, or *chesterfield* will search all three words. The pipe (|) indicates an alternative (logical OR). So, documents with any of the three synonyms will be returned for all three queries

Query Synonym Expansion	
Original query	Expanded query
sofa	sofa|couch|chesterfield
couch	sofa|couch|chesterfield
chesterfield	sofa|couch|chesterfield

people, locations, and organizations mentioned in a document and provide canonical forms for these and even links to a knowledge graph nodes for the entities. Failing to recognize entities and treat them as such can result in irrelevant search results. For example, a search for my name *tracy king* on Amazon returns many books where either the author's first or last name is Tracy and the title of the book contains the word *king*. If you include my middle name *tracy holloway king*, more of the results are relevant, but many are not. Clearly I need to work on my popularity as an author in order to rank more highly.

These entities can be indexed as special fields so that the fact that they are entities and the type of entity are recorded. This is similar to how words in the document title are indexed distinctly from those in the rest of the document so that ranking can put more weight on matching words in the title. These fields may be text fields that allow for standard inverted index retrieval (e.g., brand names in eCommerce, author names for books and documents). Some entities are not text and so are stored and searched differently, generally in ways similar to databases. Examples of such entities include prices in eCommerce and dates for documents. By having these in entity-specific formats, it is possible to reason over them, as required for queries like *dresses over $100* or *19th century poems*.

3.3 Negation and Syntactic Structure

The last category of issues affecting the quality of search results from inverted indices are for queries where finding relevant results requires an understanding of syntactic or semantic structure.

In the realm of semantics, negation is particularly complex for search. Search with inverted indices works by finding documents which contain combinations of specific words. Determining whether a document does not contain a word, much less the semantic concept that corresponds to the word in the query is much less efficient. In the *snacks without nuts* example (Fig. 2), search has to interpret this as a query for the ingredients of the snack not including nuts. Simply excluding documents without the word *nut* can exclude relevant documents: A document might contain the word *nut* if it is in a phrase like *nut free*, *no nuts*, or even *my kids are nuts about this snack* in a review of the product. If the product contains a special ingredients field, then the negation can take scope only over that field. However, checking that the word *nut* does not occur in the ingredients field is not enough since different types of nuts also should not appear there (e.g., words like *almonds*, *walnuts*, or *pecans* also have to be absent).[5] Creating the product data, query understanding [1], and search facets (e.g., left rail filters for various attributes) to handle negation requires detailed domain knowledge and systems. As a result, negation is currently rarely addressed systematically in search engines.

[5] See [5, 6] on entailment and contradiction detection more generally and [14] on hybridizing natural language inference systems.

As opposed to the semantics required for negation, information from syntactic structure is easier to capture in search [25, 28, 29]. Modifier-head relationships, such as those in the *chocolate milk* vs. *milk chocolate* example (Sect. 2.4) are of this type. If these occur frequently enough in queries, they can be treated as MWE even if they are compositional or the modification structure can be stored as additional data in the index. When processing the documents to find the words and structures to index, simply scanning for the particular phrase (referred to as an ngram) is not always enough. The ngram may occur in contexts where it does not refer to the modifier-head relationship. In addition, this simple string adjacency will miss instances where the modifier is not strictly adjacent to the head (e.g., *milk and dark chocolate assortment* or *plain, chocolate, and strawberry milk*) [22]. More complex syntactic analysis is needed to find documents which answer queries such as *who acquired PeopleSoft* versus *who did PeopleSoft acquire*. The words *PeopleSoft* and *acquire* (and its inflected forms and possibly words like *acquisition*) will match the same documents. The ranking can treat these differently using word order, but syntactic understanding is necessary as search comes closer to being question answering and not just document retrieval.

3.4 Robustness Through Embeddings

Several of the techniques discussed above involve improving recall, i.e., improving the robustness of the system to mismatches between user queries and documents. Text normalization and synonyms are of this type. In addition to the methods described above to handle these, with the advent of scalable deep-learning (DL) systems, instead of using traditional word-based inverted indices, some search engines integrate embedding-based search. DL models map text and images to arrays of numbers, referred to as embeddings, in an abstract semantic space. Words that have identical or similar meanings (e.g., *sofa, couch, chesterfield*) are close to one another in this space. Words that are related to one another (e.g., *woman, girl*) are close, but less close than the synonymous words. Words that are unrelated (e.g., *girl, chesterfield*) are far apart. Search can map the documents to embeddings and search over those.[6] The user query is mapped into the same embedding space and the closest documents are retrieved. The score that indicates how close the documents are in the semantic embedding space is used in ranking and as a threshold to decide which documents should be in the result set and which not. The remainder of this section outlines where embedding-based search works well and where it does not.

Where Embeddings Work Well DL embeddings often work well to bridge vocabulary mismatches between user queries and documents [4, 9, 18]. Instead of explicitly using synonyms via normalization or expansion, the semantic space puts words that are similar close together. A strength and weakness of this approach is that the line between true synonyms and related words is blurred with no clear cut-off

[6] Indexing and retrieving embeddings efficiently is a major challenge. It will not be discussed here.

Fig. 4 Results for the query *cougar* using embedding based search. The queries *puma* and *mountain lion* have almost identical results

between the two. This provides robustness in finding documents but can also result in related but non-exact documents being returned. Many embedding models also handle misspellings. Although embedding-based search can return relevant results for misspelled queries, it is not a replacement for a spell corrector, which can be used to message the user for confirmation and correct queries into forms that will have more accurate results and improved ranking.

Some of the greatest potential for embedding-based search is for text-based search over images. Several recent DL models map text and images into the same semantic space [12, 21]. Images are mapped into embeddings, which are indexed. The user's text query is then mapped into an embedding and the semantically closest images are returned by the search engine. Ranking features other than the semantic similarity can be used to determine the final ranking. This ensures that high-quality or popular images are returned and helps provide more visual diversity in the images. An example of a search for *cougar* is shown in Fig. 4. Results are almost identical for the queries *puma* and *mountain lion*, which are synonyms for *cougar*, even though no synonyms were added to the system.

In many cases the DL embedding models are trained on large, general domain data. This provides broad coverage, but can have issues for more specialized queries and unusual domains. The embeddings can be customized, referred to as fine-tuning, for a particular domain such as search on a fashion eCommerce site.

Where Embeddings Fail There are situations where embedding-based search does not work well. Some of these are classes discussed above which require more structure, including negation and syntactic structure. However, structure in the form of ngrams (e.g., *milk chocolate* vs. *chocolate milk*) are captured well in many embedding models. So, in these cases embeddings are not worse than inverted index-based search, but they are not always improvements.

However, embedding-based models perform much worse than inverted indices when words distribute very similarly but have meanings that are crucially distinct for search. Model numbers in eCommerce search are an example of these. When a model number occurs in a query, users want exactly that model, but embedding-based search will often return a seemingly random set of model number results. In fact, numbers in general behave in this way since they occur in similar environments (e.g., modifying the same nouns with the same adjectives and verbs surrounding those nouns) and so the embeddings treat them like synonyms. This means that results for queries like *dresses over $100* will not be relevant unless techniques as described in Sect. 3.2 are used. Names of people can also behave this way where names of the same gender and ethnicity incorrectly distribute like synonyms for one another.

4 Towards a Solution: Incorporating Structure

Section 3 outlined where neither inverted indices nor embedding-based search solve issues with search result quality. In this section we first discuss how more structured data can be used to solve these (Sect. 4.1). We then show how robustness techniques can be integrated to further enhance the results (Sect. 4.2). Finally we discuss a detailed example of combining these techniques, using color-object queries to search over images as an example (Sect. 4.3) and then sketch techniques for addressing negation in search (Sect. 4.4).

4.1 Structure Where It Matters

In the examples in Sect. 1, we saw how structured data was need to provide accurate search results for certain classes of queries. Why isn't structured data of this type used more pervasively? The first reason is that although there are established techniques for creating and searching over structured data, models for query and document understanding have to be build for the specific domain and use case. This means that effort is focused on the most important uses cases. The second is that searching over structured data is generally slower than search over inverted indices: The query understanding models are run and then the search over the structured data, often in addition to the standard search. Finally the ranking models have to take these new structure-matched features into consideration.

In general domain search, including web search, special data around entities is commonly used, including information about relations between entities [3]. The entities are identified in the documents and linked to canonical forms. Documents with matches to the query entities are ranked highly. Information learned about the entities can be used to create search page features such as entity boxes, answer boxes, highlighting in search result captions, and autocomplete [8] suggestions that map directly to the entity, thereby avoiding spurious matches (e.g., so that *tracy king* does not retrieve books about kings by authors with Tracy as one of their names). eCommerce often makes use of special entity and structured data mappings and even simple reasoning (e.g., for price queries like *dresses over $100*). Identification of brands, categories and departments, sizes, and prices can all be canonicalized and mapped to structured data [15, 23]. Often these involve straight-forward matching once the type of entity and its canonical form is identified.

Syntactic structure is used less frequently in search engines. However, it can be used to identify MWEs that might otherwise be missed. For example, if *Mickey Mouse* is treated as a MWE, it is necessary to analyze queries and titles like *mickey and minnie mouse* in order to determine that the MWE *mickey mouse* is correctly included [22]. Similarly, price queries cannot be processed with simple entity detection of prices but instead require basic syntactic or at least ngram understanding because the preposition used indicates the range (e.g., *over $100* vs. *under $100* vs. *from $100 to $250*).

4.2 Robustness Where Needed

Inverted indices make use of text normalization, stemming and lemmatization, synonyms, and spell correction to improve the robustness of search, especially in mismatches between the words used in the user's query and the documents. In addition, DL embeddings can further improve robustness. A specialized example of robustness with embeddings is color matching. As highlighted by Fig. 5, the color blue and hence the images to which the word *blue* refers can occur to a broad range of shades, including ones that merge into greens and purples. By mapping the word *blue* into a color embedding, this embedding can then be matched against color embeddings of images to be searched. Images closer to blue will be closer to the core *blue* embedding and so can be ranked higher. In addition, the color wheel shown in Fig. 5 can be used to let the user select an exact shade of blue to match, something which is difficult to do with text queries.

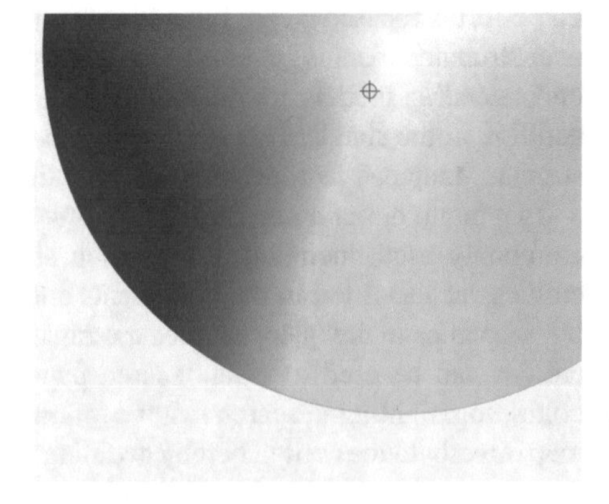

Fig. 5 Shades of blue: Color embeddings can allow the search engine to match into the blue, blue-green, and blue-purple color space

The issue with robustness is that in addition to retrieving relevant documents, it can also retrieve related and even irrelevant ones when the expansion compared to the query words is too great. This is especially the case for embeddings when the similarity score is the only control over what to match. As a result, robustness techniques are sometimes reserved for when a query has zero (referred to as null) or low results. Spell correction is often used in this way, only applying when the original query returns fewer than a fixed threshold of results. Similarly, embedding-based search may be reserved for null and low result queries [26] or for longer queries, which by definition are more likely to have fewer results because all the words have to match, thereby reducing the number of possible matches.

4.3 Color-Object Queries

In this section, we examine color-object queries when searching over stock images.[7] These queries are ones where a color word modifies an object. There may be more than one color-object pair in a query. The results for the query *white rose red background* were shown in Fig. 1. Color-object queries are relatively common in image search since users are looking for images to match their exact needs, including branding and marketing materials. When incorrectly matched results are shown, such as red roses on white backgrounds for *white rose red background*, the error is particularly jarring since the images look much different than expected. In addition, there is no way for the user to refine their query in order to see only relevant results. Finally, there is a long tail of color-object queries, including less frequent color names that overlap with non-color uses of the words (e.g., *salmon*).

The reason that irrelevant results are returned is that the inverted index contains words from the image tags and captions and there is no way to capture word order for the tags since they are just a set of words associated with the image. Among the images returned by matching words in the tags and captions, images with red roses on white backgrounds are much more common than ones with white roses on red backgrounds. So, by random selection, irrelevant results will be more common. This is exacerbated by the fact that queries for red roses are more common than for white ones and queries for images with white backgrounds are also common. This in turn results in more clicks and purchases for those images and higher popularity features in the ranking.

The treatment of color-object queries for image search requires improved document and query understanding for matching and ranking. For the images, prominent objects, including the background, have to be identified. The colors of these objects are determined. The information about the objects, colors, and the relations between the two are stored in the index as structured representations. For the queries, the color words and the objects they modify have to be identified and canonicalized. The representations for the colors in the index and the query have to be compared,

[7] The approach described here is based on techniques used in Adobe Stock search.

as do the objects and the color-object relations. Once we know which images match the query with respect to the color-object relationships, this information can be used in matching to return only images where the color-object relationships are identical or can be used in ranking to rank images with the requested color-object relationships higher in the result set. The matching approach is more aggressive, leading to higher precision results but risking excluding some relevant results. The ranking-only approach is more conservative, returning all the results the non-structured data would have returned but thereby including irrelevant results. The decisions for each of these steps can be complex. They are described in more detail below as an example of how powerful but complex incorporating structured data into search can be.

Color-object Queries For the queries, first we use a named entity recognition (NER) model to identify color phrases in queries. Depending on what types of colors are used in queries, the model can cover "kindergarten" colors (the approximately ten most basic colors, e.g., *red rose*), modified colors (e.g., *pale yellow rose*, *hot pink rose*), or the long tail of colors (e.g., *salmon rose*, *chartreuse rose*). The NER model should handle coordinated colors as well (e.g., *pink and yellow roses*), including determining their logical meaning (e.g., each rose is both pink and yellow; some roses are pink and some are yellow but there must be at least one of each). Finally, the NER must avoid detecting false positives where a color word occurs but does not refer to a color in a color-object construction. False positives include ethnicities (e.g., *black nurse*, *white nurse*), proper nouns (e.g., *snow white*), and styles (e.g., *black and white portrait*). Once a color is detected in the query, the object it modifies has to be determined. A dependency parser that is custom-trained to work on short text like queries can be used for this [25, 28]. The color NER model and the dependency parser are language dependent and so models have to be created for each language.

Color-object Images For the images, first we need to identify the prominent objects in the image and mask them. Masking determines the edges of the object so that we can extract the dominant color for each object. Each object has to be labeled as to what it is (e.g., a rose, a dog, a ball). The list of potential object labels can be extracted from the queries, in particular from the objects that occur in color-object queries.[8] The object labeling is done by an autotagger model and the confidence of the model can be used to adjust how accurate the labels are. The image background is a special case. Backgrounds are commonly referred to in color-object queries (e.g., *banana blue background*, *rose white background*). The background can be identified by masking out all the prominent objects: what is left is considered background. The background color is determined the same way that object colors are. The background label is simply as *background*; no autotagger is needed. The color-object pairs have to be stored in the index as structured data so that the specific color is associated with its object.

Color and Object Representations Consider how to represent the colors. For the index, one possibility is to use a model that maps from an image to the colors used by the query color NER model. However, this becomes difficult to manage if

[8] If the object extraction and labeling will be used for other features, then the model should label a broader set of objects.

the range of colors goes beyond the unmodified kindergarten colors. For example, for modified colors (e.g., color words modified by *light*, *pale*, *dark*, *neon*, *hot*), the index would have to list both the modified version (e.g., *dark blue*) and the simple version (e.g., *blue*) so that either type of query color could match (e.g., the queries *dark blue ball* and *blue ball* match a dark blue ball, but the query *light blue ball* does not). As discussed in Sect. 3.4, embeddings are an effective representation for colors due to their continuous nature. Instead of storing the colors as text, the embeddings can be used. These embeddings are language-independent because they represent the colors of the pixels in the image. If the colors are stored as embeddings in the index, the color words detected by the query NER have to be mapped to color embeddings. This can be done with a multi-modal text-to-color model. This model is language dependent due to the query text component.

Next consider how to represent the object labels. These could be English words. If this is the case, then for search in other languages the object words have to be translated into English. This is similar to synonym normalization (Sect. 3) only between languages instead of synonyms within a language. Alternatively, the language variants for each object label could be stored in the index, which could result in many entries for each object (i.e., one for each language). This is similar to synonym expansion only across languages instead of across synonyms within a language. An alternative is to map into language-independent concepts, either to concepts in a taxonomy or knowledge graph or to an embedding representation [11]. If the concept approach is chosen, the image autotagger has to map into these concepts so that they can be stored in the index. Then the query processing maps the object words identified by the dependency parser into the concepts.

Finally, the relationship between the color and object has to be represented and associated with the image in which they occur. This can comprise a dedicated index field for the objects in the image. Each object is then associated with at least one structured attribute, namely its color.

Color-Object Example Consider the image in Fig. 6. Two objects are detected in the image: the cup and the pastry. The remainder of the image is considered the background. The objects are labeled by an image autotagger, shown in Fig. 6 as concepts associated with English words to make them readable. Each object and the background are then associated with a color embedding that is derived from the object image, shown in the figure as a color swatch but in fact represented as a vector. The object concept and its associated color embedding are in turn associated with the document id. A list of words from tags or a caption is also associated with the image.[9]

Consider the query *pastry and red cup*. Color NER detects the color word *red*. The query dependency parser determines that the color modifies the object *cup*. The text-to-color multi-modal model maps the word *red* to an embedding. This embedding is not an exact match to the one for the coffee cup in Fig. 6 because the word *red* maps to the most canonical representation of the color, while the coffee cup has shadowing

---

[9] These are shown as English words in Fig. 6 but could be concepts similar to the ones used for the color-object representation. The same holds for the word *pastry* in the query in Fig. 7.

Structured Color-object Field Image Data		
Doc. Id	**Object (Concept)**	**Color (Embedding)**
1	Concept_123_cup	:
1	Concept_456_pastry	:
1	Concept_789_background	:

Words Associated with the Image	
Doc. Id	**Words**
1	cup, saucer, pastry, croissant, coffee, yellow, red, tasty, breakfast

Fig. 6 Image with two objects (a red cup and a brown pastry) on a yellow background. The index information is shown in the table below the image. The objects are represented as concepts (represented as Concept_*num_word* for exposition) and an associated color (represented as a color swatch corresponding to a color embedding). *Image licensed from Adobe Stock*

Fig. 7 Mapping of the query *pastry and red cup* for retrieval with color-object structured data and an inverted index

User query:	*pastry and red cup*
Color-object Field:	Concept_123_cup :
Words:	pastry

on it. The object word *cup* is mapped to a concept and associated with the color embedding. The word *and* is dropped as a stop word. The word *pastry* is treated as any other word for search since it is not part of a color-object relationship. The resulting query is shown in Fig. 7. When the query is matched against the index, the word *pastry* matches against the words in document 1. The concept Concept_123_cup matches with the object concept in document 1. The color embeddings for Concept_123_cup in the query and image are then compared. These are not an exact match since the shades of red are different. However, since they are within a pre-defined similarity threshold, they are considered a match and document 1 will be returned for the query.

The treatment of color-object queries for image search shows the power behind using structured data in conjunction with embeddings and inverted indices for search.

However, it also demonstrates the complexity of creating such structured data and integrating it into the search matching and ranking.

4.4 Towards Negation

Section 3.3 discussed why negation is complex for search, using the example query *snacks without nuts* illustrated in Fig. 2. Here, we discuss potential approaches to handling negation in search.

Negation in Queries To handle queries with negation like *snacks without nuts*, search has to determine that the query contains a negation (e.g., *without*) and the scope of that negation (e.g., *nuts*). Most queries are relatively short and there are a limited number of ways to indicate negation, which makes the detection and scope of negation easier to determine than with standard long text such as that found in documents.[10] However, once the negation and its scope is determined, search then has to correctly and efficiently retrieve documents, e.g., ones which are snacks but which do not contain nuts.

An overly simplistic approach is to retrieve all documents matching *snacks* and then exclude documents which contain the word *nut* or to rank documents with the word *nut* lower than documents without it. Unfortunately, this approach still includes many results with nuts, i.e., documents which refer to specific instances of nuts (e.g., *almonds, walnuts*). Conversely, relevant documents are excluded or ranked unduly lowly if they contain the word *nut* in phrases like *nut free*, *no nuts*, or *kids are nuts about this snack*.

Given the importance of ingredients in food items, a specialized solution can be implemented, as was done for color-object search for photos (Sect. 4.3). The ingredient lists of food items can be enhanced offline with relevant hypernyms (e.g., *nut* for *pecan*, *dairy* for *milk*). These hypernyms can be treated as keywords or as part of the structured data for the ingredients, which would also enable filters for or against specific ingredients. If search correctly identifies the query *snacks without nuts* as a food query with negation, it can match all snack documents that do not contain the word *nut* in the ingredient field or structured data. If it does not identify the query as a food query, search can back off to the approach of returning snack documents which do not contain the word *nut* in any of the text fields.

The specialized solution approach requires anticipating which classes of attributes will occur with negation. However, users can apply negation to many types of attributes (e.g., *lamp without shade*, *cup no handle*, *cities without rent control*, *plastic free restaurants*). Handling negation more broadly requires a more systematic approach. Once the negation and its scope are detected in the query, the documents

[10] Negated phrases add an extra layer of complexity in determining the scope of negation and matching that against the documents. For example *dresses without red stripes* should match dresses with blue stripes or with red polka dots since the negation applies to the concept denoted by the phrase *red stripes*. See [5, 6] for more discussion with a particular focus on question answering and textual inference.

matching the non-negated parts of the query can be retrieved. Documents mentioning the negated words can then be excluded or demoted. The remaining documents can then be checked for hyponyms of the negated word (e.g., *almonds*, *walnuts* etc. for *nuts*) and documents containing these hyponyms can be excluded or demoted. Special care has to be taken to allow for phrases like *nut free* which capture the negation within the document and hence are excellent matches for the corresponding negative query. The above approach requires more calculation at query time because a potentially large list of hyponyms has to be checked against a potentially large number of documents. To avoid this, hypernyms can instead be added to the index. This is similar to the specialized ingredient solution but on a larger scale for all of the non-stop words in the document. This has the downside of adding to the size of the index, especially to the number of document ids associated with the more abstract hypernyms. A similar trade-off was discussed in Sect. 3.1 for synonyms and other vocabulary mismatches.

Negation in Documents The above discussion outlined some methods for addressing negation in queries. However, negation also exists in documents. The search document processing and indexing must handle negation in documents when providing results for queries, whether negated or not. A simple example of this is the query *nut snacks* which should not match documents that have phrases like *a nut free snack*, *contains no nuts*, or *does not contain nuts*. A more subtle example is the query *leather jacket* which often matches faux leather jackets, where the word *faux* indicates that the jacket is not made of leather. If the user is looking for jackets in the style of leather jackets, these non-leather results are fine, but if they want a jacket made of leather, these results are irrelevant and it is difficult to construct a query to eliminate them. Often the only way to exclude these non-leather results is to use a filter for *material=leather*, if available.

Inverted indices are not well designed for encoding words as negated concepts since they are optimized to provide word-document lists (Sect. 2.1). Deciding not to index negated terms runs two risks. First, these can be perfect matches for negated queries since they are explicit about not involving the negated attribute. Second, if the processing incorrectly identifies the word as being negated, then that document will never be returned for that concept since the word is not indexed. Instead, the negated concept or words need to be associated with the information that they are negated. This takes the form of a basic structured representation. This representation allows negated queries to match directly against similarly negated document information. It also allows non-negated queries (e.g., *nut snacks*, *leather jacket*) to avoid matching against documents which negate words that should be matched (e.g., *nut* in *nut free*, *leather* in *faux leather*). This information is more costly to compute, store, and match against, but is necessary to effectively handle negation in documents and corresponding queries. Once such information is available, search can then use it for matching or, more conservatively, for ranking documents with the negated words lower than ones with the non-negated words.

This section briefly explored the complexity of handling negated queries and documents in search and some potential solutions. There is no single, simple solution, especially when large numbers of documents have to be searched over quickly. So, handling negation remains an unsolved, but crucial research problem in search.

5 Conclusion

Search historically depends on inverted indices for matching user queries to relevant documents. User behavioral data provides additional signals for query-independent and query-dependent relevance of documents. These inverted indices enable rapid search over large document collections. Optimizations in text processing, normalization, and expansion have improved search's ability to return relevant documents. However, these approaches have issues around robustness when there are mismatches between users' queries and the documents, something which is highlighted when using textual queries to search for images. Moving to representations like embeddings from DL models improves robustness for some types of data and are particularly useful for image data (Sects. 4.2 and 4.3).

However, there remain classes of queries which require information beyond the representation of words and multi-word expressions stored in an inverted index or as an embedding. These are the structured representations referred to in the title of this paper. They involve searching over structured information about the content and relationships among the entities. These structured representations in turn must support reasoning (e.g., *snacks without nuts*).

Given the specialized knowledge and models needed to create and search over this structure and given the increased cost in latency and space to search and reason over the structured data, these approaches are reserved for high-value queries which have low quality results with traditional techniques. A simple example of this is identifying brands in eCommerce queries, canonicalizing them, and then matching them against the structured brand data on the products. More complex structured data is required for representing document-internal relationships. The color-object image search queries are an example of this, where the images have to have color data associated with the objects in order to avoid cross-talk bag-of-words results where the color is present in the image but on the wrong object (e.g., for the query *white rose red background* showing red roses on a white background). The step beyond the matching to structured data and relations across entities is to be able to reason about the documents in order to provide relevant results and even answers to the users' queries. Price queries in eCommerce search (e.g., *dresses over $100*) are an example of these. For price queries, not only does the price in the query have to be identified and matched to the product price, but the word *over* (or *under* or *around* or *from ... to*) has to be identified and associated with the correct reasoning (arithmetic operation).

The future of search is going to be hybrid, involving techniques from classic information retrieval, embedding-based search, and reasoning. These hybrid search techniques will be increasingly powered by improved query and document understanding via the integration of structured data into search matching, ranking, and reasoning itself. Enabling such hybrid systems will require a deep understanding of linguistic representations of meaning, of information retrieval optimization, and of the types of information encoded in the queries and documents. It is my hope that this paper inspires further collaboration across disciplines to improve search.

For readers interested in learning more about search, there are several excellent textbooks available. References [2, 3, 17] focus on the fundamentals of search, especially inverted indices and text processing (e.g., normalization, expansion, stop words, stemming). Balog [3] examines entity-based search and how deeper understanding of entities can be used to improve all aspects of search. Baeza-Yates [1] focuses on semantic query understanding. SIGIR (Special Interest Group—Information Retrieval) is the annual conference focused on search with a published proceedings and the ACM SIGIR Forum (https://sigir.org/forum/) publishes additional papers on search. There are two annual workshops focused on eCommerce search: ECOM [13] and ECNLP [16], both of which publish proceedings. Tsagkias et al. [27] provides an overview of search issues as pertain to eCommerce.

Acknowledgement I would like to thank Roussanka Loukanova for inviting me to present at Logic and Algorithms in Computational Linguistics 2021 (LACompLing2021) and to contribute to this volume. I would also like to thank the audience of LACompLing2021, four anonymous reviewers, and Annie Zaenen for insightful questions and comments.

I would like to thank the Adobe Sensei and Search team who developed the color-object search techniques discussed in Sect. 4.3: Baldo Faieta, Ajinkya Kale, Benjamin Leviant, Judy Massuda, Chirag Arora, and Venkat Barakam.

References

1. Baeza-Yates, R.: Semantic query understanding. In: Proceedings of SIGIR. ACM (2017)
2. Baeza-Yates, R., Ribeiro-Neto, B.: Modern Information Retrieval: The Concepts and Technology Behind Search, 2nd edn. Addison-Wesley (2011)
3. Balog, K.: Entity-Oriented Search. The Information Retrieval Series, vol. 39. Springer (2018)
4. Bianchi, F., Tagliabue, J., Yu, B.: Query2Prod2Vec: grounded word embeddings for eCommerce. In: Proceedings of the 2021 Conference of the North American Chapter of the Association for Computational Linguistics: Human Language Technologies: Industry Papers, pp. 154–162. Association for Computational Linguistics (2021)
5. Bobrow, D., Condoravdi, C., Crouch, R., Kaplan, R., Karttunen, L., King, T.H., de Paiva, V., Zaenen, A.: A basic logic for textual inference. In: Proceedings of the AAAI Workshop on Inference for Textual Question Answering, pp. 47–51 (2005)
6. Bobrow, D., Crouch, D., King, T.H., Condoravdi, C., Karttunen, L., Nairn, R., de Paiva, V., Zaenen, A.: Precision-focused textual inference. In: Proceedings of the ACL-PASCAL Workshop on Textual Entailment and Paraphrasing, pp. 16–21 (2007)
7. Buttcher, S., Clarke, C.L.A., Cormack, G.V.: Information Retrieval: Implementing and Evaluating Search Engines. The MIT Press (2016)
8. Cai, F., de Rijke, M.: A survey of query auto completion in information retrieval. Found. Trends Inf. Retr. **10**, 1–92 (2016)
9. Chang, W.C., Jiang, D., Yu, H.F., Teo, C.H., Zhong, J., Zhong, K., Kolluri, K., Hu, Q., Shandilya, N., Ievgrafov, V., Singh, J., Dhillon, I.S.: Extreme multi-label learning for semantic matching in product search. In: Proceedings of KDD2021 (2021)

10. Chen, Q., Li, M., Zhou, M.: Improving query spelling correction using web search results. In: Proceedings of the 2007 Joint Conference on Empirical Methods in Natural Language Processing and Computational Natural Language Learning, pp. 181–189 (2007)
11. Chen, X., Cardie, C.: Unsupervised multilingual word embeddings. In: Proceedings of the 2018 Conference on Empirical Methods in Natural Language Processing, pp. 261–270 (2018)
12. Jia, C., Yang, Y., Xia, Y., Chen, Y.T., Parekh, Z., Pham, H., Le, Q.V., Sung, Y., Li, Z., Duerig, T.: Scaling up visual and vision-language representation learning with noisy text supervision. In: Proceedings of the 38th International Conference on Machine Learning PMLR (2021)
13. Kallumadi, S., King, T.H., Malmasi, S., de Rijke, M. (eds.): Proceedings of the SIGIR 2021 Workshop on eCommerce. CEUR-WS (2021)
14. Kalouli, A.L., Crouch, R., de Paiva, V.: Hy-NLI: a hybrid system for natural language inference. In: Proceedings of the 28th International Conference on Computational Linguistics, pp. 5235–5249. International Committee on Computational Linguistics (2020)
15. Kutiyanawala, A., Verma, P., Yan, Z.: Towards a simplified ontology for better e-commerce search. In: Proceedings of ECOM2018. CEUR-WS (2018)
16. Malmasi, S., Kallumadi, S., Ueffing, N., Rokhlenko, O., Agichtein, E., Guy, I. (eds.): Proceedings of The 4th Workshop on e-Commerce and NLP. Association for Computational Linguistics (2021)
17. Manning, C.D., Raghavan, P., Schütze, H.: Introduction to Information Retrieval. Cambridge University Press (2008)
18. Mohan, V., Song, Y., Nigam, P., Teo, C.H., Ding, W., Lakshman, V., Shingavi, A., Gu, H., Yin, B.: Semantic product search. In: Proceedings of KDD2019 (2019)
19. Peters, C., Braschler, M., Clough, P.: Multilingual Information Retrieval: From Research to Practice. Springer (2012)
20. Porter, M.F.: An algorithm for suffix stripping. Program **14**, 130–137 (1980)
21. Radford, A., Kim, J.W., Hallacy, C., Ramesh, A., Goh, G., Agarwal, S., Sastry, G., Askell, A., Mishkin, P., Clark, J., Krueger, G., Sutskever, I.: Learning transferable visual models from natural language supervision (2021). ArXiv:2103.00020
22. Senthil Kumar, P., Salaka, V., King, T.H., Johnson, B.: Mickey Mouse is not a phrase: improving relevance in E-commerce with multiword expressions. In: Proceedings of the 10th Workshop on Multiword Expressions (MWE), pp. 62–66. Association for Computational Linguistics (2014)
23. Skinner, M., Kallumadi, S.: E-commerce query classification using product taxonomy mapping: a transfer learning approach. In: ECOM SIGIR Workshop. CEUR-WS (2019)
24. Sorokina, D., Cantú-Paz, E.: Amazon search: the joy of ranking products. In: Perego, R., Sebastiani, F., Aslam, J.A., Ruthven, I., Zobel, J. (eds.) Proceedings of the 39th International ACM SIGIR conference on Research and Development in Information Retrieval, SIGIR 2016, pp. 459–460. ACM (2016)
25. Sun, X., Wang, H., Xiao, Y., Wang, Z.: Syntactic parsing of web queries. In: Proceedings of the 2016 Conference on Empirical Methods in Natural Language Processing, pp. 1787–1796 (2016)
26. Trotman, A., Degenhardt, J., Kallumadi, S.: The architecture of eBay search. In: Degenhardt, J., Kallumadi, S., de Rijke, M., Si, L., Trotman, A., Xu, Y. (eds.) Proceedings of the SIGIR 2017 eCom workshop. CEUR-WS (2017)
27. Tsagkias, M., King, T.H., Kallumadi, S., Murdock, V., de Rijke, M.: Challenges and research opportunities in eCommerce search and recommendations. ACM SIGIR Forum **54**, 1–23 (2020)
28. Wang, Z., Wang, H., Hu, Z.: Head, modifier, and constraint detection in short texts. In: Proceedings of the International Conference on Data Engineering, pp. 280–291 (2014)
29. Wang, Z., Zhao, K., Wang, H., Meng, X., Wen, J.R.: Query understanding through knowledge-based conceptualization. In: Proceedings of IJCAI, pp. 3264–3270 (2015)

Rules Are Rules: Rhetorical Figures and Algorithms

Randy Allen Harris

Abstract Rhetorical figures are form/function alignments in which the form (1) serves to convey the function(s) and (2) supports their computational detection; therefore, (3) they are particularly rich for various text mining activities and other Natural Language Understanding purposes. The figures which are especially valuable in these ways are known as *schemes*, figures that are defined by their material (phonological, orthographical, morpholexical, or syntactic) form, in distinction particularly from *tropes*, which are defined by their conceptual (semantic) form. Rhetorical schemes, however, have been almost universally ignored by linguists, including computational linguists. This article illustrates form/function alignments for a small handful of rhetorical schemes, with some examples of how they communicate specific meanings. The communicative functions of rhetorical schemes rely not so much on individual schemes as on certain collocations of schemes (sometimes with tropes or other figures as well). These collocations, in turn, are coordinated by linguistic features such that the relevant expressions fit the notion of a construction as understood in the Construction Grammar framework. Examples are drawn from epanalepsis, as well as antimetabole, mesodiplosis, and parison, which collate frequently as a group and also collectively with the trope, antithesis. The communicative functions explored include Semantic-Feature Promotion, Reciprocal Specification, Reciprocal Energy, Irrelevance of Order/Rank, Subclassification, and Reject-Replace. Rhetorical figure detection is also discussed in some detail with respect to figural and grammatical collocation.

Keywords Rhetorical figures · Rhetorical schemes · Figure detection · Antimetabole · Antithesis · Chiasmus · Epanalepsis

---

R. A. Harris (✉)
English Language & Literature, and The David Cheriton School of Computer Science, University of Waterloo, Waterloo, ON, Canada
e-mail: raha@uwaterloo.ca

R. Loukanova et al. (eds.), *Logic and Algorithms in Computational Linguistics 2021 (LACompLing2021)*, Studies in Computational Intelligence 1081,
https://doi.org/10.1007/978-3-031-21780-7_10

1 Introduction

"Rules are rules" Australian Prime Minister Scott Morrison said on 5 January 2022, giving his axiom for the decision to deny number-one-ranked Men's tennis star, Novak Djokovic, official entry into the country, where he had come in order to compete in the Australian Open [88]. A month later we heard the same thing about another sports figure on another continent. Russian figure skater, Kamila Valieva, had tested positive for a banned substance before the Beijing Winter Olympics and arguments erupted over whether she should be allowed to compete while the review process was underway. Another skater in the competition, Switzerland's Alexia Paganini, said "I have a lot of empathy for her because she, regardless of everything, she did have to get on the ice and work hard"; then she added, "I feel sorry for her, but rules are rules and they should be followed" [125].

Of course rules are rules. Prime Ministers are Prime Ministers. Tennis players are tennis players. Figure skaters are figure skaters. Grass is grass, ice is ice, and so on. These are tautologies; self-evident, *a priori* truths. As such, they carry no information. They are vacuous. But we all know what Morrison's communicative point was: there was no room for Djokovic to maneuver. He had to leave the country. Djokovic was unvaccinated against COVID-19. Australia had a rule against admitting anyone into the country who had not been vaccinated against COVID-19. Not only must Djokovic not enter, there was an axiom for it. Paganini's point was identical. Valieva had tested positive for a banned substance. The Olympics have a rule about that. There was no alternative: Valieva should not be allowed to compete. It was axiomatic. Rules are rules. But apparently rules play out differently. In the end, Djokovic was prevented from officially entering Australia while Valieva was not prevented from skating. Djokovic did not compete in the 2021 Australian Open. Valieva did compete in the Beijing Olympics.

"Rules are rules" is a tautology, but it is also a rhetorical figure, known as *epanalepsis*, a figure in which the same word or words occur at the beginning and ending of the same phrase or clause. Technically, a rhetorical figure is an abstract pattern, which the expression "rules are rules" is not; so more precisely, it is a realization or an instantiation of the figure, epanalepsis. But informally we can say it *is* an epanalepsis in the same way we can say it *is* a tautology or it *is* a sentence; namely, it satisfies the formal definition of an epanalepsis. In one of its favoured grammatical formations, in copular clauses, epanalepsis conveys at least two specific rhetorical functions, one argumentative (closing off all objections and counter-arguments), the other communicative (isolating some feature(s) of the repeating terms as an explanation for some belief or course of action). In this case, the *rigidity* of rules is invoked to explain the exclusion of Djokovic on the one hand, and to call for the exclusion of Valieva on the other hand. In the linguistic theory known as *Construction Grammar*, non-compositional form/function alignments of grammatical elements are *constructions* (cxx; singular, cx) [58]. This article argues that at least some cxx also rely on extragrammatical patterns; that is, on rhetorical figures.

In this paper, I illustrate the rhetorical functions of a few figures in specific grammatical cxx, mostly the figures known as *schemes*. Not all form/function alignments are as direct as this example of copular epanalepsis, "rules are rules". Often several figural patterns collocate with one another and with grammatical features (such as lexical category) to convey their functions. Schemes, in particular, leverage iconicity to do so [48, 50]. The computational benefits of distinctive formal patterns aligned with specific functions are almost too obvious to mention: the patterns are available for simple pattern detection and the functions make that detection worthwhile because they can get at meaning in a way that is beyond current text mining and NLU tools. Accordingly, I discuss the highly promising work in figure detection, concentrating on Marie Dubremetz and Joakim Nivre's research in some detail, especially for how the various conditions of their 'chiasmus detection' algorithm implicate other figures and associated grammatical elements. Crucially, the functions are a product, in a unification-grammar sense, of extragrammatical figural patterns and grammatical features, such as (in our "rules are rules" example) epanalepsis and main verb *be*.[1]

2 Rhetorical Figures, a Primer

This section offers a very brief overview of the incredibly tangled territory of rhetorical figuration; fuller accounts of the taxonomy I adopt can be found in [53].

The first thing to know about figures is that their terminology has two inverse problems for precision, one/many mismatches and many/one mismatches. There are multiple names for the same pattern, and there are multiple patterns that have the same figural name. Epanalepsis, for instance, is sometimes called *resumptio*, sometimes *echo sound*. It is also sometimes called *repetitio*. Meanwhile, *repetitio* can label another pattern, in which the words at the beginning of different phrases or clauses are the same. But that pattern is also called *epanaphora*, or just *anaphora*, or sometimes *epembasis* or *epibole*, and so on. It's a mess and I won't make excuses for it.

I will note some of these terminological alternatives as relevant in this article, but the most immediate lesson you should take from this discussion is that you can't go to standard accounts of rhetorical figures (e.g., [76] or [11]) and expect to map my labels onto their descriptions reliably or their descriptions back onto my labels, or to map our taxonomies back and forth. There is certainly overlap, but for the most part rhetoricians and literary critics are content with a rough figurative lore and do not worry very much about tightening up their treatment of figures, using one label or another because that's what they happened upon early in their training or research, and generally avoiding taxonomies. In this article, I follow a one-to-one mapping of pattern and label. I also adopt a taxonomy (outlined in more detail in [53]) that allows us to reduce confusion about the forms and functions of figures. The main

[1] There is as yet no significantly robust formalism for figural patterns, but see [52] for some preliminary suggestions; [60] for an adaptation of those suggestions in PERL; and [54].

point here is that while there are large catalogues of distinctive patterns, devices, and moves that are all called *rhetorical figures*, they cannot all be treated the same. For our purposes it is particularly important to note that they don't all have the same level of computational tractability. This article concentrates on the most tractable class of figures, schemes, identifying the other categories largely to exclude them.

Here are four examples of figures, each from a different taxonomic class:

1. I will not fall into calling Wente a fascist and certainly not a Stalinist [99].
2. Brutus is an honourable man [111].
3. [T]his campaign has been defined by Trump parachuting in, like an Elvis impersonator in Vegas [24].
4. He said, she said (doxa).

Example 1 realizes a maneuver usually called *paralipsis* (also *parasiopesis*, *antiphrasis*, or *occultatio*, among other labels). It occurs when the speaker says something but simultaneously denies saying it; here, Steve Palmer has stuck the labels of *fascist* and *Stalinist* on Margaret Wente, but is pretending to avoid such terms. Paralipsis is categorized as a *move*, a strategic use of language that pushes away from discourse norms in one way or another but has no distinctive linguistic or cognitive features at all. For my money, these devices are not really figures, though they have long been bundled under that term with what I consider 'true' figures, which *do* have distinctive linguistic and cognitive features, like Examples 2–4. That is the last we will hear of moves in this article.

Example 2 exhibits the figure of sarcasm, a type of irony that communicates opprobrium through asserting its opposite. It comes from a speech by Mark Antony in Shakespeare's *Julius Caesar* (Act 3, Scene 2), a speech in which Antony has repeatedly demonstrated that Brutus is highly *dis*honourable. The ironic figures do not have distinctive linguistic features (though they might be enhanced by certain styles of enunciation, facial expressions, postures and/or gestures). They rely rather on theory of mind and social understandings of the speaker and the situation; in particular, the difference between the speaker's words and the speaker's intentions as they relate to the context. The expression "Brutus is an honourable man" is a simple descriptive assertion linguistically. It is only ironic when the hearer realizes, through an awareness of the situation, including the character of the speaker, that it should be interpreted as expressing the opposite of its direct compositional semantics; here, that the adjective *honourable* is intended to invoke *dishonourable*, and, indeed *despicable*. Ironic figures are categorized as *chroma*, figures whose effects rely on theory of mind and leverage intention. They are an interesting set, but that is the last we will see of them in this article as well.

Example 3 exhibits the figure, simile, one of the many analogical figures, which rely on cross-domain comparisons to draw attention to specific attributes. They involve a source (here an Elvis impersonator) and a target (US Presidential candidate in 2016, Donald Trump) to assert that the target shares some features of the source (flamboyance, outrageousness, buffoonery). The analogical figures are categorized as *tropes*, figures which work primarily by leveraging a semantic domain or multiple semantic domains (with analogical figures, two domains) in non-entrenched ways; here, the conceptually separate semantic frames of show business and politics are brought together to transfer features from the former to the latter. Similes are especially interesting because they involve morpholexical signatures in a way that most tropes do not. More often, tropes have what we might call *figural lynchpins*, subtle cues of departure from entrenched semantics ("literality") which I will discuss briefly in a moment. Tropes are rich and pervasive types of figures, a few of which (metaphor and metonymy) have received a great deal of attention among linguists. But they are not very tractable computationally; so they, too, are largely sidelined in this article, though one of them, antithesis, will get some attention.

Example 4 exhibits the figure, epiphora, one of the many figures of lexical repetition, in which the same words occur at the ends of different phrases or clauses. (*Doxa*, by the way, is the term rhetoricians have for clichés, proverbs, maxims and other ordinary-language colligations; common instances of what Construction Grammarians call *filled* and *partially filled* cxx.) Lexical repetitions belong to the category of *schemes*, a type of figure in which the material features of expressions (phonological, orthographical, morpholexical, or syntactic) shift away from the mundane norms of speech. Overwhelmingly, phrases and clauses end with *different* words or sequences of words (just check out the phrases and clauses of this paragraph). When they end with the *same* words, like "he said, she said", hearers register that little bit of novelty and the expressions increase in salience and memorability. Schemes, because they are entirely material, are computationally very tractable, so this article focusses on them almost exclusively.

It is also essential to realize that rhetorical figures frequently travel in clusters; they collocate. Example 1 has two instances of paralipsis, for instance (*fascist* and *Stalinist*), as well as an analogical framing that we might call *metaphorical* (though not a *metaphor*): "falling into", a spatial occurrence, signals the source is something like a hole, for the target of speaking in a certain unfortunate way that we might refer to as "calling someone names" or just "insulting someone". "Falling into" is a figural lynchpin in the expression. It is not a metaphor but it creates a loose metaphorical perception (i.e., that 'insulting someone is a hole', which implies it is a 'low' behaviour that speakers can 'accidentally' perform). Example 3 also contains a metaphor as well as the simile. Here the source is *to parachute* and the target is implicit, namely 'to enter a campaign'. Example 4 includes rhyme (*he* and *she*) and parison (a repetition of syntactic structure: *he said* and *she said* have the exact same clausal structure). There are also various entailment relations among figures, especially the schemes. The figure assonance, for instance, is vowel repetition; rhyme is syllabic repetition, and all syllables contain vowels, so you don't get rhyme without also getting assonance. Epiphora is lexical repetition, and all words contain syllables. So, if epiphora occurs it brings along rhyme, and therefore also assonance (e.g., the two instances of *said* rhyme with each other in 4). Such entailments, or redundancies, are generally ignored as inconsequential and the 'higher' figure takes precedence.

One more thing about rhetorical figures, the most important thing about them for computational linguistics: they evoke rhetorical functions. This is very well known about the analogical figures, which have been understood since antiquity to serve pedagogical and argumentative functions (as well as aesthetic functions). Aristotle, for instance, says that "metaphor especially brings [learning] about", offering this example: "when [Homer] calls old age 'stubble', he creates understanding and knowledge through the genus, since both old age and stubble are things that have lost their bloom" (Aristotle *Rhetoric* [2]). But what is true of the analogical figures is true of figures more broadly [34]. We have seen, for instance, that Morrison uses epanalepsis argumentatively to justify the decision about Djokovic, and Paganini uses it to claim that Valieva should be excluded from the competition. Rhetorically, the tautological dimension attempts to foreclose all objections and counter-arguments. Communicatively, the notion of rigidity is promoted as the most relevant feature of rules.

3 Figurative Functions

Rhetorical figures are form/function alignments. Ploke, for instance, the figure of lexical repetition, achieves cognitive salience for the relevant concepts (i.e., the signata of the repeated words), effects textual cohesion and thematic coherence, and serves to stabilize reference [34, 49]. That's why technical literature–engineering, scientific, and, most relentlessly, mathematical literature–is so repetitive. Our premiere theorist of rhetorical figures as form/function alliances, Jeanne Fahnestock, has a particularly brilliant example of this in her explication of Koch's postulates, which provide a heuristic for the identification of organic pathogens:

> (1) The microorganism must be found in abundance in all organisms suffering from the disease … (2) The microorganism must be isolated from a diseased organism and grown in pure culture. (3) The cultured microorganism should cause disease when introduced into a healthy organism. (4) The microorganism must be reisolated from the inoculated, diseased experimental host and identified as being identical to the original specific causative agent [90, p. 204].

There are several repetitions in this articulation of the postulates, all serving the same basic rhetorical purposes: aiding coherence in the argument and cohesion in the text and certifying stability of reference, functions perhaps so obvious as to be invisible. But they serve those functions in a range of ways; for instance, *disease* as noun and *diseased* as adjective (realizing the figure antanaclasis, the co-occurrence of different words which look or sound the same—homonyms, polysemes, or category converts) relate a reified concept and a predicated condition; the verbs *isolated* and *reisolated* (realizing the figure polyptoton, the co-occurrence of the same stems in derivationally related words) relate the same action to the same entity at separate times.

But we will confine this illustration to the ploke of *microorganism*. Notice that *microorganism* does not designate the very same physical entity in each of its occurrences. In fact, it *cannot* reference the very same entity in our example. For one thing, the word is used generically, not specifically. But more importantly, the specific microorganisms extracted from one (diseased) organism and introduced somewhat later to a different (healthy) organism must be newly grown microorganisms in a culture (though in principle of course there may be some of the original specific microorganisms in the sample used to infect the healthy organism). Exactly the same is true of the microorganisms that are then later extracted from the (once healthy but now diseased) organism. But repetition of *microorganism* throughout the passage ensures that we know something essential remains consistent (i.e., that the specific entities are members of the same species). In this way, the word functions very much like an algebraic variable. This stability-of-reference function is not some accidental or decorative effect of a 'mere' rhetorical figure, but a fundamental semiotic (and perceptual) function related to what Charles Sanders Peirce calls the iconicity *principle of identity* [50, 102].

Perhaps the following can offer a clearer case of this function for lexical repetition:

5. a. All logicians go to heaven.
   b. Emil Post was a logician.
   Therefore,
   c. Emil Post went to heaven.

Again we see lots of lexical repetitions. Again they effect coherence and cohesion and certify stability of reference. But we can now see their specific individual and collective functioning. Propositions 5a and 5c are expressed in the figure of epiphora (phrase- or clause-final lexical repetition). Meanwhile, 5b and 5c exemplify the figure of epanaphora (phrase- or clause-initial lexical repetition). We all know how such arguments work to build or reveal set-theoretic relations among entities and classes. But for our purposes, it is important to recognize that the *figures* are doing that work. Epiphora ensures that 'going to heaven' is predicated of both *logicians* and *Emil Post*, and epanaphora ensures that both 'being a logician' and 'going to heaven' are predicated of *Emil Post*. The medial repetition of *go/went* (mesodiplosis) is also critical. Again, the figures are not accidental or decorative. The figuration *creates* the syllogism. Of course, figuration can be misused. It can just as easily produce fallacious arguments:

6. a. All logicians go to heaven.
   b. J. Alfred Prufrock went to heaven.
   Therefore,
   c. J. Alfred Prufrock was a logician.

Here we have a pair of epiphora again (6a and 6b) and a pair of epanaphora (6b and 6c) mesodiplosis of *go/went*. What is criterial, as we will see again and again, is not just the figures on their own or even collated with each other, but their co-occurrence with grammatical features, like the universal quantifier here, and which predication follows the adverb, *therefore*.

The relation between rhetorical figures and deductive logic was especially well explored in the early modern period when logic (under the label *dialectic*) and rhetoric were closely associated with each other and with grammar. Philip Melanchthon, for instance, treats syllogistic forms as their own distinct figures in his *Erotemata Dialectices*, and because deduction involves the explication of semantic properties among terms, every rule of inference he describes and exemplifies involves a lexical figure of some kind. In connection with figures we will look at later on, his nineteenth rule is especially germane:

From the converse to the converted [a conversa ad convertentem], and the reverse, the inference is valid
All the elect are called,
Therefore all those outside the assembly of the called, are not elect [86, p. 314].

4 Figure Detection

The importance for computational linguistics of the form/function relationship that rhetorical figures realize cannot be overstated. Fortunately, linguists already know a little bit about rhetorical figures; unfortunately, it is a very little bit. Or, perhaps it is more fair to say linguists know a lot about a very few figures, chiefly metaphor and metonymy, though what they call metaphor and metonymy and their usage of the terms *conceptual metaphor* and *conceptual metonymy* are at significant odds with much of the rhetorical tradition. Metaphor and metonymy are indisputably pervasive in all languages, and all dialects, varieties, registers, genres, and styles of language, both in their traditional ('novel', 'creative') variants and their quotidian residues (aka their 'conceptual' variants see [51, pp. 293–299] for some discussion). Attention to the way these quotidian residues align into terminological arrays has particularly enthused linguists (e.g., the vocabulary of battle that frames much of our talk of about argumentation: "I *buttressed* my main points but she *outflanked* me and *shot down* my whole argument; I couldn't *defend* myself against her *onslaught*"). The enthusiasm has spread to computational linguistics generally and figure detection specifically (see [112, 121] for overviews).

There are very good reasons for this enthusiasm. Figures not only pervade all corners of language, they serve indispensable communicative functions. Metaphor, for instance, forges analogical associations that have been understood from antiquity to serve pedagogical, explanatory, and motivational functions in language (and they do so not *in addition* to their aesthetic dimensions, but *because* of their aesthetic dimensions). The vocabulary of battle that pervades our talk of argumentation, for instance, tells us that arguments are frequently hostile events. Metonymy forges correlational associations that profile certain elements of a domain for particular attention. To say that "On 24 February, 2022, Putin invaded the Ukraine", for instance, when that individual did not set foot in that country on that day, singles out an individual with respect to the actions of a national army, profiling that individual as to blame for all the death, damage, and trauma that results from the invasion; or, conversely, for those who disastrously believe the invasion to be a good thing, as praiseworthy for whatever they view as 'accomplishments' resulting from the invasion. So detecting figurative patterns and building a knowledge of their functions into NLU is essential for computational systems that hope to process language adequately. Minimally, such systems need to know that 'shooting down' in the context of an argument and the context of an invasion have very different implications.

What most linguists are missing, however, is what this paper is arguing: that metaphor and metonymy are not alone in their pervasiveness or their evocation of communicative functions. Even more crucially for computational linguistics, since the form of schemes can be far more precisely specified than the form of tropes (see [91] for a summative appraisal of the very limited success of metaphor detection work, despite all of the resources poured into it). Rhetorical figures are form/function alignments. Find the form, find the function. But not all forms are equal. The functions of tropes are coded almost exclusively in the semantics (a very few also have lexico-syntactic signatures, as with similes), and semantics are not a strong suit for computers. Compare 3 with 4, for instance. An algorithm to detect the figures of 3 would not only have to detect the comparative lexico-syntax but would also need access to a knowledge base that had politicians and Elvis impersonators in different semantic domains to flag it as a simile, and the parachuting metaphor would be trickier yet. For 4, the algorithm would just have to find the same word(s) at the end of different phrases or clauses to flag that pattern as epiphora, the syllabic structure of *he* and *she* for the rhyme, the syntactic structure of *he said* and *she said* for the parison. Easy peasy.

Well, not quite easy peasy, since figures often collocate both with each other and with grammatical features, and the algorithms, as we will see, need to be tuned to those elements as well. But certainly easier and peasier than detecting tropes. The success of scheme detection is far greater than the success of trope detection, despite much less and much more marginal attention. Jakub Gawryjolek is the first researcher I am aware of who did any figure mining [39, 40], developing a tool he called JANTOR (Java ANnotation Tool Of Rhetoric), which supported automated annotation of text files for eleven different figures, including epanalepsis, epiphora (called *epistrophe*), and parison (called *isocolon*). He looked for only one trope, oxymoron, a figure of semantic conflict, because its semantic signature is relatively robust (locating antonymous head-modifier pairs, such as "open secret" and "found missing", for which he used WordNet [104]). Other significant figure-detection researchers who have focussed on schemes include O'Reilly and Paurobally [97], Strommer [114], Hromada [60], Alliheedi and Di Marco [1], Java [65], Lawrence et al. [78], Tu [118], Green and Crotts [42–45], Lagutina et al. [73], and Schneider et al. [109]

And then there is Marie Dubremetz. She wrote a series of remarkable papers with Joakim Nivre on figures of lexical repetition. If Gawryjolek invented the field, Dubremetz has raised it to its highest level thus far. We will revisit her work a bit later. Very few of these researchers, however, have gone much beyond identifying instances of a few figures, telling us little if anything about what the figures are doing.

5 Subtotal

A subtotal of where we are now, then, is:

- rhetorical figures align linguistic forms with communicative functions;
- some linguistic forms are relatively easy to detect, especially the forms associated with schemes;

but
- almost no computational research into figures has concerned itself with their communicative functions.

6 Epanalepsis, Argument Foreclosure, and Semantic-Feature Promotion

Rules are multiplex concepts. There are rules for driving, for playing games, for moral conduct, for logical inference, for business, for talking. They are not all alike. Even in the same domain there can be different sorts of rules. Some rules for driving belong to a body of normative practices, for instance (e.g., shoulder checks when merging into another lane), while others are regulated by codified regimes (e.g., stopping for a red light). Some rules for talking are socially mediated (being quiet in libraries), while others are so cognitively entrenched that speakers may not even be aware they 'follow' them (e.g., the -s, -z, -əz alternation for the regular English plural). Rules may be absolute (in chess, you always lose the game if you are in checkmate), but some are optional and contingent (you can move a pawn diagonal to its file, but only if an opposition piece is in the destination square). Some rules have conditions. Some rules have exceptions. Some rules contradict other rules. Rules are multiplex.

So, when Scott Morrison tweeted "Rules are rules" about Novak Djokovic and Alexia Paganini used the same expression about Kamila Valieva they were dipping into a bag of diverse, overlapping, and not always compatible phenomena. How does this tautological claim about a heterogeneous concept work? Morrison and Paganini both elaborated. "No one is above these rules", Morrison said [88]. "[Rules] should be followed", Paganini said [125]. There is one specific and far from universal feature of rules they are evoking, rigid application. Rules always apply, the usage insists; no one is exempt. Let's consider a few more examples of epanalepsis and see if this semantic-selection-and-promotion pattern holds. We need look no further than Twitter.

7. Hillary lost to Trump because she was Hillary. Period [106].
8. Harsh realities are harsh realities. Trump and the GOP will have free reign. Russia will guide our foreign policy. These are the facts [22].
9. When people (and I hear pundits say this) that trump is just being trump when he makes offensive, even racist, comments, I don't get it. trump isn't being trump, trump is being an asshole. or…he's an asshole being an asshole. #TrumpIsAnAsshole #TrumpIsARacist [4].

We can chart the argumentation of 7 as 'Hillary is Hillary' is the reason for the proposition 'Hillary Clinton lost to Donald Trump', and in 8, we have 'Harsh realities are harsh realities' as the reason for the (predictive) propositions about Trump and the Republican party, Russia and American foreign policy. These epanalepses, in short, have the same axiomatic quality as Morrison's and Paganini's 'rules are rules' usage. Tweet 7 even gives us the finality-declaring "Period" to reinforce the axiom, and 8 makes a very similar gesture with "These are the facts", since facts are definitionally incontestable.

Tweet 9 is the most interesting of the three, and also the thickest with epanalepses. There are three: "trump is just being trump", "trump isn't being trump", and "an asshole being an asshole" (predicated of Trump). It's the most interesting because it is both an argument and meta-argument. It rejects the reason that our Tweeter perceives is routinely offered to excuse racism or misogyny or general ignorance by Trump, namely the axiom that 'trump is just being trump', which has the same kind of closed-door incontestability as the other epanalepses we have seen. That is, she is resisting the door-closing tautological character of the expression. The author of this tweet, Beth Bacheldor, doesn't quite know how to deal with this axiom, so she just rejects it outright with an antithetical epanalepsis of her own, "trump isn't being trump", and proposes her own explanation for the relevant racist, misogynist, and generally ignorant behaviours, that 'trump is an asshole', which she then brings home with her own axiom, after predicating 'being an asshole' of Trump; namely, that Trump is an 'asshole being an asshole'. Bacheldor's reasoning is flawed, of course; in particular, "trump isn't being trump" is self-contradictory, like 'a square is not being a square', which she might have avoided by recognizing that categories can apply simultaneously–that it is possible for some entity to be both Trump and an asshole. She also appears to miss that "Trump is just being Trump", depending on the context works by selecting odious semantic features–i.e., "Trump is just being Trump; namely, a racist"; "Trump is just being Trump; namely, a misogynist"; etc.

But it is very clear how Bacheldor's reasoning proceeds, and it is also significant that she recognizes the way these tautological epanalepses work; that is, to close off objections with an indisputable claim from which everything else is said to follow.[2]

Tautological epanalepses are very common in argumentation, regularly serving this foreclosure-of-argumentation function. Here are a few more, all of them highly similar to the maxim employed by Morrison and Paganini, "rules are rules":

10. Corporations aren't people. People are people [79].
11. Business is business. (doxa)
12. A deal is a deal. (doxa)
13. The law is the law. (doxa)
14. Boys will be boys. (doxa)

[2] Beth Bacheldor is right, by the way, in her diagnosis of "Trump is just being Trump" as a popular maxim. Twitter doesn't give a hit count, but I searched for the phrase on Google as well; Google's numbers are not terribly reliable, but it returns a hit count of 6,040 for "Trump is just being Trump", many of them as story headlines or pull quotes, which of course doesn't include the number of times it has shown up on TV or other video or audio sources or around the water cooler. It is a political proverb used to explain Trump's conduct. But Bacheldor has misunderstood its use for the most part. The minimizing *just* may reduce the sting a little bit, but it is overwhelmingly used as a way to assert his unfitness, not to excuse his slip-ups. Here's a typical example: "Trump is just being Trump. Everything he does is on an outrageous scale, so it only makes sense that his ignorance, like his ego and his self-delusion, would be monumental" [70].

These expressions are meant to look and feel self-evident, indisputable. But by now we can see that there is something else going on besides a simple and vacuous "x is x" tautology. Example 10 comes from a campaign speech by Barack Obama in 2012 and is meant to draw a sharp contrast with his opponent, Mitt Romney. The first clause is false (corporations *are* people, from a legal perspective), but is included in order to refute Romney. The second one is true *a priori*. Their juxtaposition works well argumentatively because while people are legal entities, they are also flesh and blood entities, which corporations are not, and Obama counts on his audience to activate the flesh-and-blood features of personhood such that the second proposition appears to prove the first proposition. All of the doxa colligations work in a very similar way.

Morrison and Paganini, of all the possible properties that rules might have, even legislated rules, are only evoking one property, rigidity. Tweet 7 (by Michael Regan) is referencing some (alleged) properties of Hillary Clinton. It's not entirely clear what they are–uppitiness, corruption, carelessness with emails–but we do know that they all come down to the composite property of unelectability. They don't include her blonde hair, the fact that she is a lawyer, that she went to Yale, certainly not that she was a senator (an elected position). The "harsh realities are harsh realities" of tweet 8, deployed by Peter Daou, could easily fit among our doxa examples (it is an epanaleptic cliché). It identifies the relevant feature adjectivally. Daou's not talking about *all* realities. He narrows it down immediately to *harsh* realities.

Clichés 11–14 bring us back to unmodified categories, but we can see how a particular semantic feature, or a subset of semantic features, often connotative rather than denotative, are foregrounded to provide the warrant accompanying the argument-foreclosure function. The foreclosure function works because the structure implies that everything inherent to the subject is carried over equally into the predication, a complete and total mapping. But deployment in a given situation only works because specific features are promoted as explanatory and/or exculpatory. These expressions are always trotted out to explain, and often to excuse, some action or behaviour, almost always something objectionable.

'A deal is a deal', for instance, exhibits clear intransigence. 'Maybe circumstances have changed, maybe you lost your job and your spouse left you and your dog died, but you still agreed to throw me a big birthday party. A deal is a deal'. The property that is foregrounded here is the obligation a deal puts on someone, even though lots of deals are malleable, and every deal can be renegotiated if the parties are willing. The speaker is projecting their own intransigence on the past agreement. "Boys will be boys" is especially pernicious in the way it attempts to kill an argument through the promotion of particular semantic features. It selects situationally salient properties conventionally associated with boyhood, such as recklessness, maybe rudeness, often aggression, and frequently libidinousness. It never means, say, that boys will be boys in the sense that they are shorter than adults, that they attend school, that they can't legally vote or drive or drink alcohol. It usually means that boys do reckless and harmful things. And the expression is often used not of boys but of men, to excuse bad conduct, like aggression or sexual harassment, so it adds a layer of triviality or childishness to the claim structure. The people identified as boys should be excused and go unpunished because they are operating under some form of endocrinal determinism. Their actions are out of their control. It's biological.

Then there is the ultimate closed-door epanaleptic cliché, which expresses almost cosmic fatalism:

15. It is what it is. (doxa)

These tautological foreclosure moves are bad-faith argumentative ploys, what Pragma-Dialecticians call "derailments", strategic maneuvers intended not to resolve a dispute but to prevent genuine resolution [120]. An argument should be defeasible in order to be reasonable; otherwise, it is just dogmatic assertion. Tautologies are not defeasible.

Not all epanalepses have this argumentative function, however, or the communicative function of selecting and promoting specific semantic features. Here are a few more expressions realizing epanalepsis:

16. One on one. (doxa)
17. Head to head. (doxa)
18. Face to face (f2f). (doxa)
19. Side by side. (doxa)
20. Little by little. (doxa)
21. Follow4follow. (doxa)
22. Day after day. (doxa)
23. Dog eat dog. (doxa)
24. A lie begets a lie. (doxa)
25. I said what I said. (doxa)

There is nothing tautological about these expressions, though 25 suggests the intransigence we see with the tautologies, perhaps echoing 15, and we find a range of lexical material between the epanaleptic bookends, mostly prepositions but also a couple of verbs (23, 24), as well as a complementizer (25). We don't even get the same category for the repeated terms (in 16–19, as well as 23, the repeaters are all nouns; 20 repeats an adjective; 21 repeats a verb; 24 a noun phrase; 25 a full clause). What these examples mostly show is how productive epanalepsis is for prefabricated colligations.

Now a few more examples, somewhat less familiar:

26. A Canadian is a Canadian is a Canadian [117].
27. [A] dark Satanic mill ought to look like a dark Satanic mill and not like a temple of mysterious and splendid gods [98].
28. In times like these, it is helpful to remember that there have always been times like these [10].

Example 26 is from a speech by Canadian Prime Minister Justin Trudeau, 27 is attributed to Aldous Huxley by George Orwell, and 28 is attributed to media commentator, Paul Harvey. They are all more obviously crafted than our doxa examples. The Trudeau example has the sloganizing feel that most of the ones we have seen have, which was certainly one Trudeau's goals, but the additional repetition ups the ante somewhat. In a sense, it grafts *two* epanalepses together. The syntax shares that middle *Canadian*; it is simultaneously the complement of "A Canadian is" and the subject of "is a Canadian". So we might categorize it as a double epanalepsis, with that central *Canadian* sharing duties as the clause-final element of the first epanalepsis and the clause-initial element of the second. The speech in which this expression occurs followed a remarkably divisive political campaign in which one party in particular sought to pit cultural groups against one another ('old stock Canadians'–that is, Anglo-Settler Canadians–against Muslims and other more recent ethnoracialized immigrants), and Trudeau wanted to cement home the idea that identity as a Canadian supercedes all other identities, unifies the populace, and erases toxic divisions. The Huxley example is also about identity, but false identity. Large, indifferent (and, for Huxley, evil) financial institutions exploit and crush the poor, but they hide their evil behind grand architecture; the epanalepsis insists that is unnatural and wrong. The Harvey example offers solace: we have always made it through difficult stretches before, so we'll make it through this one.

I won't attempt any closer analyses of these examples or the doxa colligations, except to note a few general commonalities, and to point out that these variations suggest much more research is necessary before we can make definitive claims about how epanalepses operate rhetorically; we especially need to look at n-gram similarity among epanaleptic data to see how they interact with features of grammar and consequently what sorts of cxx have what sorts of functions. Examples 26 and 28 feature copulas (*is*, *been*) like the data we started out with ("rules are rules", 7–9 and 10–14) and they have similar argument-foreclosure and Semantic-Feature Promotion functions. The Trudeau example is especially noteworthy in this regard, but even the drawn-out Harvey example clearly works by promoting the features of continuity and recurrence in the phrase "times like these".

At this stage, we can say a few things about epanalepses, including the following:

- they are common elements of ordinary language, not confined to poetry, oratory, or other specialized genres, the frequently presumed home of rhetorical figures;
- they all have a clearly recognizable formal pattern;
- some of them have equally identifiable functions that align with those formal patterns;
- others, we can't be so sure about;
- the ones that have clearly identifiable aligned functions also have characteristic grammatical features;
- at least one specific form ($NP^A\ V_{cop}\ NP^A$) aligns with two specific functions (Argument Foreclosure and Semantic Feature Promotion).[3]

  but
- we need more data and more analysis to see what is going on.

[3] The superscript just co-indexes the phrases as identical, and the subscript *cop* identifies the verb as a copula, so the formula reads 'two identical Noun Phrases with a copula verb between them'. It is meant rather loosely so as to include, e.g., "boys will be boys", which also features a modal auxiliary. As mentioned in footnote 1, a satisfactory formalism for rhetorical figures awaits development.

In sum: we have an unmistakable example of a *construction* in the Fillmore–Kay–Lakoff–Goldberg sense of the term [58], and that construction bears the unmistakable signature of a rhetorical figure, epanalepsis. Namely, we have the NP^A V_{cop} NP^A form 'paired' with the Argument Foreclosure and Semantic Feature Promotion functions. What about the heterogeneity of the other examples? What do they say about the algorithmic value of rhetorical figures in natural language?

Firstly, notice that epanalepsis *constrains* that heterogeneity. Whatever else might be going on, the beginnings and endings of the expressions are lexically the same, which governs the entire expression. Secondly, there are neurocognitive pattern biases in play: that is, our brains and minds respond favourably to repetition as well as to positionality, which helps explain why there are so many such examples of epanalepsis in ordinary language. They are more salient and more memorable than blander versions of the same propositions, which helps individuals retain them and social groups propagate them. Many instantiations of rhetorical figures are more salient and more memorable than propositionally similar 'unfigured' expressions (compare, for instance, "little by a small amount" to 20, "dog eat canus domesticus" to 23, "I said what I stated" to 25, "Rules are codified principles for possible actions", or even "Rules must be followed", to our titular example). That is, figural patterns can provide a kind of glue, to stabilize utterances into expressions that spread easily as clichés, maxims, heuristics, and the like. It is important to realize, however, that figural patterns don't generate salience and memorability automatically. Not all utterances realizing epanalepsis are equally conspicuous or memorable. Trudeau's 26, for instance, would likely score higher in both respects than Harvey's 28. Factors like brevity, proximity of repeated terms, and, especially, the presence of other figural patterns, all affect salience and memorability.[4]

Thirdly, while the rhetorical functions of 16–25, and of the many more epanaleptic utterances that you can surely come up with on your own, in English and the other languages you may speak, are not as uniform as the tautological epanalepses, even in the little bit of data we have, patterns clearly emerge. In particular, the dual positioning, initial and final, often means dual syntactic and/or semantic roles for the repeated items, but those roles seem to be over-ruled in some way by the identity of the terms. With the prepositional examples (16–21), for instance, the first one is the head of a NP to which the second one is subordinate, but the identity of the two occurrences makes the relation seem wholly equal. Compare "side by top" to 19, for instance; in the first case we have a clear landmark (LM), *top*, and a clear trajector (TR), *side*; but in 19, the semantic roles seem to recede almost entirely and the terms *feel* like they are in a mutual, fully equal relationship, that both terms are LMs and TRs simultaneously to each other. Similarly, with 23, the first *dog* is a subject (SUBJ) and an agent (AGT), the second *dog* is an object (OBJ) and a theme (TH), or patient (P), but the expression *feels* like there are two agents and two themes, in a reciprocal

[4] There is some empirical research in this area, especially around the figure of rhyme, and the phenomenon of processing fluency seems to be relevant more broadly for rhetorical figures, but the role of figural patterns for attention, memory, and aeshetics, is not well explored at all. See [66] for some discussion, centring on a figural pattern we take up later in this article, chiasmus.

relationship. What seems to be happening is the iconicity principle of identity is overlaying a rhetorical function on the semantics. But there's not really enough to go on.

What these speculations show more than anything, in other words, is the need for more research, and for more data on which to do that research. The self-evidently true/indisputability role of the tautological epanalepses we looked at earlier is the result of traditional humanities research, with largely haphazard and intuition-based data. Corpus methods would give us much more to work with, especially the figure-detection methodologies sketched out in Sects. 4 and 7. This kind of research would not only be invaluable within given languages, but across languages. Figures of lexical repetition, in particular, would be fascinating to study in languages with different canonical word orders (SVO vs. SOV, for instance) and in languages with different degrees of positional freedom (Chinese, Polish, and Warlpiri, for instance). Factors like repetition and utterance-position have more to do with information processing than with grammar *per se*, and implicate psychological phenomena like priming and the recency effect.

To see some of the ways this research might proceed, we turn again to the work of Marie Dubremetz and her favourite figure, a figure of reverse lexical repetition that she calls *chiasmus*.

7 Detecting 'Chiasmus'

Chiasmus is best understood as a suite of rhetorical figures, not as an individual figure. Its distinctive quality is reverse repetition, as in these examples:

29. It's not the men in my life, it's the life in my men. (Mae West, qtd in [13])
30. Old King Cole was a merry old soul/And a merry old soul was he. (doxa)
31. Quitters never win and winners never quit. (doxa)
32. Tous pour un, un pour tous. [30]
    (All for one and one for all.) [31].

The reverse repetitions in 29–32 are all clear enough, and most rhetoricians would categorize them all as chiasmus (or one of its synonyms or plesionyms), but they do not represent a single figure because different constituents are repeated in each one. In 29, it looks like the *words* are repeating, but it's really only the phonological form of *life* that repeats, with a different sense in each case. The first occurrence means, roughly, 'existence', while the second occurrence means 'spirit' or 'energy'. With 30, it is the reference that repeats, with 31 it is the lexical stems (*quit* and *win*) that repeat (while the suffix, *-er*, repeats in place), but not the full words, and with 32 it is the full words (taking the sense and the reference with them). There are other variations, but that's enough to give the flavour of this pattern.

Example 32 represents the most common, or at least best known, of the chiastic patterns, and rhetoricians interested in more precision than is often given to figures, use a separate label for it, *antimetabole*, which is what we will call it (other relevant

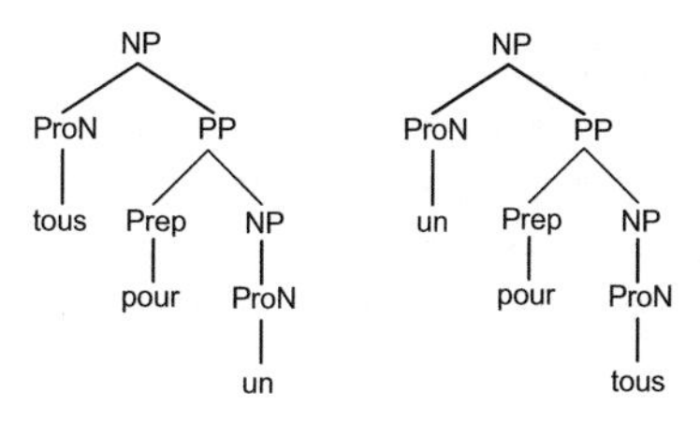

Fig. 1 The parallel syntactic structure of the two phrases of 32, realizing the rhetorical figure, parison

labels include *commutatio*, *permutatio*, and *counterchange*, though all of them are used fairly loosely in the rhetorical tradition, as is *antimetabole*). (See [81, pp. 118–120n10], for a detailed account of the terminological morass of chiastic figuration.) While Dubremetz prefers to use the more general label in her work (i.e., *chiasmsus*), she is really only interested in antimetaboles, and that is the only chiastic figure we will focus on as well.[5]

Antimetabole is not, however, the only figure we will consider in this section, just the only *chiastic* figure. As noted and illustrated in Sect. 2, figures frequently travel together; they also frequently code communicative functions together. Take example 32 as a prototypical antimetabole. That's how Dubremetz and Nivre regard 32, citing it recurrently to exemplify chiasmus. But if you look again, you'll quickly see that it also exemplifies other figures. The word *pour* repeats as well; the term for medial repetition is *mesodiplosis*. And both phrases have the same syntactic pattern (see Fig. 1); that is, they exhibit the figure parison. There are other figures realized in 32 as well–you may have noticed that it begins and ends with *tous*, for instance, which realizes epanalepsis yet again–but the three most crucial for its communicative function are antimetabole, mesodiplosis, and parison. We will return to this grouping.

Let's get back to Dubremetz. Her work took place over a few years, as she advanced toward her doctorate under Nivre, which she earned in 2017. Dubremetz and Nivre summarize this work efficiently in their 2018 paper for *Frontiers in Digital Humanities* [28], which I highly recommend.

As are most pattern-detection researchers, Dubremetz and Nivre were anxious about false positives. In one of their early attempts, for instance, they analyzed the text of a single book consisting of 130,000 words, which they found to contain 66,000 examples of reverse repetitions (which they call "criss-cross patterns"). But among them all, there was only a single instance they would call a "real chiasmus" [26, p. 24]. That's a false-positive rate of 65,999 to 1. In other words, utterly useless. I contend that the problem has more to do with their notion of a "real chiasmus" than with their initial detection methods *per se*, but their results led them in some interesting directions. In particular, they treat antimetabole as a graded phenomenon

[5] Dubremetz and Nivre are aware that chiasmus designates multiple different reverse-repetition figures–they define it as "a family of figures that consist in repeating linguistic elements in reverse order" ([26, pp. 23–24]; see also [28, pp. 2, 1])–but they choose not to refine their terminology.

with prototypical examples and controversial/borderline cases. That decision made antimetabole-detection not a matter of binary classification for them, but rather a task of extracting all of the "criss-cross" lexical occurrence patterns they could find and ranking them for their supposed antimetabolic purity.

Here, for instance, are their top and their bottom representatives of antimetabole in terms of 'purity' or 'realness'.

33. [=their #1] There are only two kinds of men: the righteous who think they are sinners and the sinners who think they are righteous.
34. [=their #3000] You hear about constitutional rights, free speech and the free press. Every time I hear these words I say to myself, "That man is a Red, that man is a Communist!"

The reverse repetitions in 33, easily spotted by you I'm sure, are *righteous* and *sinners*. But if you're still looking for them in 34 (it took me a while on my first encounter), they are *hear* and *free*. Example 33 certainly feels more antimetabolic and more rhetorical–the key repetitions pop out at you–while example 34 feels wholly bland and unfigured. We can sympathize with Dubremetz and Nivre for wanting to call it non-chiastic, a false positive; or, in any case, a piece of language with an extremely low chiastic quotient to it. But if antimetabole is, as they define it "the repetition of a pair of words in reverse order" [28, p. 1], a definition that is fully in keeping with the rhetorical tradition (see, for instance, [76, pp. 3, 13]), and is also, by the way, binary (either words repeat in reverse order or they don't), then 33 and 34 are both equally, 100%, *bona fide* antimetaboles. The difference between these two is not the presence or absence of antimetabole or the purity and the degraded nature of antimetabole in the respective instances.[6]

The way to understand the difference between them hinges not on some imaginary 'essence' of antimetabole that does not even make it into their definition; in fact, quite the opposite. In terms of figural purity, 34 is actually more singularly antimetabolic than 33. The difference between them hinges on the presence or absence of *other figures*. Example 33 has the two other figures we identified in the Dumas example: 33 has mesodiplosis (*who think they are*), and parison (the two parts of the expression are also syntactically identical; see Fig. 2). Meanwhile, 34, the allegedly adulterated example, has neither of these figures. It is fully antimetabolic, but it is not as elegant as 33 or 32; not as salient, not as memorable. More importantly, there is no tightly iconic communicative function in 34 of the sort we see in the examples Dubremetz and Nivre value for their 'purity'. The repeated terms only stand in the loosest of relations to each other in 34, while the repeated terms of 33 and 32 stand in very specific, inverse relations to each other. The reason that antimetabole, mesodiplosis, and parison collocate frequently is that (1) they effect a neurocognitively appealing utterance, recruiting attention and imprinting on memory, and (2) they evoke distinctive communicative functions. The famous 32, for instance, evokes reciprocality: the group (*tout*) is obliged to defend and foster the individual (*un*) and the individual members (*un*) are obliged to defend and uphold the interests of the group (*tout*). The

[6] See [53] for some related discussion.

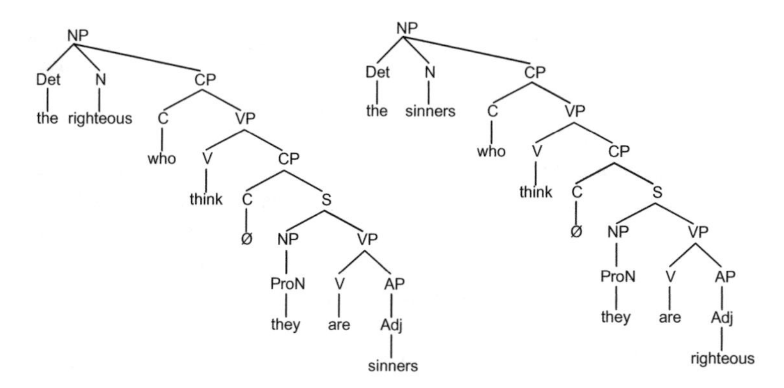

Fig. 2 The parallel syntactic structure of the two clauses of 33, realizing the rhetorical figure, parison

antimetabole conveys a reverse directionality, the movement of what John Austin calls *commissive* energy from the *tout* to the *un* and equally from the *un* to the *tout* [3, pp. 120, 150–162].[7]

The reverse directionality of 33 is somewhat different. It conveys a kind of alpha-to-omega, omega-to-alpha coverage of all men, the sinners and the righteous, who share a comprehensive confusion, each believing they belong to the other category. Logically, of course, to say there are two kinds of men in the world does not entail that there are *only* two kinds of men in the world, but the structure of 33 implies that all possibilities have been covered. The complex of antimetabole, mesodiplosis and parison frequently convey these functions (reciprocality and comprehensiveness), among a few others, and it's easy to see why. Most serious scholars of chiastic structures are aware of the supporting factors that profile the reverse repetitions, often by building one or more of those factors into their definition of chiasmus/antimetabole.[8] Where all of these approaches go wrong, however, is by treating these contributing factors to the success of chiastic instances as integral to chiastic structure itself. This is both messy, since chiastic instances often occur without these factors, and unparsimonious, since, e.g., the catalogue of rhetorical figures already includes parison. Why have an independent figure of syntactic parallelism and then include syntactic parallelism in the definition of other figures? To analogize from traditional Phrase Structure Grammar, that is like defining NP as, e.g., (Det) (AP)* N (PP), and then

[7] Strictly speaking, Austin calls it *commissive elocutionary force*, not *energy*. But since the two phenomena are materially convertible, I hereby convert force to energy metaphorically as well.

[8] Fahnestock, for instance, embeds parallelism into her definition of antimetabole [34, pp. 24, 135], and Patricia Lissner describes chiastic structures as occurring "usually within larger signifying dual platforms that are alike or related. By *platform* I have reference to phrases or clauses isogenous or nearly isogenous to one another in lexis and syntax" [81, p. 61], thereby folding in parallelism. *The Concise Oxford Dictionary of Literary Terms*, defines *chiasmus* as a device in "which the order of the terms in the first of two parallel clauses is reversed in the second", adding that this "may involve a repetition of the same words …or just a reversed parallel between two corresponding pairs of ideas" [5, p. 34].

to define PP as P (Det) (AP)* N (PP), rather than simply P NP. If we assume figural patterns collocate and we define those patterns only by their necessary and sufficient conditions, we can describe expressions like 32 and 33 quite elegantly. The antimetabole guarantees two occurrences each of two distinct elements. The mesodiplosis mediates their relation to each other. The parison stabilizes the syntax to ensure the mutuality of that relationship. With 32, for instance, first *tout* is the head of an NP to which the PP/NP/N, *un*, is attached, then *un* is the head of an NP to which the PP/NP/N, *tout*, is attached. By changing position (antimetabole) in the same syntactic structure (parison) around the same mediating predicate (mesodiplosis), *tout* and *un* are put into a reciprocal relationship.

Dubremetz and Nivre's 'purity' detection methods are very revealing in this regard, the way they end up with 33 on the top of their hit parade and 34 at the chiastic nadir. Shockingly, one would think, since they are ranking *chiastic* purity, a pattern defined in terms of reverse repetition, their sorting methods have nothing to do with reverse repetition at all. Reverse repetition of lexemes is the entry point for them, but that's also what gives them all their thousands and thousands of 'false positives', which their algorithm has to sort and filter away for their impurities.

One criterion is unsurprising, filtering via stop words. They don't want utterances like this one (where the relevant terms are coded in bold and bold-italics):

35. My government respects the **application** of ***the*** European directive and ***the*** **application** of the 35-hour law [26, p. 25].

So, they eliminate all 'criss-crossing' articles, pronouns, auxiliaries, very-high-frequency content words, and so on. This is a perfectly reasonable strategy—not for finding "true antimetaboles", I would argue, but for eliminating (or downgrading) uninteresting phenomena.[9] It is worth noting, however, that both *all* and *one* are in their stop list, so they couldn't even find 32 (more precisely, the equally famous English version of 32). It is the parade example of antimetabole, universally regarded as the most prototypical chiastic structure of all, including by Dubremetz and Nivre [26, p. 25], but their stop-word criterion would eliminate it from consideration[10]

More relevant to rhetorical figuration is their use of n-gram context in their purification algorithm. This criterion would very clearly differentiate 33 from 34, since the similarity of lexical environments for both "the righteous who think they are sinners" and "the sinners who think they are righteous" phrases is very high indeed, virtually identical, while the n-gram similarity between the word string "hear about constitutional rights, free speech" and the word string "free press. Every time I hear" is nil after the primary terms (*free* and *hear*). It would also, by the way, be kind to 32, if 32 could make it past the stop-word filter.

---

[9] I see no reason to declare 35 to be non-chiastic, for instance, though we can certainly note that its chiastic features are accidental and without consequence. We wouldn't declare a word like *apprehend* to be non-metaphorical in an expression like "Fred apprehends Wilma's meaning", though we may well want to note that its metaphoricity is utterly mundane and invisible to anyone who doesn't know the etymology (i.e., from L., *adprehendĕre*, 'to lay hold of, seize').

[10] Their stop-word list is available at http://snowball.tartarus.org/algorithms/english/stop.txt.

And they use syntactic role as a filtering criterion, what they describe as tuning their "algorithm [for] the symmetrical swap of syntactic roles" [26, p. 29]. You can see that this part of their algorithm would also be really happy with 33, where *righteous* and *sinners* do swap roles. First, *righteous* is the subject and *sinners* is a relativized complement; then *sinners* is the subject and *righteous* is a relativized complement. On the flipside, the algorithm would be really unhappy with 34 where *hear* and *free*, a verb and an adjective, don't even have grammatical roles. (This criterion heavily biases their algorithm for NPs.)

Two other factors are worth considering with respect to whether or not a linguistic instance exemplifies a 'real' rhetorical figure or some kind of spurious figure, intentionality and proximity, both of which notions participate in Dubremetz and Nivre's reasoning (and both of which, again, trace easily to the rhetorical tradition).[11] On the relevance of intentionality, consider this headline from *The Guardian* about Boris Johnson's lockdown misconduct during the COVID-19 pandemic, for a story which reports on growing discontent in the Conservative Party:

36. More Tory MPs call for PM to go as No 10 tries to limit Partygate report fallout [113]

There is no reason to believe that the *MP/PM* inversion was deliberate here, or even that it was noticed by the headline writer. But lack of intentionality does not somehow nullify their mutual chiasticity. When I was a child and I said something like "I'm going to the park but I'll be home before dark", my mother would inevitably say "You're a poet but don't know it" (a practice that continues to this day). The words *park* and *dark* rhyme whether one intends them to rhyme or not. They each have the phonological character of 'sharing' their final (and only) syllable, of having a final vowel-consonant sequence that the other has as well. So, will you, nill you, they rhyme. *MP* and *PM* have the same constituents in the inverse sequence. Will you, nill you, they are mutually chiastic. You may also have noticed that I used the word *Party* in my set up of 36 and that the word *Partygate* appears in 36. Whether you did or you didn't, those two words exhibit polyptoton, the rhetorical figure where at least two words share the same stem but also exhibit different morphology. Will you, nill you, they realize polyptoton. A figure is a pattern, irrespective of the intentions

[11] See, for instance, their talk of "rhetorical purpose" for intentionality as a defining criterion [27, p. 46]. The matter of proximity is more complicated. Dubremetz is clearly aware that high proximity is not criterial for figuration–as in her discussion of epanaphora, the figure in which words repeat at the beginning of clauses or phrases, where she acknowledges that it can occur in non-successive clauses or phrases, with such units as paragraphs or verses intervening. But she adopts high proximity methodologically. "To keep the problem simple", she says, "we will consider only epanaphora relying on immediately successive sentences" [25, p. 46]. (This approach is in contrast with Claus Strommer's work on epanaphora [114]). And it may be that she has a similar view of chiastic figures, that proximity is only a methodological convenience not a defining feature; certainly it is both reasonable and a long-established practice to rely on proximity in corpus research. But proximity is a common feature in traditional definitions of many schemes, including chiastic figures (e.g., antimetabole as "Repeating two words in successive phrases, but in reverse order" [82, p. 329]).

of the person who produces instances of that pattern, or, in the doubly epanaleptic slogan of our research group, "A figure is a figure is a figure".

As for proximity, while it factors into many definitions of antimetabole, chiastic figures more broadly, and schemes overall, it is misguided. A proximity-based definition of antimetabole, for instance, eliminates one of the most famous antimetaboles in all of modern literature. The first six words of T. S. Elliot's poem, "East Coker", are

37. In my beginning is my end [32, p. 123].

After this line, we get 213 more lines, several pages, and more than 1,600 words, before we reach the last six words:

38. In my end is my beginning [32, p. 129].

A definition of *antimetabole* that can't account for 37 and 38 is just not doing its job. What is relevant for an antimetabole to have a communicative or rhetorical effect is simply that it be processed as such. Proximity is a powerful factor to ensure that the relations among stimuli are perceived, like identity and inverse ordering that can evoke reciprocality or, as in the Eliot example, a sense of cyclical return. Strommer [114] even shows that proximity can be a significant predictor of rhetorical intent. But proximity is not criterial for antimetabole, or for very many figures. Eliot does without proximity by leveraging other factors with his "East Coker" antimetabole–the primacy effect for 37 and the recency effect for 38, for instance, and the semantic relevance of the words *beginning* and *end* at the beginning and end of a piece of discourse [49, pp. 29–30]. Also, of course, there are additional figures in play (parison, mesodiplosis, antithesis), enhancing their salience and memorability.

Proximity plays an obvious role in Dubremetz and Nivre's algorithm, and an important role because, in addition to its methodological convenience, proximity plays a significant role in figuration. While is not criterial for many figures, proximity very frequently plays a major role in amplifying the salience generated by repetition and relative position. But we do need to note that proximity is an 'additional' component of their algorithm, beyond the basic reverse-repetition desideratum of chiasticity.

Meanwhile, intention does not play a role in their algorithm, but it does inform their whole idea of what they are searching for; that is, what it means for a piece of text to be a 'true' chiasmus or a 'false' chiasmus (or, more precisely, for what it means for one piece of text to be a truer chiasmus than another, since their classifier assigns rankings to instances, not mutually exclusive categorizations). The notion of 'better' or 'worse' antimetaboles, or figures of any kind, is misguided, and largely based on a systematic ambiguity of category labels. Dubremetz and Nivre's ranking assignments would be like ranking 39 as a better equation than 40 on the basis of its formal properties:

39. $e = mc^2$.
40. $mc = e^2$.

In one sense, of course, 39 *is* a better equation than 40. It maps better onto the observable facts than 40, when we know the context and especially what the variables stand for. But it is not a better example of an equation *per se*. In categorial terms, an equation is an equation is an equation. So, 39 is more *true* than 40, understood as an expression characterizing features of the phenomenal world. We might also apply criteria like elegance to its expressivity and scope to its implications, to judge its aesthetics and its value relative to other equations, including 40. But neither 39 nor 40 is somehow a purer realization of the category, equation. They are both equations, in a way that, say, 33 and 34 certainly are not. What we mean when we say 33 is a better antimetabole than 34, or what we *should* mean when we say such things, is that it is a more effective nugget of language and that antimetabole is a figure that makes an important (but not exclusive) contribution to its effectiveness.

Here, in short, is the takeaway about Dubremetz and Nivre's algorithm in figurative terms: they are unwittingly looking for the co-presence of parison, which swaps the syntactic (and semantic) roles of the relevant terms when it combines with antimetabole, and they are just as unwittingly looking for the co-presence mesodiplosis. As we've seen, the n-gram context of *righteous* and *sinners* in both phrases of 33 is *who think they are*. It is completely and utterly coextensive with an instance of mesodiplosis. While Dubremetz and Nivre clearly do not realize it, their ranking criteria are largely tuned to the co-presence (or absence) of *other figures*. The more additional figures are present, the higher they rank the instance.[12]

So, there is very good news about Dubremetz and Nivre's work; some not exactly bad news, but less good news; and some great news. The very good news is that their work is highly reliable at finding at least some figural collocations, and figural collocations is where the form/function action is. The less good news is that they aren't fully aware of what they are doing, though I do want to make it clear that this is not their fault. They are relying on the rhetorical tradition, and rhetoricians

---

[12] Schneider et al. [109] expand Dubremetz's work in interesting ways that suggest further approaches to the search for figural collocations. They detect what they call "general chiasmus" rather than antimetabole and report considerable success (improving "the average precision from 17 to 28% and the precision among the top 100 results from 13 to 35%"–p. 96). Their 'general chiasmus' includes e.g., utterances like 32 and 33, but they exemplify their expansion of candidates over Dubremetz and Nivre by this utterance (from Schiller), which they can detect but Dubremetz and Nivre cannot: "**Eng** ist die **Welt**/und das **Gehirn** ist **weit** (**Narrow** is the **world**,/and the **brain** is **wide**".) This example is a reversal of syntactic structure (the first clause has the structure, NP V_{cop} AP, and the second clause has the structure, AP V_{cop} NP), which is given additional salience by the antithesis of *eng* and *weit*.

Schneider et al. use broader search criteria than Dubremetz and Nivre, looking only for reverse repetition of part-of-speech, which brings back way more candidates. Then they filter the candidates by augmenting "the Dubremetz features" in two ways. Firstly, they sort all possible pairs of the relevant tokens (i.e., in the sequence $A_1B_1A_2B_2$, they sort A_1B_1, A_1B_2, A_1A_2, B_1B_2, B_1A_1, and B_2A_2) for whether or not their lemmata are identical. Secondly, they sort them on the basis of the "semantic relationships between the supporting tokens" (p.97).

It is unfortunate that Schneider et al. do not approach their data in collocational terms, nor in terms that would allow them to differentiate, for instance, syntactic chiasmi from lexical chiasmi (i.e., antimetaboles), but they show the possibility of refining the "Dubremetz features" further in exactly these directions.

have done a really bad job of representing and understanding figural collocations. What one finds when one looks into a rhetoric handbook to try and understand figuration is firstly, a definition, like this one by rhetorician Michael Mills: "two or more words are repeated in a successive clause, but in reverse order" [87, p. 144]. That definition is highly typical as a definition of *antimetabole*. The author goes on to identify antimetabole as a type of chiasmus. But secondly what we get for an example, invariably, is just like the one Mills gives, if not identical with it: "All for one and one for all" (i.e., our 32). It is indeed antimetabolic, as we have seen, but we also know there are other figures here as well, and Mills not only fails to mention them, but curates this example, and one other which also features mesodiplosis and parison ("absence of evidence is not evidence of absence"), as typical antimetaboles and *only* as antimetaboles. In other words, these are the prototypes that Dubremetz and Nivre encountered in the rhetorical literature and therefore that serve as models for what they are trying to find and rank highly as the purest of antimetaboles. The rhetorical tradition, for all its length and multiplicity of theoretical perspectives (or, more likely, *because* of its length and multiplicity of perspectives), is incomplete and misguided. As a rhetorician, it's embarrassing, but we have been really shoddy in these respects.

The other less good news about their work is that Dubremetz and Nivre don't have any interest in the functions that their figures (or, rather, their figural collocations) serve. This, too, is something that the rhetorical tradition has been somewhat sloppy about. There are a few important exceptions [19, 34, 101, 103, 116], but much of the work on figures focusses on aesthetics and wittiness and warnings against over use; occasionally a communicative purpose is noted, but never in terms of collocation.

As for the really great news, Dubremetz and Nivre have laid an excellent foundation for detecting figural collocations and for charting out the functions they serve. It's a side effect of their ranking strategies, but an incredibly important one.

8 Antimetabole, Mesodiplosis, and Parison; AMP Constructions

A few of the figure detection researchers have looked beyond simple detection. James Java looked at figures as diagnostics for authorship attribution [65], Claus Strommer investigated correlations of affect with the figure of epanaphora [114]. Nadezhda Lagutina, Ksenia Lagutina, and their colleagues explored certain figures they associate with rhythm as correlates of author- and genre-styles [73]. The only study to look for specific form/function correlations, however, is by John Lawrence and his associates at the Centre for Argument Technology, the leading site for argument mining. They ran a fascinating pilot study on form/function correlations, focussing on "five schemes of [lexical] repetition: anadiplosis, epanaphora, epistrophe [synonymous with *epiphora*], epizeuxis, and polyptoton" [78, p. 294]. Two other studies are worth mentioning here as well, Jakub Gawryjolek's early broad-spectrum detection research [39] and Katherine Tu's more targeted detection research [118]. Gawryjolek's work is important because he detected all the figural patterns in a given text (that is, all the patterns JANTOR was built to detect; James Java's tool did this as well [65]). Among other things, that allowed him to find passages with particular figural densities. Tu's work is important because she specifically looked at the collocations of a small range of figures. None of these studies looked for grammatical features in combination with figures, except in the incidental way we have noted for Dubremetz.

What is needed, if we are to incorporate rhetorical figures seriously into computational linguistics, is a programme that brings together the detection of rhetorical figures, in isolation but more importantly in collocations, with theories of (1) form/function alignments, and (2) Construction Grammar. This section and the following one explore what such a programme would look like, centring on antimetabole and its frequent collocates, mesodiplosis, parison, and antithesis.

Antimetabole, mesodiplosis, and parison function so frequently together that we can refer to them collectively as *AMP*. There are always variations in language, so the AMP designation is not watertight. For instance, there are instances like the following:

41. The brain was made for religion and religion for the human brain [123, p. 149].
42. Plato is philosophy, and philosophy, Plato [33, v. 4, p. 40].
43. Ambition stirs imagination nearly as much as imagination excites ambition [20, p. 26].[13]
44. I had neither enough power to compensate for my lack of speed nor enough speed to make up for my lack of power [85, p. 193].
45. It seemed fitting that my former tormentor had gone from a boarding school regulated like a prison to a prison run like a boarding school [72].

Instances 41 and 42 are both missing mesodiplosis. The second clauses in both instances exhibit what Transformationalists dubbed *Gapping* (e.g., [63]) and rhetoricians have known for millennia as *prozeugma*. In both traditions, the verb (or verb sequence) in the first clause is understood to apply not only to its own direct complement (or object) but to the one in the subsequent clause as well. With instances 43, 44, and 45, mesodiplosis is again missing but this time not because of 'omission' or (in the Standard Theory Transformational paradigm) 'deletion'; rather, through the figure of synonymia (i.e., the use of synonyms): the pairs *stir* and *excite*, *compensate* and *make up for*, and *regulated* and *run* each share the same core meanings, at least in the context of the relevant clauses here. Parison is present in all three of 43, 44 and 45, but not in 41 or 42. Despite these differences, 41–45 function communicatively pretty much like AMP collocations; in a sense, the mesodiplosis is 'understood' in 41 and 42, and the meaning equivalences in 44 and 45 perform a similar role.

[13] This example, incidentally, is the one 'true' antimetabole Dubremetz and Nivre report from their 66,000 "criss-cross" patterns in a 130,000 word book, Winston Churchill's *The River War* [26, p. 24].

In fact, in cases where one or more figures are not present, one can impose them on a text for such tasks as argument summarization. Fahnestock points this out. She proposes that one can convert "loose" or "lax" arguments into the figurative relations they express to render their rational epitomes. For instance, the causal relations inherent in 46a can be rendered more clearly by 46b, which realizes a figure known as *gradatio* (a series of repetitions of words at the ends and beginnings of contiguous clauses or phrases), rendering the causal structure of the argument explicit.[14]

46. a. In a hypothetical example given by Dr. Mech, a wolf kills a moose. The remains slowly disintegrate and add minerals and humus to the soil, making the area more fertile. Lush vegetation grows, which attracts snowshoe hares, which in turn draw foxes and other small predators, which coincidentally eliminate many of the mice that live nearby. A weasel that used to hunt the mice moves to another area and in so doing is killed by an owl. The chain could be extended indefinitely See [34, p. 109].
    b. A wolf kills a moose, creating a moose carcass. The moose carcass creates enriched soil. The enriched soil grows lush vegetation. Lush vegetation attracts snowshoe hares. Snowshoe hares attract foxes and other small predators. Foxes and other small predators cause the elimination of most mice. The elimination of most mice displaces a weasel. The weasel is killed by an owl.

Or, if we move a little closer to home, the kernel of 47a might be realized as 47b, which has an AMP structure combined with epiphora ("have faults"):

47. a. I hope my Friends will pardon me, when I declare, I know none of them without a Fault; and I should be sorry if I could imagine, I had any Friend who could not see mine [36, p. 108].
    b. I know my friends have faults and my friends know I have faults.

One might, in other words, 'reduce' a text like 46a or 47a to its figurative core, 46b or 47b. Such a move is eerily reminiscent of Zellig Harris's Discourse Grammar, which used transformations to produce sentences in more parsable and semantically precise "kernel" forms [55]. In a similar way, one might render, say, 42 as "Plato is philosophy, and philosophy is Plato" (which is perhaps more reminiscent of Standard Theory Transformational Grammar [17]), or 44 as "I had neither enough power to COMPENSATE for my lack of speed nor enough speed to COMPENSATE for my lack of power" (assuming COMPENSATE to be something like a semantic predicate, in the sense of Generative Semantics (e.g., [74])). Much of the theory of sentential syntax in the Bloomfieldian tradition was primarily about what rhetoricians call *style*, though few of its practitioners (with the partial exception of Harris) seem to have been aware of that fact.

There are other variations of the basic AMP structure, usually involving synonyms or paraphrases, most of which can also be 'transformed' into AMPs, but 41–45 illustrate the main points.

---

[14] See [34, pp. 109–110, 205n11]. I have extrapolated her discussion somewhat and 46b is my own rendering.

We will now look at the AMP constructions by way of three rhetorical functions: Irrelevance of order or rank, Reciprocality, and Comprehensiveness.

8.1 Irrelevance of Order or Rank

If you recall the quotation from Melanchthon on p.8, he uses antimetabole to explicate complementarity, and Christine Grausso's fascinating dissertation outlines an eight-way taxonomy of chiastic functions from logical principles [41].[15] But the most obvious use of the AMP complex for logicians and mathematicians, perhaps for anyone, can be seen in examples like the following:

48. [T]he expression "men and women" is, conventional meanings set aside, equivalent with the expression "women and men". Let x represent "men", y, "women"; and let + stand for "*and*" and "*or*", then we have

$$x + y = y + x$$

an equation which would equally hold true if x and y represented *numbers*, and + were the sign of arithmetical addition [7, p. 33].
49. Gretzky and McNall, McNall and Gretzky, joined at the hip, joined at the wallet [8, p. 196].
50. We're fric and frac. ...We're frac and fric. Always together [81, p. ii].
51. Imagine you and me, me and you [6].
52. Claudius: Thanks, Rosencrantz and gentle Guildenstern.
    Gertrude: Thanks, Guildenstern and gentle Rosencrantz [111].

The first instance, George Boole's description of how natural language conjunctions and disjunctions exhibit the principle of commutation (48), gives the game away. The principle of sequential iconicity [48] requires that the default syntactic order reflects the temporal order of the events it expresses. So, there is no more efficient way to express the irrelevance of temporal order than to juxtapose two inverse orders, crystalized in the illustrative equation Boole highlights, $x + y = y + x$. It doesn't matter if you add x to y or add y to x. The order of operation is irrelevant. You get the same sum. As Fahnestock notes, "the commutative laws provide a blueprint for a particular class of arguments epitomized by the antimetabole" [34, p. 134]. Instances 49–52 are much the same. What is conveyed about the order of operations with some AMPs (e.g., 48) is also conveyed about order of temporal encounters (e.g., 49, which suggests that if you see Gretzky, you will soon see McNall, and if you see McNall, you will soon see Gretzky; ditto for fric and frac, frac and fric) and rank of importance (e.g., 52, from *Hamlet*, Act 2, Scene 2, which indicates that Claudius and

[15] I don't fully buy all of her categorizations, and she also misses some that I regard as clear and important, as well as largely ignoring grammatical factors and figural collocations. But there are some very promising overlaps between her work and mine, especially on the meaning side of the form/meaning alliances of the chiastic figures.

Gertrude see Rosencrantz and Guildenstern as utterly interchangeable, reinforced by the interchangeable application of the adjective, *gentle*). Indeed, sequence and rank differences are often both negated in this cx; apparent especially in 50, which also includes minimally distinct nonce words for the referents.

We have seen that the reciprocality function comes into play with AMP constructions which reverse the NP-Semantic Role assignments (i.e., in 32), and we will look at reciprocality more fully in the next subsection. But when the relevant terms reverse positions around conjunctions and disjunctions (in natural language or in predicate calculus), as with addition and multiplication operators, they retain their roles (or lose and gain the same role), so order is foregrounded. With the reminder that we need more data to make such claims very strongly, it would appear that Irrelevance-of-order-or-rank AMP constructions are characterized by mesodiplosis of conjunctions and disjunctions. In English at least, we might get the same function when the relevant antimetabolic terms are immediately juxtaposed, as in this example:

53. Church, cult, cult, church–so we get bored somewhere else every Sunday [96].

Church and *cult* are utterly interchangeable in 53.

8.2 Reciprocality

AMP constructs in natural language regularly express reciprocality. We have seen this function with 32 already, the commissive reciprocality of *un* and *tout*. By changing position (antimetabole) in the same syntactic structure (parison) around the same mediating predicate (mesodiplosis), the relevant terms are put into a reciprocal relationship, either of energy flow or of identity-specification. Other examples include:

Reciprocal Energy

54. Women are changing the universities and the universities are changing women [46, p. 629].
55. Wenn du lange in einen Abgrund blickst, blickt der Abgrund auch in dich hinein [94, p. 105]
    (When you look into the abyss, the abyss also looks into you [95, p. 89]).
56. [Y]ou can communicate with your pet dog, and your pet dog can communicate with you [18].
57. Si equus lapidem funi allegatum trahit, retrahetur etiam & equus aequaliter in lapidem: nam funis utrinq; distentus eodem relaxandi se conatu urgebit Equum versus lapidem, ac lapidem versus equum] [92, p. 13].
    (If a horse draws a stone tied to a rope, the horse … will be equally drawn back towards the stone: for the distended rope, by the same endeavour to relax or unbend itself, will draw the horse as much towards the stone as it does the stone towards the horse [93, p. 83]).

Reciprocal Specification

58. In America, race is class and class is race [62].
59. Gay rights are human rights, and human rights are gay rights [21].
60. For Y am sorwe, and sorwe ys Y.
    (For I am sorrow, and sorrow is me.) [14, p. 597]
61. La vie, c'est le germe et le germe, c'est la vie. [100, p. 328].
    (Life is the germ and the germ is life.) [34, pp. 139–140]

I've given the two basic varieties of AMP Reciprocality constructions, and they help to highlight the way in which grammar is equally as important to the functions of these expressions as is rhetoric. Instances 54–57, with some variation, especially in portions of the Newton instance, feature transitive verbs and (like our prototype, 32) prepositions. These are AMP examples with the communicative function of Reciprocal Energy. 'Partial' AMP cxx can also serve this function, depending on how they are configured. Example 43, with the *stir/excite* synonymia, for instance, is a Reciprocal Energy AMP, with causative energy flowing from ambition to imagination and ("nearly as much") from imagination to ambition. Reciprocal Energy AMPs activate what Ron Langacker calls *flow*. Langacker describes transitivity as "the flow of energy along an action chain" [75, p. 293], and these AMPs double up that energy flow with two equal and opposite action chains. With 32 obligation flows from the all to the one and equally from the one to the all (this is perhaps more apparent when one considers the 'conceptual metaphor', OBLIGATION IS ENERGY). The prototype here is 57, especially in the final AMP of the original Latin which might be a slogan for Reciprocal Energy, *equum versus lapidem, ac lapidem versus equum*.

The other examples here, 58–61, exhibit Reciprocal Specification, which concerns not energy but identification.[16] Here the mesodiplosis is not of transitive verbs or prepositions but uniformly of copular verbs. Reciprocal Specification is effectively bi-implication. All things to do with race, 58 says, equally implicate class (in America), and all things to do with class equally implicate race. And so on. As with Reciprocal Energy constructions, Reciprocal Specification constructions can include 'partial' AMPs. Example 42, for instance, is Ralph Waldo Emerson's total identification of philosophy with Plato (and vice versa).

8.3 Comprehensiveness

You likely did not notice that my complaint about the problems plaguing rhetorical terminology back in Sect. 2 was chiastic. I hadn't defined the term yet when I said the vocabulary of rhetorical figures suffers from both "one/many mismatches and many/one mismatches", and then explicated that chiastic expression with another one, "[t]here are multiple names for the same pattern, and there are multiple patterns

[16] This function is closely related to Grausso's Equalization chiastic type [41, 81 et passim].

that have the same figural name". But you surely recognized the point that phrasing was steering you towards: rhetorical terminology is comprehensively sloppy. Dubremetz and Nivre's highest scoring chiasmus, 33, also serves this Comprehensiveness function: all men are comprehensively included and while we get two categories both are comprehensively confused, believing themselves to be in the other category. More Comprehensive AMP examples follow:

62. I meant what I said and I said what I meant. An elephant's faithful–one hundred per cent [110, passim].
63. A place for everything and everything in its place. (Doxa; often attributed to Benjamin Franklin)
64. Whether we bring our enemies to justice or bring justice to our enemies, justice will be done [12].
65. I would argue that insofar as the Four-Color Proof has been checked, it is not rigorous and that insofar as it is rigorous, it has not been checked [119, p. 131].

With 62 we have an utterly and completely sincere Horton (the elephant in the story who is asserting his reliability) who declares an utterance he has just made to be a comprehensive representation of his intentions and that his intentions are comprehensively revealed by that utterance. With the maxim of obsessive housekeepers, 63, there is no item that does not have an allocated space which it is occupying and there are no allocated spaces which are unoccupied. The occupation of places by things is full and complete. President George W. Bush's declaration of Comprehensive vengeance after the 911 terrorist attack, 64, says that all enemies will be punished, no matter the effort required. Philosopher of mathematics, Thomas Tymoczko is arguing that the Four-Colour Proof is utterly uncertain in 65, Comprehensively untrustworthy. He later sums up,

> if mathematicians know the Four-Color Theorem on the basis of the Four-Color Proof–and I believe they do–then the two fundamental principles of proof which we began with are jointly false of the Four-Color Proof [119, p. 135].

I don't have a linguistic explanation for when Comprehensiveness comes to the front of the stage as an AMP function in instances like 62–65; it is certainly more complex than a mesodiplosis of a transitive or copular verb. Examples 64 and 65 are both conditional; example 62 includes a complement; example 63 is actually not an AMP, since there is no mesodiplosis, and the central terms (*for* and *in*) are not very close synonyms, not even in this context.

But it may just be that Comprehensiveness is a kind of default function, when the other functions are not activated by grammatical features. Perhaps subordination plays a role. Antimetabole is frequently discussed as "creat[ing] a closed loop" [35, p. 98], noted for its "period-like" quality [77, p. 322], framed as "a figure which mimes circularity" [64, p. 113], something which we saw earlier in the 37 and 38 instances from Eliot's "East Coker". And–we should add–all of these accounts of antimetabole are accompanied by examples that include mesodiplosis and parison. The tag for the headline in an Australian newspaper makes this 'closed loop' iconicity as clear as possible. The headline is chiastic:

66. Greater Victoria builders say they can't find workers to build new homes, because they can't find homes for the workers [29].

And here is the tag: "The 'vicious circle' has deepened a persistent worker shortage during a period of record demand". This example is not a full AMP so it is not reciprocal, which reflects the ontological (and grammatical) asymmetry between humans and artefacts. Workers can build homes (as AGENTS) but homes cannot build workers–not without a heavy overlay of conceptual figuration. But the rhetorical pattern of 66 does enact the comprehensive futility of a vicious circle. In another example–this time an activist biologist trying to protect an ecosystem in Canada from catastrophic infestations introduced by salmon farming–we see this circularity as a bureaucratic boondoggle, reminding us that sometimes balance can signal an obstructive stalemate:

67. [W]hen I was writing these letters, trying to find someone to take responsibility for this circular mess, it was like watching a tennis match. The province sent my concerns to the feds and the feds lobbed them back to the province. [89, p. 28].

The closed-loop evocation of chiastic expressions informs most of the functions they serve, but it seems especially prevalent when the AMP complex collocates with epanalepsis, so that the beginning and ending of the instance are identical. The Reciprocal Energy AMPs, in particular, suggest a cyclic flow; e.g., from the all to the one and then back to the all again in our prototype, 32. But this evocation is perhaps nowhere more prominent than with the Comprehensiveness function, which suggests the complete seal-the-deal coverage of a domain (vengeance, for instance, in 64, where all possible ways to visit 'justice' upon enemies are exhausted). Frequently (but certainly not always) this function capitalizes on binaries: two types of men in 33; saying and meaning in 62; the reverse directionalities of 64. More data should tell.

9 Antithesis, Antimetabole, Mesodiplosis, and Parison; AAMP Constructions

When I discussed Dubremetz and Nivre's purity algorithm, I neglected one of their important criteria. A "chiasmus candidate" was ranked higher if it "contains one of the negative words 'no', 'not', 'never', 'nothing'" [28, p. 6]. This may seem like an odd condition on the surface, but they did their homework, citing nineteenth century rhetorician, Pierre Fontanier [37, p. 160] for his observation "that prototypical chiasmi are used to underline opposition" and noting that chiasmus scholar Harald Horvei characterizes them as often appearing "in [a] general atmosphere of antagonism" [26, p. 27]; in fact, Horvei goes somewhat beyond that, noting that in such an atmosphere, chiasmi can take "on the formal characteristics of the antithesis" [59, p. 49]. The AB and BA clauses of antimetabole frequently realize antitheses.[17]

We can see the presence of AMPs in an atmosphere of antagonism clearly in a little patch of dialogue from the resolutely anti-figural Ernest Hemingway, when he has Anselmo answer his moral opposite, Pablo, antimetabolically in *For Whom the Bell Tolls*:

> "Coward", Pablo said bitterly. "You treat a man as coward because he has a tactical sense. Because he can see the results of an idiocy in advance. It is not cowardly to know what is foolish".
> "Neither is it foolish to know what is cowardly", said Anselmo, unable to resist making the phrase [56, p. 54].

"Unable to resist making the phrase", Hemingway tells us of Anselmo's words here. The atmosphere of antagonism between the two men is so dense that the gravitational pull of opposition causes Anselmo to reverse Pablo's words, which beautifully illustrates the iconicity here, moral opposition expressed in syntactic reversal. Ward Farnsworth notes that chiastic patterns are "often used to refute" [35, p. 99], and Fahnestock provides this double-refutation exemplum from Richard Lewontin:

68. Just as there can be no organism without an environment, so there can be no environment without an organism [80, p. 48], [34, p. 154].[18]

Lewontin here is countering both Lamarck (whom he characterizes as claiming that organisms incorporate the environment into their biology) and Darwin (characterized as claiming the environment is independent of organisms) by arguing that neither concept makes sense without the other.[19]

Fahnestock calls this AMP function "corrective", because it very often works by juxtaposing the denial or downgrading of some relationship or "energy flow" next to a claim of the accuracy, authenticity or desirability of the opposite relationship or flow [34, pp. 150–155], as in these examples:

69. The fire [=electrical charge] does not proceed from touching the finger to the wire, as is supposed, but from the wire to the finger (Benjamin Franklin, in [38, p. 182]; [34, p. 124]).
70. Men need not trouble to alter conditions, conditions will so soon alter men [15, p. 26].

---

[17] Grausso has a category she calls *One-way effect chiasmus*, which she notes includes "a negation involving the term 'not' in the sentence or some similar phrasing" [41, p. 40]; that is, the one-way chiasmi would seem to involve antithesis. This category overlaps with the AAMP constructions I investigate in this paper, and may subsume all of them. This is yet one more area where further research is needed.

[18] I have used a slightly different quotation than does Fahnestock, from a different source.

[19] While the argumentative intention of 68 is a double refutation, and it therefore partakes in an atmosphere of antagonism, the primary communicative function of the AMP is Reciprocal Specification: organisms and environments mutually define (and determine) each other.

Sometimes, the energy flow can be as low-wattage as nominal modification, when the linear order and the parison do all the work on their own (i.e., no mesodiplosis or synonymous mediation), as in this exchange from the American TV comedy show, *Seinfeld*, where Elaine corrects the 'weighting' of George's two-term designation of Elaine's roommate:

71. Elaine: I'm getting a break from my roommate.
    George: Oh, the actress-waitress.
    Elaine: No, the waitress-actress [23].

While we are largely confining our discussions to antimetabole, the iconicity of reversal for evoking opposition is widespread throughout the chiastic suite; for instance, in these examples:

72. We didn't land on Plymouth Rock. Plymouth Rock landed on us [84, p. 232].
73. Blanche: You're not as gallant as when you were a boy.
    John: You're not as buoyant as when you were a gal [47, pp. 215–216].

Example 72 is an expression Malcolm X used often, so the *us* refers to African Americans. It is not an antimetabole because the two uses of *Plymouth Rock* have different senses (as does the relevant term in 29 above, *life*). The first usage is 'literal', referring to a geographical object, though it certainly has historical and cultural resonance. The second usage is metonymic, referring primarily to that cultural resonance, in fact to the whole cultural apparatus of American mythology. The expression is clearly not making a factual claim that some historical physical object fell on all African Americans, but *Plymouth Rock* is well chosen as a metonym because *rock* conveys the heaviness and bluntness of oppression. This type of chiastic figure, an antanametabole, is a trope because it relies on a conceptual shift (in this case metonymy; in 29, paronomasia, a pun). Antanametabole is important for figure mining since its functions are less stable than antimetabole (so we need to be able to tell them apart), but such structures are never distinguished from antimetabole/chiasmus in the rhetorical tradition. Minimally, one needs to know when a given word in some expression is 'literal' or 'figural'. Example 73 is a commutatio, in which the chiastic reversal is of phonological units (phonemes, syllables, as in the 73 example, and even phonological features). Example 73 is from an old radio sitcom called *The Bickersons*, and even though John's response is almost entirely unconnected from Blanche's remark semantically (*buoyant* and *gallant* are only very tangentially related) it is clearly offered as a rebuttal in some way.[20]

Notice that all of these examples, 68–73 (with the possible exception of 72, if the algorithm can't handle clitic negations) would score well on Dubremetz and

[20] Both *antanametabole*, a neologism, and *commutatio*, an adaptation of one plesionym of *antimetabole*, are unique to my research project. *Antanametabole* is a blend of *antimetabole* and *antanaclasis*, a figure in which a signans is used at least twice with at least two different signata ("we must … all hang together, or most assuredly we shall all hang separately", reputedly said by Benjamin Franklin to his co-signers after signing the American Declaration of Independence, that they needed to remain unified or they would all be executed [61, p. 52]). *Commutatio* is the Latin term for *chiasmus*. There is an excess of terminology for rhetorical figures, so we just repurposed this one for phonological chiasmi.

Nivre's chiasmus-ranking procedure because of the presence of negation along with high n-gram numbers and syntactic role swapping, along with the criterial reverse repetition.

Horvei's remark that antimetaboles often exhibit "the formal characteristics" of antithesis when contention is in the air is only part of his story. Much of his monograph, *The Changing Fortunes of a Rhetorical Term: The History of the Chiasmus* [59], is concerned with the interactions between chiasmi and the trope, antithesis. Antithesis is a complex figure that requires closer study and taxonomizing, but for now we can define it as opposing predications, often realized by antonyms or by clausal negations. Both antonyms and clausal negations are frequent travelling companions with AMP constructions, to the extent that Quintilian, for instance, regarded antimetabole as a variant of antithesis [105, 9.3.85]. Others regard antithetical antimetaboles as a distinct figure, called *antimetathesis* and defined as occurring when "a repetition is made in an inverse order by the members of an antithesis" [107, p. 74].

It is not hard to see why antithesis and chiastic structures have a mutual affinity. Reversal and opposition are closely related concepts. The chiastic figures are defined by material reversal. Antithesis is defined by semantic opposition. This collocation, along with mesodiplosis and parison (with parison 'relaxed' enough to include negation) is common enough that we can give it a unique label as well, *AAMP*. AAMP cxx express two basic functions, Subclassification and Reject-Replace.

9.1 Subclassification

The subclassification function is very straightforward. Here are some relevant data:

74. All compounds are molecules (since compounds consist of two or more atoms), but not all molecules are compounds (since some molecules contain only atoms of the same element). [122, p. 7]
75. Ultrabooks are laptops after all, but not all laptops are ultrabooks. In essence, Ultrabooks are just a special type of laptops. [124]
76. Every finite state language is a terminal language, but there are terminal languages which are not finite state languages. [16, p. 30]
77. I have come to the personal conclusion that while all artists are not chess players, all chess players are artists. (Marcel Duchamps qtd in [69, p. 14])

These are clear variations of the reciprocal-specification AMPs with negation (and consequently, the other *A*, antithesis) added. If we think of reciprocal-specification AMPs in set-theoretic terms, we have total amalgamation. All things that implicate race also implicate class, and vice versa, according to 58; all rights that concern Gays concern humans more broadly, according to 59, and vice versa. They all belong in the same set. Negation denies one of those identifications, which leaves some molecules out of the set of compounds, for instance (74), but no molecules out of the set of compounds, some laptops out of the set of ultrabooks, but no ultrabooks out of the set of laptops (75), and so on.

9.2 *Reject-Replace*

The Reject-Replace cxx are similarly variations of a reciprocal AMP, namely the reciprocal-energy AMPs, with negation added:

78. Es ist nicht das Bewußtsein der Menschen, das ihr Sein, sondern umgekehrt ihr gesellschaftliches Sein, das ihr Bewußtsein bestimmt [68, p. 21].
    (It is not the consciousness of men that determines their existence, but their social existence that determines their consciousness [67, p. v].
79. [B]ehavior built language in the brain; the brain did not build language in behavior [115].
80. And so, my fellow Americans: ask not what your country can do for you–ask what you can do for your country [71].
81. We don't build services to make money; we make money to build better services (Mark Zuckerberg, qtd in [83]).
82. Non ut edam, vivo, sed ut vivam, edo.
    (I do not live to eat, but eat to live.) (Quintilian 9.3.85, [105])
83. Plain statement must be defined in terms of metaphor, not metaphor in terms of plain statement [9, p. 69].

Again, it is fairly easy to see what is going on. We have the basic lexical repetition in reverse around a central term with the same basic syntactic structure such that the roles swap as the positions swap. With 74–77 the central term is a copula, so the overall construction is very close to Reciprocal Specification AMPs, but one specification is blocked. The bi-implication is disabled and we end up with a subset relation. The Reject-Replace AAMPs are more varied and the label is perhaps a bit too absolute.[21] Mitigate-Replace or Downgrade-Replace might be more accurate for the milder expressions, such as weaker versions of Fahnestock's "corrective" function, like 70, in which G. K. Chesterton doesn't utterly discard the notion of trying to alter conditions, just says it is rather pointless. (The sentence that follows this, by the way, is a gem. One need not bother to try and alter conditions, he says, since "The head can be beaten small enough to fit the hat" [15, p. 26].)

But most of these AAMPs are more strongly moral or factual so the Reject-Replace function seems correspondingly more absolute. Benjamin Franklin's "corrective" about the flow of electricity, for instance, just asserts that the assumption that it moves from finger to wire is factually incorrect, wrong, and should therefore be rejected and replaced by the correct belief, that it moves from wire to finger (69;

[21] There is a mutual-exclusion aspect of this function that is similar to Grausso's Exclusion chiastic type [41, p. 89 et passim], though of course it is not a neutral exclusiveness. One option is promoted as correct, the other as incorrect and even reprehensible.

[38, p. 182]). I think all of the "corrective" AAMPs can be subsumed by the Reject-Replace AAMPs, but we need more data.[22] In any case, both 78 and 79 have this same kind of factual status, where mistaken beliefs need to be rejected and replaced with the correct opposite beliefs. The reasoning evoked here follows the principle of non-contradiction, with the semantic opposition of the two options being amplified by the syntactic reversal. With 80–83 the moral dimension arises. Example 80 is among the most famous and widely propagated antimetaboles, especially among Americans, rivalling 32. One of Fahnestock's great contributions is her observation that figural patterns often "epitomize lines of argument" [34, p. xii; see especially, pp. 3–44], and the Kennedy instance is as clear an example of this as one might hope. It is an exhortation to reject an ethos of entitlement and embrace an ethos of duty or responsibility, which epitomizes the entire argument of Kennedy's inaugural address. Mark Zuckerberg completely rejects the suggestion that Facebook's associated applications and widgets are in any way financially motivated (81); rather, Facebook only raises money to provide those (allegedly needed and beneficial) services. The maxim of 82 was historically one of the most robust and widely cited examples of chiasmus, purporting to utterly oppose one way of living (hedonism) with the morally upright one (Stoicism), and 83 brings the notion of obligation ("must be") to understanding metaphor against 'literal' language. What all of these examples have in common are two opposite 'action chains' signalled by transitive verb mesodiploses, but one chain is negated with the effect that it is utterly discredited and rejected and its converse is promoted as the only possibility.

10 Conclusion

There isn't any symbolism in the book, Ernest Hemingway once wrote of his novella, *The Old Man and the Sea*, which is famous as a kind of parable for nobility in the face of declining faculties in an indifferent world. "The sea is the sea", he wrote, "The old man is an old man. The boy is a boy and the fish is a fish. The sharks are all sharks". He also adds, just so the point is not overlooked, "All the symbolism that people say is shit" [57, p. 780]. But, of course, "the sea" is many things, most of them thick with cultural resonances, so are old men, boys, fish, and sharks. Hemingway makes the naive assumption here that literary symbolism, which is really just a subset of meaning more broadly, is 'put into' language by authors. Meaning is interactional, with speakers/authors and hearers/readers both building it on an utterance-by-utterance basis with resources they share as a socially entrenched 'code'. So, the sea is not 'just' some singular concept of a large body of salty water. The old man is not just another random [+old, +male, +human] entity that happens to go fishing one day.

[22] Fahnestock identifies at least some Subclassification AAMPs (not her terminology) as also "corrective", with examples like "All pro-lifers are conservative, but not all conservatives are pro-lifers", though these do not seem to have the moral force of the Reject-Replace cxx. They seem rather to be simple factual claims.

What, then, is Hemingway communicating with these epanaleptic tautologies? What function is he contributing to the meaning interaction by deploying this construction? Just what we have seen. He is selecting some specific features of "the sea", "old man", "boy", etc., and deflecting others. By his "no symbolism" framing, he emphasizes the brute materiality of the sea, the individuality of his characters, the simple biology of the fish and the sharks.[23] What this reveals is that, even in the denial of rhetorical function, Hemingway deploys the functions of this figural construction.

This article has scratched the surface of the claim that rhetorical figures link linguistic patterns with semiotic functions in systematic ways that implicate grammatical features, illustrating it with examples from the figure epanalepsis and two related figural collocations, Antimetabole-Mesodiplosis-Parison (AMP) and that same grouping with Antithesis (AAMP). But this scratch reveals the promise beneath, warranting a research programme that is highly suggestive for NLU and advanced text mining practices, like argument mining. This article has also outlined how such a research programme would need to proceed; in particular, by merging with Construction Grammar. It is not the rhetorical figure, epanalepsis, that conveys the rhetorical functions we have explored. It is the combination of that figural pattern with specific grammatical features; i.e., that the repeated term is first a Subject and then a Complement with copular predication. Similarly, it is not the figural collation of antimetabole with mesodiplosis and parison (including variations that might involve prozeugma and/or synonymia) that conveys the rhetorical functions we discussed, but that collocation with specific grammatical features; in particular, the functions are notably different if the mesodiplosis is copular or transitive, but prepositions play a role as well. The further collocation with antithesis changes the game again, but in systematic ways.

It is also worth noting that rhetorical functions cannot always be fully isolated. There is a sense of Comprehensiveness, for instance, with both the Irrelevance of Order/Rank and Reciprocal Specification AMP constructions; in the first case because two inverse orders suggests that all possibilities are exhausted ($x < y$ and $y < x$), and in the second case because bi-implication suggests that each term comprehensively specifies the other. But in each case, a different rhetorical function is foregrounded, significantly eclipsing Comprehensiveness. This situation is not unlike lexical functioning in tropic extensions. The term *mumble*, for instance, denotes talking in a low, mostly inaudible way, and a background component of that meaning comes from onomatopoeic association of the nasal and mid vowel sounds that blend

[23] By the way, the person to whom he wrote this passage (it appears in a 13 September 1952 letter) was Bernard Berenson, a literary critic who realized a bit more about how language works than Hemingway seems to. He supplied the following lines as promotional copy for *The Old Man and The Sea*: "No real artist symbolizes or allegorizes–Hemingway is a real artist–but every real work of art exhales symbols and allegories. So does this short but not small master-piece" [108, p. 518]. The critic manages to make both Hemingway ("real artist", "master-piece") and the audience ("symbols and allegories") happy here, but more importantly he's saying it doesn't matter what Hemingway imagines he did or did not put into the novella, non-material meanings certainly factor into the communicative event of reading it.

together in a low register and the embodied iconicity of lips barely parting for a predominantly bilabial articulation, but it is not fully onomatopoeic in the way that *meow* or *beep-beep* or *cock-a-doodle-doo* is onomatopoeic. The term *run* for executing a computer programme denotes specific electromechanical actions but retains the notion of 'rapid energy expenditure' from its metaphorical extension of animal locomotion. Meaning is layered rather than encapsulated. With AMP cxx especially, the closed-loop iconicity, which evokes Comprehensiveness, seems always to be one of the layers.

And, just like lexical semantics, in order to understand more fully what the form/function correspondences are when rhetorical patterns and grammatical features amalgamate into constructions, the appropriate research programme will need to rely heavily on corpus methods. This article has argued for the importance of that research programme and sketched some of the ways it can profitably proceed.

Acknowledgements This article is based on my keynote address to LACompLing 2021. I would like to thank the organizers for that opportunity and the participants for their generous discussion of these ideas, singling out in particular Roussanka Loukanova and Larry Moss. I also thank Christine Grausso for discussion and commentary on an earlier draft of this article, Ritika Puri for a highly conscientious edit; and, most especially, Ramona Kühn, Claus Strommer, Katherine Tu, and Yetian Wang also for comments on earlier drafts of this article but more so for rich discussion of these ideas as they have developed over the years (and also for riding shotgun on my LaTeX adventure, with major extra shotgun props for Yetian); as well as the thoughtful and helpful but anonymous reviewers for this volume.

References

1. Alliheedi, M., Di Marco, C.: Rhetorical figuration as a metric in text summarization. In: Proceedings, Canadian Artificial Intelligence Conference (2014)
2. Aristotle: On Rhetoric: A Theory of Civic Discourse. Oxford University Press (1991/c350 BCE)
3. Austin, J.L.: How to Do Things with Words. William James Lectures, 2nd edn. (1955). Clarendon Press (1975/1962)
4. Bacheldor, B.: When people (and i hear pundits say this) that Trump is just being Trump ... [Tweet later deleted] (2017)
5. Baldrick, C.: The Concise Oxford Dictionary of Literary Terms. Oxford University Press (1990)
6. Bonner, G., Gordon, A.: Happy Together (1967)
7. Boole, G.: An Investigation of the Laws of Thought: On Which Are Founded the Mathematical Theories of Logic and Probabilities. Walton and Maberly (1854)
8. Brunt, S.: Gretzky's Tears: Hockey,Canada, and the Day Everything Changed. Knopf (2009)
9. Buck, G.: Metaphor: A Study in the Psychology of Rhetoric. Inland Press (1899)
10. Bukrate: Paul Harvey quotes (n.d.). https://bukrate.com/author/paul-harvey-quotes
11. Burton, G.O.: Silva Rhetorica (2022). http://rhetoric.byu.edu/. Accessed 02 Jan 2023
12. Bush, G.W., Frum, D.: Address before a joint session of the congress on the state of the union. https://www.presidency.ucsb.edu/documents/address-before-joint-session-the-congress-the-state-the-union-23
13. Chandler, C.: There's nothing better in life than diamonds. The Guardian (2007). https://www.theguardian.com/theguardian/2007/sep/21/greatinterviews. Accessed 02 Jan 2023

14. Chaucer, G.: The Riverside Chaucer. Oxford University Press (1986/c1380)
15. Chesterton, G.K.: What's Wrong with the World. Dodd, Mead (1910)
16. Chomsky, A.N.: Syntactic Structures. Mouton (1957)
17. Chomsky, A.N.: Aspects of the Theory of Syntax. The MIT Press (1965)
18. Chomsky, A.N.: Beyond fascism [Noam Chomsky Interviewed by John Holder and Doug Morris]. ZNet (2013). https://chomsky.info/20130524/. Accessed 02 Jan 2023
19. Christiansen, N.: Figuring Style: The Legacy of Renaissance Rhetoric. Studies in Rhetoric/Communication. The University of South Carolina Press (2013)
20. Churchill, W.: The River War: An Account of the Reconquest of the Sudan. Dover (2006)
21. Clinton, H.: LBGT rights are human rights (transcript-audio-video). https://www.americanrhetoric.com/speeches/hillaryclintonintllbgthumanrights.htm
22. Daou, P.: Harsh realities are harsh realities. Trump and the GOP ... [tweet later deleted.] (2016). https://twitter.com/peterdaou/status/809854708844232704. Accessed 02 Jan 2023
23. David, L., Seinfeld, J., Goldman, M.: The robbery. Seinfeld (1990). https://www.seinfeldscripts.com/TheRobbery.htm
24. Dowd, M.: Fall of the house of Bush. The New York Times (2015). https://www.nytimes.com/2015/11/01/opinion/sunday/fall-of-the-house-of-bush.html. Accessed 02 Jan 2023
25. Dubremetz, M.: Detecting rhetorical figures based on repetition of words: chiasmus, epanaphora, epiphora. Ph.D. thesis, Uppsala University, Uppsala Sweden (2017). https://www.kalendarium.uu.se/Evenemang?eventId=31460
26. Dubremetz, M., Nivre, J.: Rhetorical figure detection: The case of chiasmus. In: NAACL-HLT Fourth Workshop on Computational Linguistics for Literature, pp. 23–31. Curran Associations (2015). https://doi.org/10.3115/v1/W15-0703
27. Dubremetz, M., Nivre, J.: Syntax matters for rhetorical structure: The case of chiasmus. In: Proceedings of the Fifth Workshop on Computational Linguistics for Literature, NAACL-HLT, pp. 47–53 (2016). https://doi.org/10.18653/v1/W16-0206, https://aclanthology.org/W16-0206
28. Dubremetz, M., Nivre, J.: Rhetorical figure detection: chiasmus, epanaphora, epiphora. Front. Digit Humanit. **5**, 10 (2018). https://doi.org/10.3389/fdigh.2018.00010, https://www.frontiersin.org/article/10.3389/fdigh.2018.00010
29. Ducklow, Z.: Greater Victoria builders say they can't find workers to build new homes, because they can't find homes for the workers. Capital Daily (2022). https://www.capitaldaily.ca/news/greater-victoria-construction-labour-shortage. Accessed 02 Jan 2023
30. Dumas, A.: Les Trois Mousquetaires. Dufour & Mulat (1849)
31. Dumas, A.: The Three Muskateers. Penguin Classics (2006/1849)
32. Eliot, T.S.: The Complete Poems and Plays, 1909–1950. Houghton Mifflin Harcourt (1952)
33. Emerson, R.W.: The Prose Works of Ralph Waldo Emerson. Fields, Osgood, & Company (1870)
34. Fahnestock, J.: Rhetorical Figures in Science. Oxford University Press (1999)
35. Farnsworth, W.: Farnsworth's classical english rhetoric. David R. Godine (2011). https://doi.org/10.1017/ccol0521856965
36. Fielding, H.: The History of Tom Jones, a Foundling. J. J. Tourneisen (1796)
37. Fontanier, P.: Les Figures du discours. Flammarion (1827)
38. Franklin, B., Cohen, I.: Benjamin Franklin's Experiments: A New Edition of Franklin's Experiments and Observations on Electricity. Harvard University Press (1941)
39. Gawryjolek, J.: Automated annotation and visualization of rhetorical figures. Master's thesis, University of Waterloo, Waterloo ON Canada (2009). https://uwspace.uwaterloo.ca/handle/10012/4426
40. Gawryjolek, J., Di Marco, C., Harris, R.A.: Automated annotation and visualization of rhetorical figures. In: 9th International Workshop on Computational Models of Natural Argument, International Joint Conference on Artificial Intelligence (2009)
41. Grausso, C.M.: Chiasmus: a phenomenon of language, body and perception. Ph.D. thesis, University of Edinburgh, Edinburgh UK (2020). https://era.ed.ac.uk/handle/1842/37291

42. Green, N.L.: Recognizing rhetoric in science policy arguments. Argum. & Comput. **11**(3), 257–268 (2020). https://doi.org/10.3233/AAC-200504, https://content.iospress.com/articles/argument-and-computation/aac200504
43. Green, N.L.: Some argumentative uses of the rhetorical figure of antithesis in environmental science policy articles. In: Proceedings of the Workshop on Computational Models of Natural Argument, pp. 85–90. CEUR-WS.org (2021)
44. Green, N.L.: The use of antithesis and other contrastive relations in argumentation. Argum. & Comput. 1–16 (2022). https://content.iospress.com/articles/argument-and-computation/aac210025
45. Green, N.L., Crotts, L.J.: Towards automatic detection of antithesis. In: Computational Models of Natural Argument, pp. 69–73 (2020)
46. Greer, G.: The proper study of womankind. TLS (1988-06-03)
47. Grothe, M.: Never Let a Fool Kiss You or a Kiss Fool You. Random House (1999)
48. Haiman, J.: The iconicity of grammar. Language (Baltimore) **56**(3), 515–540 (1980)
49. Harris, R.A.: Ploke. Metaphor. & Symb. **35**(1), 23–42 (2020). https://doi.org/10.1080/10926488.2020.1712781
50. Harris, R.A.: Chiastic iconicity. In: Fischer, O., Ljungberg, C., Tabakowska, E., Lenninger, S. (eds.) Iconicity in Cognition and Across Semiotic Systems. John Benjamins (2023)
51. Harris, R.A.: The Linguistics Wars: Chomsky Lakoff, and the Battle over Deep Structure, second edn. Oxford University Press (2022)
52. Harris, R.A., Di Marco, C.: Constructing a rhetorical figuration ontology. In: Symposium on Persuasive Technology and Digital Behavior Intervention, Convention of the Society for the Study of Artificial Intelligence and Simulation of Behaviour (AISB), pp. 47–52 (2009)
53. Harris, R.A., Di Marco, C.: Rhetorical figures, arguments, computation. Argum. & Comput. **8**(3), 211–231 (2017). https://doi.org/10.3233/AAC-170030
54. Harris, R.A., Di Marco, C., Ruan, S., O'Reilly, C.: An annotation scheme for rhetorical figures. Argum. & Comput. (2018). https://doi.org/10.3233/AAC-180037, https://content.iospress.com/articles/argument-and-computation/aac037
55. Harris, Z.S.: Discourse analysis. Language **28**(1), 1–30 (1952)
56. Hemingway, E.: For Whom the Bell Tolls. Charles Scribner's Sons (1968/1940)
57. Hemingway, E., Baker, C.: Selected Letters, 1917-1961. Charles Scribner's Sons (1981)
58. Hoffman, T., Trousdale, G. (eds.): The Oxford Handbook of Construction Grammar. Oxford University Press (2013). https://doi.org/10.1093/oxfordhb/9780195396683.013.0008
59. Horvei, H.: The Changing Fortunes of a Rhetorical Term: The History of the Chiasmus. Self published (1985)
60. Hromada, D.: Initial experiments with multilingual extraction of rhetoric figures by means of PERL-compatible regular expressions. In: Proceedings of the Second Student Research Workshop Associated with RANLP, pp. 85–90. Incoma Ltd (2011). https://aclanthology.org/R11-2013/
61. Huang, N.S.: Benjamin Franklin in American thought and culture, 1790–1990. Am. Philos. Soc. (1994). https://doi.org/10.2307/2082233
62. Hunt, D.: Race in America, Sunday Edition CBC Radio (2014)
63. Jackendoff, R.: Gapping and related rules. Linguist. Inq. **2**(1), 21–35 (1971)
64. Jacoff, R.: Introduction to Paradiso. In: Jacoff, R. (ed.) The Cambridge Companion to Dante, second edn. Cambridge University Press (2007). https://doi.org/10.1017/CCOL0521844304.007
65. Java, J.: Characterization of prose by rhetorical structure for machine learning classification. Ph.D. thesis, Nova Southeastern University College of Engineering and Computing, Fort Lauderdale, FL, USA (2015). https://nsuworks.nova.edu/gscis_etd/347/
66. Kara-Yakoubian, M., et al.: Beauty and truth, truth and beauty: chiastic structure increases the subjective accuracy advance of online statements (online version cited, ahead of print). Can. J. Exp. Psychol./Revue canadienne de psychologie experimentale (2022). https://doi.org/10.1037/cep0000277
67. Karl, M.: Zur Kritik der politischen Oekonomie. F. Duncker (1859)

68. Karl, M.: A Contribution to the Critique of Political Economy. Progress Publishers (1977)
69. Kasparov, G., with Greengard, M.: Deep Thinking: Where Machine Intelligence Ends and Human Creativity Begins. PublicAffairs (2017)
70. Kemmick, E.: Prairie lights: we shouldn't forget what Trump doesn't know. Missoula Current (2017). https://missoulacurrent.com/opinion/2017/05/montana-trump-history-kemmick/?print=print. Accessed 02 Jan 2023
71. Kennedy, J.F., Sorensen, T.: Inaugural address (1961). https://www.presidency.ucsb.edu/documents/inaugural-address-2. Accessed 02 Jan 2023
72. Kurzweil, A.: Whipping boy. The New Yorker (2014). https://www.newyorker.com/magazine/2014/11/17/whipping-boy. Accessed 2 Jan 2023
73. Lagutina, N.S., Lagutina, K.V., Boychuk, E.I., Vorontsova, I.A., Paramonov, I.V.: Automated rhythmic device search in literary texts applied to comparing original and translated texts as exemplified by English to Russian translations. Autom. Control Comput. Sci. **54**(7), 697–711 (2020). https://doi.org/10.3103/S0146411620070147
74. Lakoff, G.: Irregularity in Syntax. Holt, Rinehart, & Winston (1970)
75. Langacker, R.W.: Cognitive Grammar: A Basic Introduction. Oxford University Press (2008). https://doi.org/10.1093/acprof:oso/9780195331967.001.0001
76. Lanham, R.A.: A Handlist of Rhetorical Terms. University of California Press (1999)
77. Lausberg, H., Orton, D.E., Anderson, D.: Handbook of Literary Rhetoric. Brill (1998)
78. Lawrence, J., Visser, J., Reed, C.: Harnessing rhetorical figures for argument mining. Argum. & Comput. **8**(3), 289–310 (2017). https://doi.org/10.3233/AAC-170026, https://content.iospress.com/articles/argument-and-computation/aac026
79. Leighton, K.: Obama in Ohio: 'corporations aren't people. People are people.' (2012). https://talkingpointsmemo.com/livewire/obama-in-ohio-corporations-aren-t-people-people-are-people. Accessed 02 Jan 2023
80. Lewontin, R.C.: The Triple Helix: Gene, Organism, and Environment. Harvard University Press (2001)
81. Lissner, P.A.: Chi-thinking: chiasmus and cognition. Ph.D. thesis, University of Maryland, College Park, MD, USA (2007). https://drum.lib.umd.edu/handle/1903/7687
82. Mack, P.: A History of Renaissance Rhetoric 1380-1620. Oxford University Press (2011)
83. Magid, L.: Zuckerburg claims "we don't build services to make money". Forbes (2012). https://www.forbes.com/sites/larrymagid/2012/02/01/zuckerberg-claims-we-dont-build-services-to-make-money/?sh=77ae8e692b11. Accessed 02 Jan 2023
84. Malcolm, X., with Haley, A.: The Autobiography of Malcolm X. Random House (2015)
85. Mandela, N.: The Long Walk to Freedom. Little, Brown and Company (1994)
86. Melanchthon, P.: The Dialectical Questions. Brill (2021/1547)
87. Mills, M.S.: Concise Handbook of Literary and Rhetorical Terms. Estep-Nichols Publishing (2010)
88. Morrison, S.: Mr Djokovic's visa has been cancelled. ... (2022). https://twitter.com/ScottMorrisonMP/status/1478848008363991049. Accessed 02 Jan 2023
89. Morton, A.: Not on My Watch. Random House (2021)
90. Murphy, F.: The historical perspective: what constitutes discovery (of a new virus). In: Kielian, M., Maramorosch, K., Mettenleiter, T. (eds.) Advances in Virus Research, pp. 197–220. Academic Press (2016/1953)
91. Neidlein, A., Wiesenbach, P., Markert, K.: An analysis of language models for metaphor recognition. In: Proceedings of the 28th International Conference on Computational Linguistics, pp. 3722–3736. Association for Computational Linguistics, Barcelona, Spain (Online) (2020). https://doi.org/10.18653/v1/2020.coling-main.332, https://aclanthology.org/2020.coling-main.332
92. Newton, S.I.: Philosophiæ Naturalis Principia Mathematica. Samuel Pepys (1687)
93. Newton, S.I.: The Mathematical Principles of Natural Philosophy (In Three Volumes), vol. 1. Printed for H.D. Symonds (1803/1687)
94. Nietzsche, F.: Jenseits von Gut und Böse: Vorspiel einer Philosophie der Zukunft. C. G. Naumann (1886)

95. Nietzsche, F.: Beyond Good & Evil: Prelude to a Philosophy of the Future. Knopf Doubleday (2010/1886)
96. O'Donnell, S.(Writer), Moore, S.D.(Director): The Joy of Sect, The Simpsons [Episode 191] (1998)
97. O'Reilly, C., Paurobally, S.: Lassoing rhetoric with OWL and SWRL. Master's thesis, Westminister University, London, UK (2010). https://www.academia.edu/2095469/Lassoing_Rhetoric_with_OWL_and_SWRL
98. Orwell, G.: The Road to Wigan Pier. Victor Gollancz (1937)
99. Palmer, S.: A response to Margaret Wente, or why congress is good for Canada (2015). http://canuckhm.ca/a-response-to-margaret-wente-or-why-congress-is-good/
100. Pasteur, L., Vallery-Radot, P.: Fermentations et Generations dites Spontanees. Oeuvres de Pasteur, vol. II. Masson (1922)
101. Peacham, H.: The Garden of Eloquence. Scholar's Facsimiles (1954/1593)
102. Peirce, C.: Collected Papers of Charles Sanders Peirce. The Belknap Press of Harvard University Press (1931)
103. Perelman, C., Olbrecht-Tyteca, O.: The New Rhetoric : A Treatise on Argumentation. University of Notre Dame Press (2008/1969)
104. Princeton University: About wordnet. (2010). https://wordnet.princeton.edu/
105. Quintillian, M.F.: The Orator's Education. Harvard University Press (2002)
106. Reagan, M.: Hillary lost to Trump because she was Hillary. Period (2017). https://twitter.com/ReaganWorld/status/820499615443079169. Accessed 02 Jan 2023
107. Ruffin, J.D.: Forms of Oratorical Expression and their Delivery: or, Logic and Eloquence Illustrated. Simkin, Marshall, Hamilton, Kent & Co. (1920)
108. Samuels, E., Samuels, J.: Bernard Berenson, the Making of a Legend. Harvard University Press (1987)
109. Schneider, F., Barz, B., Brandes, P., Marshall, S., Denzler, J.: Data-driven detection of general chiasmi using lexical and semantic features. In: Proceedings of the 5th Joint SIGHUM Workshop on Computational Linguistics for Cultural Heritage, Social Sciences, Humanities and Literature, pp. 96–100. Association for Computational Linguistics, Punta Cana, Dominican Republic (online) (2021). https://doi.org/10.18653/v1/2021.latechclfl-1.11, https://aclanthology.org/2021.latechclfl-1.11
110. Seuss, D.: Horton Hatches the Egg. Random House (1940)
111. Shakespeare, W.: The Complete Works of William Shakespeare. http://shakespeare.mit.edu/
112. Shutova, E., Sun, L., Gutiérrez, E.D., Lichtenstein, P., Narayanan, S.: Multilingual metaphor processing: experiments with semi-supervised and unsupervised learning. Comput. Linguist. **43**(1), 71–123 (2017). https://doi.org/10.1162/COLI_a_00275, https://aclanthology.org/J17-1003
113. Staff: More Tory MPs call for PM to go as No. 10 tries to limit Partygate report fallout. The Guardian (2022). https://www.theguardian.com/politics/2022/may/26/more-tory-mps-call-for-boris-johnson-to-go-as-no-10-tries-to-limit-sue-gray-partygate-report-fallout. Accessed 02 Jan 2023
114. Strommer, C.: Using rhetorical figures and shallow attributes as a metric of intent in text. Ph.D. thesis, University of Waterloo David Cheriton School of Computing, Waterloo ON Canada (2011). https://uwspace.uwaterloo.ca/handle/10012/5972
115. Studdert-Kennedy, M., Terrace, H.: To the editors: in response to "at the birth of language". The New York Review of Books (2016). https://www.nybooks.com/articles/2016/09/29/berwick-chomsky-birth-language/. Accessed 02 Jan 2023
116. Tindale, C.W.: Rhetorical Argumentation: Principles of Theory and Practice. Sage Publications (2004). https://doi.org/10.4135/9781452204482
117. Trudeau, J.: We beat fear with hope [transcript]. MacLean's (2015). https://www.macleans.ca/politics/ottawa/justin-trudeau-for-the-record-we-beat-fear-with-hope/. Accessed 02 Jan 2023
118. Tu, K.: Collocation in rhetorical figures: a case study in parison, epanaphora and homoioptoton. Major research project, University of Waterloo, Waterloo ON Canada (2019). http://uwspace.uwaterloo.ca/handle/10012/14778

119. Tymoczko, T.: Computers, proofs and mathematicians: a philosophical investigation of the four-color proof. Math. Mag. **53**(3), 131–138 (1980). https://doi.org/10.1080/0025570X.1980.11976844
120. Van Eemeren, F.H., Houtlosser, P.: Strategic maneuvering: a synthetic recapitulation. Argumentation **20**, 381–392 (2006)
121. Veale, T., Shutova, E., Klebanov, B.B.: Metaphor: a computational perspective. Synth. Lect. Hum. Lang. Technol. **9**(1), 1–160 (2016). https://direct.mit.edu/coli/article/44/1/191/1586/Metaphor-A-Computational-Perspective
122. Volpe, P.E.: Man, Nature, and Society: An Introduction to Biology. W.C Brown Company (1975)
123. Wilson, E.: The Meaning of Human Existence. Liveright Publishing Corporation (2014)
124. Yadav, M.: Ultrabooks vs laptops (2013). http://java-maheshyadav.blogspot.com/2013/01/ultrabooks-vs-laptops.html. Accessed 02 Jan 2013
125. Yang, N.: "Rules are rules": figure skating community argues Kamila Valieva should not be allowed to compete. The Boston Globe (2022). https://www.bostonglobe.com/2022/02/15/sports/rules-are-rules-figure-skating-community-argues-kamila-valieva-should-not-be-allowed-compete/. Accessed 02 Jan 2023

Integrating Deep Neural Networks with Dependent Type Semantics

Daisuke Bekki, Ribeka Tanaka, and Yuta Takahashi

Abstract Recent studies in computational semantics have explored how the advantages of both type-logical semantics and deep neural networks can be combined to develop a theory or system of natural language understanding that can perform complex inferences while being learnable. In this study, we propose a theory of language understanding that fuses neural language models and dependent type semantics (DTS), a proof-theoretic semantics of natural language based on dependent type theory, by replacing names and predicates in DTS with tensor-based distributional representations and neural classifiers over them. Under this unified view, interesting correspondences are found between proof-theoretic and machine learning notions. For instance, the value of the loss function provides proof for an atomic predicate, and the neural parameters are regarded as proof-theoretic assumptions about the real world.

Keywords Dependent type semantics · Deep neural network · Symbolic and soft reasoning

1 Introduction

1.1 Symbolic and Soft Reasoning

When classifying and comparing the inference systems in computational semantics, reasoning with type-logical semantics, or logic in general, is often described as *symbolic*, while reasoning with deep neural networks is described as *soft*. This

---

D. Bekki (✉) · R. Tanaka · Y. Takahashi
Ochanomizu University, Tokyo, Japan
e-mail: bekki@is.ocha.ac.jp

R. Tanaka
e-mail: tanaka.ribeka@is.ocha.ac.jp

Y. Takahashi
e-mail: takahashi.yuta@is.ocha.ac.jp

R. Loukanova et al. (eds.), *Logic and Algorithms in Computational Linguistics 2021 (LACompLing2021)*, Studies in Computational Intelligence 1081,
https://doi.org/10.1007/978-3-031-21780-7_11

dichotomy reflects their complementary advantages: symbolic reasoning can represent complex linguistic phenomena, such as negation, quantification, and anaphora/presupposition, regardless of how deep they get embedded in syntactic structures, while soft reasoning can perform tasks such as similarity computation, classification, sentence generation, and translation, regardless of the size and diversity of the real texts.

With the prospect that these two types of reasoning complement with each other, it is natural to seek a way to fuse them. This has been explored in various ways in multiple fields, including, at least, natural language processing, linguistics, and functional programming languages. Previous studies on the fusion of symbolic and soft reasoning have taken one of the following approaches[1]:

1. Emulating symbolic reasoning by embedding knowledge graphs [16, 19, 34], SAT problems [33], first-order logic [17, 37].
2. Introducing a similarity measure between symbols using distributional representations instead of symbols [24].
3. Using neural networks to control the direction of the proof search [32, 38].
4. Embedding neural networks in symbolic reasoning [12, 23].
5. Choosing between symbolic and soft reasoning module depending on the characteristics of problems [22].

In this paper, we pursue the fourth approach. While the work of Cooper [12] and Larsson [23] is an attempt to extend their model-theoretic semantics based on a type system called TTR [11], we adopt Dependent Type Semantics (DTS) [6, 7] as a framework of *proof theoretic* semantics based on Martin-Löf type theory (MLTT) [27], and *implement* neural networks therein. In other words, we will construct a component that performs soft reasoning within the framework of symbolic reasoning by using MLTT.

Since the fourth approach has not yet been fully investigated, the question remains as to whether the fourth approach, namely, embedding neural networks in a type system, qualifies as a fusion of symbolic and soft reasoning in the first place. Therefore, before describing how to materialize such an approach (which will be done in Sect. 2), it would be helpful to clarify the criteria for a reasoning system to be considered soft reasoning, as well as the advantageous aspects of soft reasoning.

1.2 Criteria for a Soft Reasoning System

Generally, a soft reasoning system is one whose representation possesses the following properties.

Continuous truth values: The truth value (or its counterpart) of a proposition (under a given interpretation) must be defined as a real number, not as an ele-

[1] See Sect. 5 for a detailed comparison of the different approaches.

ment of a discrete set such as $\{1, 0\}$, thus allowing inferences under uncertain premises, with shaky credibility.

Comparable representations: Words, phrases, and sentences must be represented so that their similarity is defined using real numbers: the more similar two expressions are, the larger the value will be. This provides a natural way to perform tasks centered around text classification, question and answering, and translation.

Learnable representations: Representations must be parameterized and learnable, which is often achieved when they give rise to, for any input, differentiable functions from the parameters to the losses. This allows a reasoning system to learn from and inductively fit to data.

Whether these criteria are sufficient conditions for an inference system to be considered a soft reasoning system is left open here, but if we take them to be necessary conditions, they are useful in the following sense: if the system that we construct (as an extension of DTS) satisfies the three criteria above, then it is a symbolic reasoning system that has the advantages of a soft reasoning system.

Let us first confirm that the current version of DTS [7] satisfies none of the criteria above and why this is so.

1.3 Names and Predicates in DTS

DTS uses names and n-place predicates for its representations. An n-place predicate is a term of type $\mathbf{entity}^n \to \mathsf{type}$. The **entity** type is a type for finite entities, which is defined as an enumeration type[2] and plays the role of *the domain of entities* in model-theoretic semantics.

The behavior of the enumeration types, following MLTT, is specified by the following set of three rules: the formation rule $(\{\}F)$, the introduction rule $(\{\}I)$, and the elimination rule $(\{\}E)$ (together with their accompanying β-equality rules).

$$\frac{}{\{a_1, \ldots, a_n\} : \mathsf{type}}\ (\{\}F) \qquad \frac{}{a_i : \{a_1, \ldots, a_n\}}\ (\{\}I)$$

$$\frac{M : \{a_1, \ldots, a_n\} \quad P : \{a_1, \ldots, a_n\} \to \mathsf{type} \quad N_1 : P(a_1) \quad \ldots \quad N_n : P(a_n)}{\mathsf{case}_M^P(N_1, \ldots, N_n) : P(M)}\ (\{\}E)$$

According to the $(\{\}I)$ rule, a term of type **entity** is one of $a_1, \ldots, a_n$. Now, suppose that $\mathsf{john} \equiv a_i$ and **dog** is a unary predicate of type $\mathbf{entity} \to \mathsf{type}$. Then the meaning of the sentence *john is a dog* is represented as the type $\mathbf{dog}(\mathsf{john})$. Via Curry-Howard correspondence, types are propositions. Therefore, $\mathbf{dog}(\mathsf{john})$ is a proposition, which is *true* if and only if a proof inhabits the type $\mathbf{dog}(\mathsf{john})$, namely, that there is a proof for John being a dog. This conception of truth in DTS follows that of intuitionistic traditions: in these traditions, truth is the inhabitance of a proof

[2] In the notation of the enumeration type, the curly braces $\{\}$ are used following the notation of Martin-Löf, but note that the enumeration type is not a set of constructors.

that the proposition holds, and the truth-value of a proposition A is determined by the existence of a proof for A or a proof of contradiction from the assumption that A holds. Since proofs are discrete objects, the conception of truth in DTS does not satisfy the property *Continuous truth values* explained above.

Moreover, neither the predicate **dog** nor the name john are comparable representations. In the real world, (at least some aspects of) puppies are similar to (some aspects of) dogs. In DTS (and in most symbolic logics), however, the predicates **dog** and **puppy** are two different constant symbols, and their similarity can be captured only by the entities that satisfy them. Types in DTS are also not learnable.

Thus, we confirm that the current version of DTS does not meet any of the criteria for a soft reasoning system, and neither does any other system that is based on type-logical semantics. Now, it is worth asking whether—and if so, how—the non-soft behaviors of names and predicates in DTS, or type-logical semantics in general, are intrinsic to symbolic reasoning. In other words, are the criteria listed in Sect. 1.2 incompatible with symbolic reasoning in the first place?

1.4 Are Symbolic and Soft Reasoning Incompatible?

We can immediately see that continuous truth values exist in some logical systems such as fuzzy logic so they must not be inconsistent with symbolic reasoning. How about comparable and learnable representations?

In deep neural networks, distributional representations are employed in a metric linear space. Thus the representations are comparable because the distance between them is defined via their inner product. Representations become learnable when the loss function under a given input and ground truth is a differential function from a metric linear space to a real line.

Turning to DTS, we would like to emphasize that there is no reason for their representations not to be comparable and learnable, contrary to what is widely perceived in the field. In most type-logical semantics, symbols are defined as elements of discrete sets rather than as elements in a metric linear space. We should, however, be aware that this is not a requirement that type-logical semantics must fulfill.[3] In other words, there is no intrinsic reason for why symbolic reasoning cannot also be soft. No problem arises when, for instance, names and predicates in DTS or type-logical semantics are replaced by distributional representations.

Of course, in the realm of quantification, the essential requirements are that different elements must be distinguishable from each other and that countable elements are, indeed, able to be counted, which is not satisfied when only distributional representations are used.

---

[3] What constitutes type-logical semantics, or logic in general, is controversial. For example, some answers might include having harmony between introduction and elimination rules of logical operators when they are displayed in terms of natural deduction, adjunction between conjunction and implication, or adjunction between universal quantification, substitution, and existential quantification. Note that none of these requires that symbols be defined as elements in a discrete set.

Therefore, in our integration of DTS and deep neural networks, we should be able to distinguish representations for different entities and compute the similarity between them at the same time, a seemingly impossible task. However, this can be achieved within the framework of DTS by using the following steps to redefine names and predicates.

2 Neural DTS

We claim that we can implement these concepts, continuous truth values, and comparable and learnable representations within the proof-theoretic setting of DTS.

We are going to replace the names and predicates of DTS with tensor-based distributional representations and neural classifiers over them: a classifier for n-place predicates takes as input an n-tuple of terms of enumeration type and embeds them in a metric linear space, and a neural network is used to obtain the probability distribution for the extent to which a given entity belongs to the class in question. If we consider the difference between the output probability distribution and the distribution that assigns the maximum probability for the class in question, we can simultaneously obtain the proof term for a predicate and the loss function for training the neural network.

In the following sections, we will describe the procedure for achieving soft reasoning in DTS.

2.1 Real Numbers and Quotient Types

In standard set theory such as ZF/ZFC, the quotient set is generally used to construct the real numbers: the set of real numbers is constructed from the set of Cauchy sequences by the equivalence relation on Cauchy sequences. On the other hand, MLTT, which DTS is based on, is not equipped with quotient sets, thus how to express the mathematical concepts that are realized by using quotient sets has been a challenging issue for MLTT. Note that this remains an issue for the Calculus of Inductive Constructions (CIC), and that proof assistants, such as Agda [35], Coq [36], and LEGO [26], inherit this issue from MLTT and CIC.

There are two major approaches for resolving this issue: one is to introduce quotient types and the other is to use *setoids*. The introduction of the quotient type, which is the proof-theoretic counterpart of the quotient set, has been studied [20, 21].

To show that Neural DTS can be constructed from DTS based on the standard intensional MLTT, we will take a setoid approach rather than adopting a new type, such as the quotient type. One reason for choosing this approach is that, to the best of our knowledge, incorporating quotient types into the standard intensional MLTT introduces non-canonical elements (see, for example, Hofmann [20]). On the other hand, the setoid approach preserves the canonicity of the standard intensional MLTT.

Roughly speaking, a type system is canonical if any closed proof term in the system is reducible to a canonical form, and a proof term is in a canonical form if the last rule in the construction of the term is an introduction rule. Since the setoid approach preserves canonicity, it also preserves the properties which follow from canonicity, such as consistency.

2.2 Setoids in DTS

Setoids, an alternative to sets in ZF/ZFC, originated with the notion of sets in Bishop's constructive mathematics [8], and since then, various definitions of setoids have been proposed [2–5, 15]. The Coq Repository at Nijmegen (CoRN), which is a mathematical library of Coq, adopted setoids to formalize analysis, and setoids have become a standard tool for developing mathematics in Coq and Agda that involve real numbers. The sections that follow give an overview of how the real numbers are constructed in DTS using setoids, which is mainly based on Palmgren's formulation of total setoids [28, 29].

Definition 1 A *setoid* is a pair $(\underline{X}, \sim_X)$ consisting of a type $\underline{X}$ and an equivalence relation $\sim_X$ on $\underline{X}$.[4]

Infix notation is often used for binary relations. For the sake of simplicity, we represent the proposition that the binary relation $\sim_X$ is an equivalence relation as $equiv(\sim_X)$ (Note that $\sim_X$ is not a binary relation on X but on $\underline{X}$). The exact definition (using Σ-type[5]) is:

$$equiv(\sim_X) \stackrel{def}{\equiv} \begin{bmatrix} (x : \underline{X}) \to (x \sim_X x) \\ (x, y : \underline{X}) \to (x \sim_X y) \to (y \sim_X x) \\ (x, y, z : \underline{X}) \to (x \sim_X y) \times (y \sim_X z) \to (x \sim_X z) \end{bmatrix}$$

X *setoid* indicates that X is a setoid. With these notations, we can rewrite Definition 1 in the form of a formation rule.

---

[4] $\underline{X}$ is called a *preset* in the terminology of Bishop [8] and *carrier* of X in the terminology of Barthe et al. [5].

[5] Following Bekki and Mineshima [7], we adopt DTS-style notations for Π-types and Σ-types, illustrated as follows:

DTS-notation	MLTT-notation
$(x : A) \to B$	$(\Pi x : A)B$
$(x : A) \times B$ or $\begin{bmatrix} x : A \\ B \end{bmatrix}$	$(\Sigma x : A)B$

See Definitions 18 and 19 for the further details.

Definition 2 (*Setoid formation*)

$$\frac{\underline{X} : \mathsf{type} \quad \sim_X : \underline{X} \times \underline{X} \to \mathsf{type} \quad equiv(\sim_X)\ true}{(\underline{X}, \sim_X)\ setoid}$$

Note that, a setoid of the form $(\underline{X}, \sim_X)$ is not a type in the predicative setting of DTS. We also define the binary relation $\in$ between an element and a setoid as follows.

Definition 3 (*Setoid membership*)

$$\frac{(\underline{X}, \sim_X)\ setoid \quad x : \underline{X}}{x \in (\underline{X}, \sim_X)}$$

This definition tells us that x can be an element of several different setoids.

Remark 1 As an example of the simplest kind of setoid, we generate a setoid in which any type A is embedded with its intensional equality. It is obvious that $=_A$ is an equivalence relation.

$$\mathbb{A} \stackrel{def}{\equiv} (A, =_A)$$

Definition 4 (*Setoid function*) A *setoid function* f from a setoid X to a setoid Y is a pair $(\underline{f}, ext_f)$ consisting of a function $\underline{f} : \underline{X} \to \underline{Y}$ and a proof term ext_f that proves the extensionality of $\underline{f}$:

$$ext_f : \left(x, y : \underline{X}\right) \to (x \sim_X y) \to (\underline{f}x \sim_Y \underline{f}y)$$

Definition 5 (*Exponential of setoids*) The exponential of setoids $X \equiv (\underline{X}, \sim_X)$ and $Y \equiv (\underline{Y}, \sim_Y)$ is $(\underline{X \to Y}, \sim_E)$ (notation: $X \to Y$), where $\underline{X \to Y}$ is a type defined as:

$$\underline{X \to Y} \stackrel{def}{\equiv} \left(\underline{f} : \underline{X} \to \underline{Y}\right) \times \left(x, y : \underline{X}\right) \to (x \sim_X y) \to (\underline{f}x \sim_Y \underline{f}y)$$

and $\sim_E$ is a binary relation defined as:

$$(\underline{f}, \_) \sim_E (\underline{g}, \_) \stackrel{def}{\equiv} \left(x : \underline{X}\right) \to \underline{f}x \sim_Y \underline{g}x$$

A function application operator ev is defined for each domain-codomain pair of setoids.

$$\frac{(\underline{f}, \_) \in X \to Y \quad a \in X}{ev_{X,Y}((\underline{f}, \_), a) \stackrel{def}{\equiv} \underline{f}a \in Y}$$

Definition 6 (*Product of setoids*) The product of setoids $X \equiv (\underline{X}, \sim_X)$ and $Y \equiv (\underline{Y}, \sim_Y)$ is the pair $(\underline{X \times Y}, \sim_P)$ (notation: $X \times Y$), where $\underline{X \times Y}$ is a type defined as:

$$\underline{X \times Y} \stackrel{def}{\equiv} \underline{X} \times \underline{Y}$$

and $\sim_P$ is a binary relation defined as:

$$(x, y) \sim_P (u, v) \stackrel{def}{\equiv} (x \sim_X u) \times (y \sim_Y v)$$

Projections work as expected.

$$\frac{p \in X \times Y}{\pi_1(p) \in X} \qquad \frac{p \in X \times Y}{\pi_2(p) \in Y}$$

Definition 7 (*Binary Relation on setoids*) A *binary relation* R between setoids X and Y is a pair $(\underline{R}, ext_R)$ consisting of a binary relation $\underline{R} : (x : \underline{X}) \to (y : \underline{Y}) \to \textsf{type}$ and a proof term ext_R that proves the extensionality of $\underline{R}$:

$$ext_R : \left(x, x' : \underline{X}\right) \to \left(y, y' : \underline{Y}\right) \to (x \sim_X x') \to (y \sim_Y y') \to \underline{R}xy \to \underline{R}x'y'$$

Definition 8 (*Quotient setoids*) Let $X \equiv (\underline{X}, \sim_X)$ be a setoid and $\sim : (x : \underline{X}) \to (y : \underline{X}) \to \textsf{type}$ be a binary relation on $\underline{X}$. If $\sim$ is an equivalence relation on $\underline{X}$, then we define a *quotient setoid* $X/\sim$ as:

$$X/\sim \stackrel{def}{\equiv} (\underline{X}, \sim)$$

with a setoid function $q : X \to X/\sim$ defined by identity function on $\underline{X}$ and its extensionality.

Remark 2 Let $X \equiv (\underline{X}, \sim_X)$ be a setoid, and $X/\sim$ be a quotient setoid. Due to the extensionality of the identity function, the following holds for any $x, y \in X$:

$$x \sim_X y \to x \sim y$$

Thus, the equivalence relation $\sim_X$ is *finer* than $\sim$ on the type $\underline{X}$.

Definition 9 (*Subsetoids*) Let X be a setoid. A *subsetoid* of X is a pair $(\partial S, i_S)$, where ∂S is a setoid and $i_S : \partial S \to X$ is an injective setoid function.

When we define a product of setoids or a quotient setoid by using subsetoids, we ignore their injective setoid functions and consider their presets only. But we write, e.g., $S/\sim$ for a subsetoid S by abuse of notation.

Definition 10 (*Subsetoid membership*) Let X be a setoid. An element $a \in X$ is a *member* of the subsetoid ∂S of X if there exists an element $s : \partial S$ such that $a \sim_X i_S(s)$. Namely,

$$a \in_X (\partial S, i_S) \stackrel{def}{\equiv} (s : \partial S) \times (a \sim_X i_S(s))$$

Note that $a \in_X (\partial S, i_S)$ is a type. If A and B are subsetoids of X, then

$$A \subseteq_X B \overset{def}{\equiv} (x : X) \rightarrow x \in_X A \rightarrow x \in_X B$$

Definition 11 (*Separation of subsets*) Let $X = (\underline{X}, \sim_X)$ be a setoid and A a type. A subsetoid $\left\{ x \in X \middle| A \right\}$ of X is defined as:

$$\left\{ x \in X \middle| A \right\} \overset{def}{\equiv} \left(\left((x : \underline{X}) \times A, \sim_S \right), i \right)$$

$$\text{where } (x, \_) \sim_S (y, \_) \overset{def}{\equiv} x \sim_X y$$
$$i(x, \_) \overset{def}{\equiv} x$$

2.3 Setoids of Natural Numbers, Integers, Rationals, and Reals

Definition 12 The setoid of *natural numbers* $\mathbb{N}$ is obtained by embedding the natural number type N with its intensional equality:

$$\mathbb{N} \overset{def}{\equiv} (N, =_N)$$

Definition 13 The setoid of *integers* $\mathbb{Z}$ is defined as:

$$\mathbb{Z} \overset{def}{\equiv} (\mathbb{N} \times \mathbb{N}) / \sim_{\mathbb{Z}}$$

where $\sim_{\mathbb{Z}}$ is defined as $(m, n) \sim_{\mathbb{Z}} (p, q) \overset{def}{\equiv} m + q =_N p + n$.

Definition 14 The setoid of *rational numbers* $\mathbb{Q}$ is defined as[6]

$$\mathbb{Q} \overset{def}{\equiv} (\mathbb{Z} \times \left\{ z \in \mathbb{Z} \,\middle|\, \neg(z \sim_{\mathbb{Z}} 0_{\mathbb{Z}}) \right\}) / \sim_{\mathbb{Q}}$$

where $\sim_{\mathbb{Q}}$ is defined as $(a, (b, \_)) \sim_{\mathbb{Q}} (c, (d, \_)) \overset{def}{\equiv} a \cdot d =_{\sim_{\mathbb{Z}}} c \cdot b$.

Definition 15 (*Cauchy sequence*) $\mathsf{Cauchy}(\mathit{seq})$ is a proposition that a sequence (a setoid function) $\mathit{seq} \in \mathbb{N} \rightarrow \mathbb{Q}$ is a *Cauchy sequence* and is defined as[7]

---

[6] $0_{\mathbb{Z}}$ is defined as $(0, 0) : \underline{\mathbb{Z}}$.

[7] The function $\mathsf{asRational}^*$ is defined as a compostion:

$$\mathsf{asRational}^* \overset{def}{\equiv} \mathsf{asRational} \circ \mathsf{asInteger} : \underline{\mathbb{N}} \rightarrow \underline{\mathbb{Q}}$$

$$\mathsf{Cauchy}(seq) \stackrel{def}{\equiv} (i : \mathbb{N}) \to (i \neq 0) \to (k : \mathbb{N}) \times (j_1, j_2 : \mathbb{N}) \to (j_1 > k) \\ \to (j_2 > k) \to |(\underline{seq}(j_1)) - (\underline{seq}(j_2))| < \frac{1}{\mathsf{asRational}^*(i)}$$

Definition 16 The setoid of *real numbers* $\mathbb{R}$ is defined as

$$\mathbb{R} \stackrel{def}{\equiv} \left\{ seq \in \mathbb{N} \to \mathbb{Q} \middle| \mathsf{Cauchy}(seq) \right\} / \sim_{\mathbb{R}}$$

where $\sim_{\mathbb{R}}$ is

$$s_1 \sim_{\mathbb{R}} s_2 \stackrel{def}{\equiv} (i : \mathbb{N}) \to (i \neq 0) \to (k : \mathbb{N}) \times (j : \mathbb{N}) \to (j > k) \\ \to |\underline{\pi_1(s_1)}(j) - \underline{\pi_1(s_2)}(j)| < \frac{1}{\mathsf{asRational}^*(i)}$$

The setoids of n-dimensional real vector spaces $\mathbb{R}^n$ are definable as the product setoids, together with the standard distance function:

$$\mathbb{R}^n \stackrel{def}{\equiv} \overbrace{\mathbb{R} \times \cdots \times \mathbb{R}}^{n}$$

We may then prove, within the type theory of DTS, that $\mathbb{R}^n$ are linear spaces. We denote the standard basis for $\mathbb{R}^n$ as $e_1, \ldots, e_n \in \mathbb{R}^n$.

2.4 Embedding of an Entity into a Vector Space

Now, we consider obtaining a distributional representation for each entity. Since the **entity** type is defined as the enumeration type $\{a_1, \ldots, a_n\}$, which consists of n constructors, the *embedding function* $\underline{\mathsf{onehot}}$ for entities is defined by using the enumeration elimination constructor.

(1) $\vdash \underline{\mathsf{onehot}} \stackrel{def}{\equiv} \lambda x.\mathsf{case}_x^{\lambda\_.\mathbb{R}^n}(e_1, \ldots, e_n) : \mathbf{entity} \to \underline{\mathbb{R}^n}$

Recall that, as discussed in Remark 1, any type can be embedded into a setoid. There is an *entity setoid* $\mathbb{ENT} \stackrel{def}{\equiv} (\mathbf{entity}, =_{\mathbf{entity}})$. Once we prove the extensionality of $\underline{\mathsf{onehot}}$, which is a routine exercise, we may lift $\underline{\mathsf{onehot}}$ to a setoid function

where the *casting* functions are defined as follows: put $1_{\mathbb{Z}}$ as $(1, 0) : \underline{\mathbb{Z}}$, and $1'_{\mathbb{Z}}$ as a fixed pair of $1_{\mathbb{Z}}$ and a proof term $\mathsf{peano4}(0) : \neg(1_{\mathbb{Z}} \sim_{\mathbb{Z}} 0_{\mathbb{Z}})$. Then,

$$\mathsf{asInteger} \stackrel{def}{\equiv} \lambda n.(n, 0) : \underline{\mathbb{N}} \to \underline{\mathbb{Z}}$$

$$\mathsf{asRational} \stackrel{def}{\equiv} \lambda z.(z, 1'_{\mathbb{Z}}) : \underline{\mathbb{Z}} \to \underline{\mathbb{Q}}$$

onehot from the entity setoid to the n-dimensional real vector space, as in (2) (Recall Definition 3 for $\in$).

(2) $\vdash (\underline{\mathsf{onehot}}, ext_{\mathsf{onehot}}) \in \mathbb{ENT} \to \mathbb{R}^n$

Then, consider an $n \times m$ matrix $W_{\mathsf{emb\_ent}}$ that is parametrized by $p_{\mathsf{emb\_ent}} \in \mathbb{R}^{n \times m}$. This gives rise to a linear setoid function from $\mathbb{R}^n$ to $\mathbb{R}^m$.

(3) $p_{\mathsf{emb\_ent}} \in \mathbb{R}^{n \times m} \vdash W_{\mathsf{emb\_ent}} \in \mathbb{R}^n \to \mathbb{R}^m$

These functions can be composed as follows.

(4) $p_{\mathsf{emb\_ent}} \in \mathbb{R}^{n \times m} \vdash W_{\mathsf{emb\_ent}} \circ \mathsf{onehot} \in \mathbb{ENT} \to \mathbb{R}^m$

In words, a composite function $W_{\mathsf{emb\_ent}} \circ \mathsf{onehot}$ embeds each entity (a term of type **entity**) into an m-dimensional real vector.[8] Note that this function is represented by using parameters that are elements of the $n \times m$-dimensional real vector space.

2.5 Neural Classifiers as Predicates

Next, we are going to replace predicates in DTS with neural classifiers. For this purpose, we may adopt any architecture for the neural classifiers. We take a linear regression using multilayer perceptron with only one hidden layer as the simplest example, but the point is to demonstrate that we may implement any neural classifiers within DTS by using setoids.

Suppose that the signature of DTS contains k-many one-place predicates (namely, terms of type **entity** $\to$ type). Suppose that we denote them by $P_1, \ldots, P_k$. Let $e_1, \ldots, e_k$ be the standard basis for $\mathbb{R}^k$ and define a $k \times l$ matrix $W_{\mathsf{emb\_pred}}$ that is parametrized by $p_{\mathsf{emb\_pred}} \in \mathbb{R}^{k \times l}$, giving rise to a linear setoid function from $\mathbb{R}^k$ to $\mathbb{R}^l$. Then, we assign each one-place predicate P_j the following vector in $\mathbb{R}^l$.

(5) $p_{\mathsf{emb\_pred}} \in \mathbb{R}^{k \times l} \vdash W_{\mathsf{emb\_pred}}(e_j) \in \mathbb{R}^l$

Now, consider the classifier that is applied to the entity a_i for a given one-place predicate p_j. First, we concatenate their distributional representations by using the concatenation operator $\oplus$.

(6) $p_{\mathsf{emb\_ent}} \in \mathbb{R}^{n \times m}, p_{\mathsf{emb\_pred}} \in \mathbb{R}^{k \times l}$
$\vdash W_{\mathsf{emb\_ent}}(\mathsf{onehot}(a_i)) \oplus W_{\mathsf{emb\_pred}}(e_j) \in \mathbb{R}^{m+l}$

Second, an $(m+l) \times o$ matrix W_{hidden} that is parametrized by $p_{\mathsf{hidden}} \in \mathbb{R}^{(m+l) \times o}$ is applied to (6) as a hidden layer, and then a non-linear setoid function broadcasts to each dimension. Here, we choose the $\mathsf{sigmoid}$ function as an example of a non-linear function, but any will do.

[8] This sort of embedding of entities into a vector space could be regarded, in Searle's sense, as giving a *cluster of descriptions* for each entity, especially when their values are learned by fitting to data.

(7) $p_{\mathsf{emb\_ent}} \in \mathbb{R}^{n \times m}, p_{\mathsf{emb\_pred}} \in \mathbb{R}^{k \times l}, p_{\mathsf{hidden}} \in \mathbb{R}^{(m+l) \times o}$
$\vdash \mathsf{sigmoid}(W_{\mathsf{hidden}}(W_{\mathsf{emb\_ent}}(\mathsf{onehot}(a_i)) \oplus W_{\mathsf{emb\_pred}}(e_j))) \in \mathbb{R}^o$

Third, a $o \times 1$ matrix W_o that is parametrized by $p_o \in \mathbb{R}^{o \times 1}$ is applied to (7) as an output layer and then another sigmoid function is broadcasted.

(8) $p_{\mathsf{emb\_ent}} \in \mathbb{R}^{n \times m}, p_{\mathsf{emb\_pred}} \in \mathbb{R}^{k \times l}, p_{\mathsf{hidden}} \in \mathbb{R}^{(m+1) \times o}, p_o \in \mathbb{R}^{o \times 1}$
$\vdash \mathsf{sigmoid}(W_o(\mathsf{sigmoid}(W_{\mathsf{hidden}}(W_{\mathsf{emb\_ent}}(\mathsf{onehot}(a_i)) \oplus W_{\mathsf{emb\_pred}}(e_j)))))$
$\in \mathbb{R}$

The output, given as a real number between 0 and 1, represents the degree to which the entity a_i applies to the predicate P_j. Let us call this deep neural classifier $\mathsf{pred}(a_i, j)$, where a_i is an entity and j is an index of the given predicate.

(9) $\mathsf{pred}(a_i, j) \stackrel{def}{\equiv}$
$\mathsf{sigmoid}(W_o(\mathsf{sigmoid}(W_{\mathsf{hidden}}(W_{\mathsf{emb\_ent}}(\mathsf{onehot}(a_i)) \oplus W_{\mathsf{emb\_pred}}(e_j)))))$
$\in \mathbb{R}$

Let us set a $\mathsf{threshold} \in \mathbb{R}$ as a hyperparameter between 0 and 1 that is large enough to consider that an entity a_i belongs to P_j when $\mathsf{pred}(a_i, j)$ is larger than the threshold. Then, the one-place predicate P_j is *defined* as follows.

(10) $P_j \stackrel{def}{\equiv} \lambda x.\mathsf{pred}(x, j) \geq \mathsf{threshold} : \mathbf{entity} \to \mathsf{type}$

We assume that the inequality $\geq$ is a setoid relation defined by

(11) $m \in \mathbb{R}, n \in \mathbb{R} \vdash m \geq n \stackrel{def}{\equiv} (d : \underline{\mathbb{R}}) \times (m \sim_{\mathbb{R}} n + d) : \mathsf{type}$

and is accompanied by its extensionality proof. By the definition of a setoid relation, for any entity a_i, $\mathsf{pred}(a_i, j) \geq \mathsf{threshold}$ is a type of DTS. An entity a_i belongs to P_j if $\mathsf{pred}(a_i, j) \geq \mathsf{threshold}$ has a proof, namely, if the real number $\mathsf{pred}(a_i, j)$ is sufficiently large.

Therefore, we may successfully replace every one-place predicate with a neural classifier that is represented as a type in DTS. It is straightforward to extend this definition to the more general cases of n-ary predicates, where we only concatenate all the arguments.[9]

[9] In the case of a two-place predicate p_j, for instance, applied to the two entities a_i and a_k, the corresponding two-place classifier is constructed as follows, by introducing a new parameter p_{hidden_2}:

$p_{\mathsf{emb\_ent}} \in \mathbb{R}^{n \times m}, p_{\mathsf{emb\_pred}} \in \mathbb{R}^{k \times l}, p_{\mathsf{hidden}_2} \in \mathbb{R}^{(2m+l) \times o}$
$\vdash \mathsf{sigmoid}(W_{\mathsf{hidden}_2}(W_{\mathsf{emb\_ent}}(\mathsf{onehot}(a_i)) \oplus W_{\mathsf{emb\_ent}}(\mathsf{onehot}(a_k)) \oplus W_{\mathsf{emb\_pred}}(e_j))) \in \mathbb{R}^o$

We may define the three-place classifiers in the same manner by introducing yet another parameter $p_{\mathsf{hidden}_3} \in \mathbb{R}^{3m+l}$.

3 Does Neural DTS Provide Soft Reasoning?

Can we say that names and predicates are soft symbols in the Neural DTS that was defined above? In this section, we will show that the Neural DTS satisfies the three requirements for soft symbols described in Sect. 1.2.

3.1 Continuous Truth Values

In the previous section, we defined the type **entity** as an enumeration type $\{a_1, \ldots, a_n\}$. Then, we represented the degree to which an entity a belongs to the predicate P_j by the value $\mathsf{pred}(a, j)$, where pred is a neural classifier that is implemented in DTS.

One can see that the Neural DTS satisfies the first property of Sect. 1.2 in the following way. For each entity a_i and each predicate P_j, the value $\mathsf{pred}(a_i, j)$ is a real number between 0 and 1 and, if normalized, can be regarded as the probability that a_i is P_j.

3.2 Comparable Representations

For two entities a_i and a_j, $W_{\mathsf{emb\_ent}}(\mathsf{onehot}(a_i))$ and $W_{\mathsf{emb\_ent}}(\mathsf{onehot}(a_j))$ are the embedded vectors represented in $\mathbb{R}^m$, respectively. Since $\mathbb{R}^m$ is a metric vector space if given the standard inner product, the similarity between a_i and a_j can be expressed by the cosine similarity.

(12) $$\frac{W_{\mathsf{emb\_ent}}(\mathsf{onehot}(a_i)) \cdot W_{\mathsf{emb\_ent}}(\mathsf{onehot}(a_j))}{\| W_{\mathsf{emb\_ent}}(\mathsf{onehot}(a_i)) \| \| W_{\mathsf{emb\_ent}}(\mathsf{onehot}(a_j)) \|}$$

Similarly, for the two predicates P_i and P_j, $W_{\mathsf{emb\_pred}}(i)$ and $W_{\mathsf{emb\_pred}}(j)$ are the embedding vectors represented in $\mathbb{R}^l$. So, as in the case of entities, the similarity between P_i and P_j can be expressed by the cosine similarity.

(13) $$\frac{W_{\mathsf{emb\_pred}}(i) \cdot W_{\mathsf{emb\_pred}}(j)}{\| W_{\mathsf{emb\_pred}}(i) \| \| W_{\mathsf{emb\_pred}}(j) \|}$$

3.3 Learnable Representations

Recall that our neural classifier is parametized by the following.

(14) $$p_{\mathsf{emb\_ent}} \in \mathbb{R}^{n \times m},\ p_{\mathsf{emb\_pred}} \in \mathbb{R}^{k \times l},\ p_{\mathsf{hidden}} \in \mathbb{R}^{(m+l) \times o},\ p_o \in \mathbb{R}^{o \times 1} \\ \vdash \mathsf{pred}(a_i, j) \in \mathbb{R}$$

Now, as an example of a loss function, we define the mean square error for the ground truth of each entity a_i.

(15) $$\mathsf{mse}(a_i, j) \stackrel{def}{\equiv} \begin{cases} \lambda p.(1 - \mathsf{pred}(a_i, j))^2 & \text{if } a_i \text{ is } P_j \\ \lambda p.(0 - \mathsf{pred}(a_i, j))^2 & \text{otherwise} \end{cases}$$

Parameters are learnable from the ground truth if $\mathsf{mse}(a_i, j)$ is a differentiable function of $p_{\mathsf{emb\_ent}}$, $p_{\mathsf{emb\_pred}}$, p_{hidden} and p_o for every input pair (a_i, j).

We have thus shown that it is possible to construct a deep neural network inside DTS and to embed entities and predicates in distributional representations, thereby allowing us to replace the predicates of DTS with neural classifiers. Although we have presented only a very simple deep neural network, it is clear that the method can be applied to almost any neural network currently being studied.

4 Verification Conditions for Predicates

So far, we have presented a very rough idea of how to integrate DTS and deep neural networks. Even so, our work raises some essential questions regarding the meaning of language.

First, an n-place predicate in DTS forms a type when combined with n-many terms, but the type formed in this way is not canonical (it is not constructed by a formation rule), so no introduction rule is given to it. For example, **dog**(john) is a type and, therefore, a collection of proofs (that john is a dog), but since **dog**(john) is not a canonical type, there is no introduction rule for the type **dog**(john). Introduction rules are verification conditions that give meaning to the type in proof-theoretic semantics. Therefore, a type without introduction rules is a proposition whose meaning is not defined.

We should recognize that this problem is not unique to proof-theoretic semantics. In model-theoretic semantics, the one-place predicate **dog** is interpreted as a set of entities, but an interpretation may assign any set to **dog**. This means that the semantic theory does not say anything about what the one-place predicate **dog** means. This is the same situation as in DTS, which does not say anything about what inhabits the type **dog**(john).

From this perspective, replacing predicates with the neural networks as presented in this paper takes on a different connotation. Specifically, non-canonical types, such as **dog**(john), are decomposed and replaced by setoid relations, such as (10). This makes it possible to describe all predicates using only canonical types, which further suggests that it is not just the semantics of mathematics that can be described by canonical types, but the semantics of natural language as well. Under this fusion, the meaning of every natural language sentence is given verification conditions.

However, there may be objections to such a fusion. Neural networks model the mechanisms by which the human brain perceives things. One of the features of neural networks is that they allow an agent (namely, the possessor of the neural network) with limited or partial information about the external world to make inferences, sort of maximum likelihood predictions about the rest of the external world. If this view is also applied to the neural classifiers embedded in Neural DTS, it falls short of a semantic theory of natural language, because it is unlikely that the meaning of natural language is fully explained by the neural classifiers that models the limited cognitive capacity of agents.

One possible response to this objection is to assume that for most (if not all) predicates P_j there is a *true* classifier, which is a map that sends each e_i to $\mathbb{R}$, namely, takes as input the elements of the standard basis corresponding to each entity and returns the probability that the element belongs to that predicate. The uniqueness of the *true* classifier for each predicate can be relativized to the extent that there is a true classifier for each smaller community of native speakers who use the predicate. It would not be too optimistic to assume that these functions are differentiable.

Given that neural networks can approximate any differentiable real-valued function, DTS, as a theory of meaning, need only replace predicates with corresponding *true* classifiers (rather than by neural classifiers).

On the other hand, when we want to consider a theory of natural language understanding, then we can use neural classifiers as approximations of the true classifiers. The current formulation of Neural DTS has neural parameters as free variables. Updating it through learning from data reflects how a speaker updates their own knowledge and, consequently, brings it closer to the true meaning.

These considerations on the meaning of predicates are a step beyond the scope of formal semantics, including model-theoretic semantics. The decomposition of the meaning of predicates is in the domain of lexical semantics rather than formal semantics. Thus, a comparison between the representation of word meanings in lexical semantics and those in Neural DTS will be a topic for future studies.

5 Related Work

There are several methods that combine reasoning and neural networks. We have outlined a method for combining symbolic reasoning with neural networks by *implementing* the latter in the former: a neural network model can be constructed within a system of symbolic reasoning. Specifically, we have presented a method for constructing neural classifiers as proof terms of a dependent type system, and these classifiers describe the meanings of predicates in natural language (see (9) and (10) above). Our semantic framework, which we call Neural DTS, provides a proof-theoretic explanation for the meanings of predicates in natural languages.

This section discusses some of the existing approaches for combining reasoning and neural networks and contrasts them with our work. Note that, for some approaches, reasoning is not always restricted to symbolic reasoning, such as those for recognizing textual entailment or natural language inference. We start with a brief discussion on the existing approaches that are found in logic-oriented work.

5.1 Logic and DNN

Some methods that combine symbolic reasoning and neural networks centered around *emulating* the former by means of the latter. First, we focus on research into knowledge base completion. Knowledge base completion is the task of predicting whether an edge belongs to a given knowledge graph, and this task can be treated as symbolic reasoning on knowledge graphs. To mention a few, Guu et al. [19], Das et al. [16], and Takahashi et al. [34] deal with knowledge base completion by embedding knowledge graphs in vector spaces. Guu et al. [19] proposed a technique for extending any composable vector space model to a compositional model. Not only can the resulting neural network model can answer queries about edges on graphs, but it can also answer ones about paths on graphs. Das et al. [16] presented a compositional model that also considers entity types and multiple paths between a given pair of entities. Takahashi et al. [34] discussed the problem of how to make dimension reduction compatible with compositional constraints on matrix representations of relations, and proposed an approach that trains relation parameters jointly with an autoencoder.

Demeester et al. [17] formulated a method for imposing first-order logical rules, representing commonsense knowledge, on vector space models to achieve knowledge base completion. During training, the first-order rules are imposed in a lifted form; this is a form independent of any entity-tuples. This approach is able to impose a large number of first-order rules as commonsense knowledge, preserving computational efficiency. First-order logical rules have also been used to define neural networks. Šourek et al. [37] introduced Lifted Relational Neural Networks, which are represented as sets of weighted first-order rules. These weighted first-order rules can be transformed into a feed-forward neural network via Herbrand model construction. The resulting neural network can be used to perform some symbolic inference tasks that are similar to knowledge base completion.

In trying to emulate symbolic reasoning, research has also been conducted to see if neural network models can be used to solve SAT problems. Selsam et al. [33] presented a message-passing neural network called NeuroSAT that was designed to solve SAT problems automatically. To reproduce several features of truth-value semantics, SAT problems were rendered as formulas in conjunctive normal form and translated into undirected graphs with two kinds of edges. These graphs were then embedded into a vector space model.

There are also methods of *equipping* symbolic reasoning with neural networks. Wang et al. [38] proposed an approach that used neural networks for premise selection in automated theorem proving. Premise selection is the task of selecting relevant statements that are useful for proving a given mathematical conjecture. In their approach, statements and conjectures are given as formulas in higher-order logic and are then translated into graphs that capture the syntactic and semantic features of those formulas. Next, the resulting graphs are embedded into vectors by performing graph convolution. The embeddings of a statement and a conjecture are then concatenated and sent to a classifier to predict how useful the statement is for that conjecture.

In addition to premise selection, proof search in automated theorem proving has also been equipped with neural networks. Rocktäschel and Riedel [32] introduced Neural Theorem Provers, which are end-to-end differentiable provers for knowledge base completion. As a part of their search procedure, these provers have a differentiable unification operation that uses vector representations of symbols. They are able to learn the similarity between vector representations and to induce useful logical rules for predicting edges in knowledge graphs.

5.2 Natural Language and DNN

Research on how DNN models can be utilized to achieve reasoning in natural language has been conducted in the field of natural language processing. The task is known either as *recognizing textual entailment* (RTE) or *natural language inference* (NLI), and it determines whether a pair of input sentences (premises and hypothesis) possesses a relation of entailment, contradiction, or neither. It was proposed as a benchmark task for evaluating systems designed around natural language understanding (cf. Cooper et al. [13]) and has gained more attention with the recent developments of DNN models and large-scale RTE datasets.

There are two primary approaches for performing NLI with DNNs. The first approach is to train end-to-end neural models on NLI datasets (e.g., Rocktäschel et al. [31]). The other approach is to pre-train language representation models and then fine-tune them on the specific NLI task (e.g., Devlin et al. [18]). Both can be viewed as approaches that aim to have DNNs learn reasoning in natural language directly from a large number of inference problems. Both of these approaches have exhibited good performance on general-domain large-scale NLI datasets, such as SNLI [9] and MNLI [39]. They have even attained performance that is comparable to that of human [25], which demonstrates the sufficiency of these approaches.

However, it has been shown that DNNs have not necessarily succeeded in learning symbolic reasoning in natural language. The limitations of DNN models have been explored by carefully observing their performance on more organized and controlled data sets that focus on specified reasoning types in natural language.

Richardson et al. [30] systematically generated NLI datasets in which each problem targeted a specific semantic phenomenon, such as negation, counting, and comparatives. Their experiment demonstrates that state-of-the-art models, including BERT, that are pre-trained on large NLI datasets, perform poorly on these logical inference problems unless they are re-trained with a subset of the generated data. Yanaka et al. [40] examined whether a DNN model could learn monotonicity inference in natural language, which would allow it to handle unseen combinations of patterns. They developed an evaluation protocol that was designed to measure the systematic generalization ability of DNN models for several aspects of monotonicity, such as predicate replacements and structural recursiveness. A series of experiments on three DNN models, including BERT [18], showed that the models' generalization ability was limited to cases where sentence structures were similar to the training data.

Overall, these studies demonstrated that it is still difficult for DNN models to emulate symbolic reasoning in natural language by learning inferences directly from input-output data.

A *hybrid* approach is one possible method for achieving soft and symbolic reasoning. Kalouli et al. [22] proposed a hybrid NLI system that identifies whether an input NLI pair is linguistically challenging or not and uses either symbolic or deep learning components to make a final entailment judgment. The symbolic component consists of an inference engine based on Natural Logic, while the deep learning component adopts the state-of-the-art language representation model BERT. The system performs well on existing NLI data sets that consist of pairs requiring either symbolic or soft reasoning. In a more practical setting, however, it must be able to handle more complex cases where both soft and symbolic reasoning are needed. It is not trivial to break down complex inference problems into simpler ones that require either soft or symbolic reasoning.

Another method is to embed the DNN in the symbolic approach. Lewis and Steedman [24] proposed a pioneering approach that replaced predicate symbols by distributed representations that were composed via Combinatory Categorial Grammar (CCG) syntactic structures and then used them in inferences.

The replacement of predicate symbols by deep neural classifiers rather than distributed representations goes back to Cooper [12] and Larsson [23]. In particular, Larsson replaced a predicate by a classifier that was represented as a record. Specifically, the classifier was a witness for a (dependent) record type in TTR. Such a record typically consists of a weight vector, a threshold, and a function that classifies a given situation by using the weight and the threshold. Larsson's method differs from our approach in that a classifier is defined *outside* the type system. We will give a detailed comparison of our approach with Larsson's in future work.

5.3 Programming Languages and DNN

In their study of reverse derivative categories, Cockett et al. [10] attempted to axiomatize differential operators in terms of Cartesian differential categories (CDCs) consisting of seven axioms. Cruttwell [14] succeeded in formulating the back-propagation of neural networks in CDCs. Additionally, Abadi and Plotkin [1] used CDCs to design a functional programming language equipped with a differential operator. This line of work has some affinity with Neural DTS in the sense that it represents neural networks in a certain functional programming language.

6 Conclusion and Future Work

The neural classifiers in Neural DTS, as introduced in this paper, are functions of type **entity** $\rightarrow$ type, following the previous versions of DTS but with internal structures, namely, real-valued parameters and loss values. Each **entity** and predicate symbol are also embedded in a vector space, making it possible to define soft symbols that are both comparable and differentiable. The parameters can be trained on the data, by which similar expressions will obtain similar representations that enable us to compute similarities between them.

Several challenges lie ahead for Neural DTS. One is to simulate Neural DTS through implementation and to explore its effectiveness for computational semantics. This will rely on a combination of existing DTS implementations and neural network technology. A possibly more challenging question is how to obtain the ground truth about predicates. One idea is to parse texts using CCG parsers, convert their syntactic structures into representations in DTS, and then deduce knowledge about the atomic predicates by using an automatic theorem prover. Although there are some technical problems involved in exploring this approach, it may open up the prospect of a new field that combines type theory, formal semantics, lexical semantics, cognitive science, and natural language processing.

Acknowledgements We sincerely thank the anonymous reviewers and the participants of LACompLing 2022, especially Zhaohui Luo, Tracy Holloway King, and Koji Mineshima, for their insightful comments. We also owe a lot to the discussion on the earlier version of this work with Kentaro Inui, Sadao Kurohashi, and Minao Kukita. This work was partially supported by the Japan Science and Technology Agency (JST) CREST Grant Number JPMJCR20D2.

Appendix: Dependent Type Theory (DTT)

Definition 17 (*Alphabet*) An *alphabet* is a pair $(\mathcal{V}ar, \mathcal{C}on)$ where $\mathcal{V}ar$ is a collection of variables and $\mathcal{C}on$ is a collection of constant symbols.

Definition 18 (*Preterms*) The collection of preterms of DTT (notation Λ) under an alphabet $(\mathcal{V}ar, \mathcal{C}on)$ is defined by the following BNF grammar, where $x \in \mathcal{V}ar$ and $c \in \mathcal{C}on$.

$$
\begin{aligned}
\Lambda := {} & x \mid c \mid \mathsf{type} \\
& \mid (x : \Lambda) \rightarrow \Lambda \mid \lambda x.\Lambda \mid \Lambda\Lambda \\
& \mid (x : \Lambda) \times \Lambda \mid (\Lambda, \Lambda) \mid \pi_1(\Lambda) \mid \pi_2(\Lambda) \\
& \mid \Lambda \uplus \Lambda \mid \iota_1(\Lambda) \mid \iota_2(\Lambda) \mid \mathsf{unpack}_L\,(M, N) \\
& \mid \{a_1, \ldots, a_n\} \mid a_1 \mid \ldots \mid a_n \mid \mathsf{case}^{\Lambda}_{\Lambda}\,(\Lambda, \ldots, \Lambda) \\
& \mid \Lambda =_{\Lambda} \Lambda \mid \mathsf{refl}_{\Lambda}(\Lambda) \mid \mathsf{idpeel}^{\Lambda}_{\Lambda}\,(\Lambda) \\
& \mid \mathbb{N} \mid 0 \mid \mathsf{s}(\Lambda) \mid \mathsf{natrec}^{\Lambda}_{\Lambda}(\Lambda, \Lambda)
\end{aligned}
$$

Free variables, substitutions, and β-reductions are defined in the standard way. The full version of DTT also employs well-ordered types and universes, as adopted in [27], the details of which I omit here for the sake of space.

Definition 19 (*Vertical/Box notation for Σ-type*) $\begin{bmatrix} x : A \\ B \end{bmatrix} \stackrel{def}{\equiv} (x : A) \times B$

Definition 20 (*Implication, Conjunction and Negation*)[10]

$$\begin{aligned} A \to B &\stackrel{def}{\equiv} (x : A) \to B \quad \textit{where } x \notin \mathit{fv}(B) \\ \begin{bmatrix} A \\ B \end{bmatrix} &\stackrel{def}{\equiv} \begin{bmatrix} x : A \\ B \end{bmatrix} \quad \textit{where } x \notin \mathit{fv}(B) \\ \bot &\stackrel{def}{\equiv} \{\} \\ \neg A &\stackrel{def}{\equiv} A \to \bot \end{aligned}$$

Definition 21 (*Signature*) A collection of signatures (notation σ) for an alphabet $(\mathcal{V}ar, \mathcal{C}on)$ is defined by the following BNF grammar:

$$\sigma ::= () \mid \sigma, c : A$$

where () is an empty signature, $c \in \mathcal{C}on$ and $\vdash_\sigma A : \mathsf{type}$.

Definition 22 (*Context*) A collection of contexts under a signature σ (notation Γ) is defined by the following BNF grammar:

$$\Gamma ::= () \mid \Gamma, x : A$$

where () is an empty context, $x \in \mathcal{V}ar$ and $\Gamma \vdash_\sigma \mathsf{type}$.

Definition 23 (*Judgment*) A judgment of DTT is the following form

$$\Gamma \vdash_\sigma M : A$$

where Γ is a context under a signature σ and M and A are preterms, which states that there exists a proof diagram of DTT from the context Γ to the type assignment $M : A$. The subscript σ may be omitted when no confusion arises.

Definition 24 (*Truth*) A judgment of the form Γ' *true* states that there exists a term M that satisfies $\Gamma \vdash M : A$.

Definition 25 (*Structural Rules*)

$$\frac{A : \mathsf{type}}{x : A}\,(VAR) \qquad \frac{}{c : A}\,(CON) \quad \textit{where } \sigma \vdash c : A$$

$$\frac{M : A \quad N : B}{M : A}\,(WK) \qquad \frac{M : A}{M : B}\,(CONV) \quad \textit{where } A =_\beta B$$

[10] $\bot$ is defined as an empty enumeration type. $\mathit{fv}(M)$ is a set of free variables in the term M.

Definition 26 (Π-*types*)

$$\dfrac{A : \mathsf{type} \quad \begin{array}{c}\overline{x : A}^{\,i} \\ \vdots \\ B : \mathsf{type}\end{array}}{(x : A) \to B : \mathsf{type}}\ (\Pi F), i \qquad \dfrac{A : \mathsf{type} \quad \begin{array}{c}\overline{x : A}^{\,i} \\ \vdots \\ M : B\end{array}}{\lambda x.M : (x : A) \to B}\ (\Pi I), i$$

$$\dfrac{M : (x : A) \to B \quad N : A}{MN : B[N/x]}\ (\Pi E)$$

Definition 27 (Σ-*types*)

$$\dfrac{A : \mathsf{type} \quad \begin{array}{c}\overline{x : A}^{\,i} \\ \vdots \\ B : \mathsf{type}\end{array}}{(x : A) \times B : \mathsf{type}}\ (\Sigma F), i \qquad \dfrac{M : A \quad N : B[M/x]}{(M, N) : (x : A) \times B}\ (\Sigma I)$$

$$\dfrac{M : (x : A) \times B}{\pi_1(M) : A}\ (\Sigma E) \qquad \dfrac{M : (x : A) \times B}{\pi_2(M) : B[\pi_1(M)/x]}\ (\Sigma E)$$

Definition 28 (*Enumeration Types*)

$$\dfrac{}{\{a_1, \ldots, a_n\} : \mathsf{type}}\ (\{\}F) \qquad \dfrac{}{a_i : \{a_1, \ldots, a_n\}}\ (\{\}I)$$

$$\dfrac{M : \{a_1, \ldots, a_n\} \quad P : \{a_1, \ldots, a_n\} \to \mathsf{type} \quad N_1 : P(a_1) \quad \ldots \quad N_n : P(a_n)}{\mathsf{case}_M^P\,(N_1, \ldots, N_n) : P(M)}\ (\{\}E)$$

Definition 29 (*Disjoint Union Types*)

$$\dfrac{A : \mathsf{type} \quad B : \mathsf{type}}{A \uplus B : \mathsf{type}}\ (\uplus F) \qquad \dfrac{M : A}{\iota_1(M) : A \uplus B}\ (\uplus I) \qquad \dfrac{N : B}{\iota_2(N) : A \uplus B}\ (\uplus I)$$

$$\dfrac{L : A \uplus B \quad P : (A \uplus B) \to \mathsf{type} \quad M : (x : A) \to P(\iota_1(x)) \quad N : (x : B) \to P(\iota_2(x))}{\mathsf{unpack}_L^P\,(M, N) : P(L)}\ (\uplus E), i$$

Definition 30 (*Intensional Equality Types*)

$$\dfrac{A : \mathsf{type} \quad M : A \quad N : A}{M =_A N : \mathsf{type}}\ (=F) \qquad \dfrac{A : \mathsf{type} \quad M : A}{\mathsf{refl}_A(M) : M =_A M}\ (=I)$$

$$\dfrac{E : M =_A N \quad P : (x : A) \to (y : A) \to (x =_A y) \to \mathsf{type} \quad R : (x : A) \to Pxx(\mathsf{refl}_A(x))}{\mathsf{idpeel}_E^P\,(R) : PMNE}\ (=E)$$

Definition 31 (*Natural Number Types*)

$$\frac{}{\mathbb{N} : \mathsf{type}} (\mathbb{N}F) \qquad \frac{}{\mathsf{0} : \mathbb{N}} (\mathbb{N}I) \qquad \frac{n : \mathbb{N}}{\mathsf{s}(n) : \mathbb{N}} (\mathbb{N}I)$$

$$\frac{n : \mathbb{N} \quad P : \mathbb{N} \to \mathsf{type} \quad e : P(\mathsf{0}) \quad f : (k : \mathbb{N}) \to P(k) \to P(\mathsf{s}(k))}{\mathsf{natrec}_n^P(e, f) : P(n)} (\mathbb{N}E)$$

Definition 32 (*First Universe*)

$$\frac{}{\mathsf{U} : \mathsf{type}} (\mathsf{U}F) \qquad \frac{M : \mathsf{U}}{\mathsf{dec}\,(M) : \mathsf{type}} (\mathsf{U}E)$$

$$\frac{A : \mathsf{U} \qquad \begin{matrix}\overline{x : \mathsf{dec}\,(A)}^{\,i} \\ \vdots \\ B : \mathsf{U}\end{matrix}}{(x : A)\,\underline{\to}\, B : \mathsf{U}} (\mathsf{U}I),i \qquad \frac{A : \mathsf{U} \qquad \begin{matrix}\overline{x : \mathsf{dec}\,(A)}^{\,i} \\ \vdots \\ B : \mathsf{U}\end{matrix}}{(x : A)\,\underline{\times}\, B : \mathsf{U}} (\mathsf{U}I),i$$

$$\frac{}{\underline{\{}a_1, \ldots, a_n\underline{\}} : \mathsf{U}} (\mathsf{U}I) \qquad \frac{A : \mathsf{U} \quad B : \mathsf{U}}{A \,\underline{\uplus}\, B : \mathsf{U}} (\mathsf{U}I)$$

$$\frac{A : \mathsf{U} \quad M : \mathsf{dec}\,(A) \quad N : \mathsf{dec}\,(A)}{M \underline{=}_A N : \mathsf{U}} (\mathsf{U}I) \qquad \frac{}{\underline{\mathbb{N}} : \mathsf{U}} (\mathsf{U}I)$$

References

1. Abadi, M., Plotkin, G.D.: A simple differentiable programming language. Proc. ACM Program. Lang. **4**(POPL), 38:1–38:28 (2020). https://doi.org/10.1145/3371106
2. Aczel, P.H.: Galois: a theory development project (1993)
3. Bailey, A.: Representing algebra in LEGO. Thesis, University of Edinburgh (1993)
4. Barthe, G.: Formalising mathematics in UTT: fundamentals and case studies. Report (1995)
5. Barthe, G., Capretta, V., Pons, O.: Setoids in type theory. J. Funct. Program. **13**(2), 261–293 (2000). https://doi.org/10.1017/S0956796802004501
6. Bekki, D.: Proof-theoretic analysis of weak crossover. In: Logic and Engineering of Natural Language Semantics 18 (LENLS18), pp. 75–88 (2021)
7. Bekki, D., Mineshima, K.: Context-passing and underspecification in dependent type semantics. In: Modern Perspectives in Type Theoretical Semantics. Studies of Linguistics and Philosophy, pp. 11–41. Springer (2017)
8. Bishop, E.: Foundations of Constructive Analysis. McGraw-Hill (1967)
9. Bowman, S.R., Angeli, G., Potts, C., Manning, C.D.: A large annotated corpus for learning natural language inference. In: The 2015 Conference on Empirical Methods in Natural Language Processing, pp. 632–642. Association for Computational Linguistics (2015). https://doi.org/10.18653/v1/D15-1075, https://aclanthology.org/D15-1075
10. Cockett, R., Cruttwell, G., Callagher, J., Pacaud Lemay, J.S., MacAdam, B., Plotkin, G., Pronk, D.: Reverse derivative categories, 15 October 2019 (2019). ArXiv:1910.07065v1 [cs.LO]
11. Cooper, R.: Records and record types in semantic theory. J. Logic Comput. **15**(2), 99–112 (2005)

12. Cooper, R.: Representing types as neural events. J. Logic Lang. Inf. **28**, 131–155 (2019). https://doi.org/10.1007/s10849-019-09285-4, https://link.springer.com/article/10.1007/s10849-019-09285-4
13. Cooper, R., Crouch, D., van Eijck, J., Fox, C., van Genabith, J., Jaspars, J., Kamp, H., Milward, D., Pinkal, M., Poesio, M., Pulman, S., Briscoe, T., Maier, H., Konrad, K.: Using the framework. Report (1996)
14. Cruttwell, G., Gavranović, B., Ghani, N., Wilson, P., Zanasi, F.: Categorial foundations of gradient-based learning, 2 March 2021 (2021). ArXiv:2103.01931v1 [cs.LG]
15. Cruz-Filipe, L.: Constructive real analysis: a type-theoretical formalization and applications. Thesis, University of Nijmegen (2004)
16. Das, R., Neelakantan, A., Belanger, D., McCallum, A.: Chains of reasoning over entities, relations, and text using recurrent neural networks. In: Proceedings of the 15th Conference of the European Chapter of the Association for Computational Linguistics, vol. 1 (Long Papers), pp. 132–141. Association for Computational Linguistics (2017). https://aclanthology.org/E17-1013
17. Demeester, T., Rocktäschel, T., Riedel, S.: Lifted rule injection for relation embeddings. In: Proceedings of the 2016 Conference on Empirical Methods in Natural Language Processing, pp. 1389–1399. Association for Computational Linguistics (2016). https://doi.org/10.18653/v1/D16-1146, https://aclanthology.org/D16-1146
18. Devlin, J., Chang, M.W., Lee, K., Toutanova, K.: Bert: Pre-training of deep bidirectional transformers for language understanding. In: The 2019 Conference of the North American Chapter of the Association for Computational Linguistics: Human Language Technologies (NAACL-HLT 2019), vol. 1 (Long and Short Papers), pp. 4171–4186. Association for Computational Linguistics (2019). https://doi.org/10.18653/v1/N19-1423, https://aclanthology.org/N19-1423
19. Guu, K., Miller, J., Liang, P.: Traversing knowledge graphs in vector space. In: Proceedings of the 2015 Conference on Empirical Methods in Natural Language Processing, pp. 318–327. Association for Computational Linguistics (2015). https://doi.org/10.18653/v1/D15-1038, https://aclanthology.org/D15-1038
20. Hofmann, M.: Extensional concepts in intensional type theory. Thesis, University of Edinburgh (1995)
21. Hofmann, M.: A simple model for quotient types. In: Dezani-Ciancaglini, M., Plotkin, G. (eds.) TLCA'95. Lecture Notes in Computer Science 902, pp. 216–234. Springer (1995)
22. Kalouli, A.L., Crouch, R., de Paiva, V.: Hy-nli: a hybrid system for natural language inference. In: The 28th International Conference on Computational Linguistics, pp. 5235–5249. International Committee on Computational Linguistics (2020). https://doi.org/10.18653/v1/2020.coling-main.459, https://aclanthology.org/2020.coling-main.459
23. Larsson, S.: Discrete and probabilistic classifier-based semantics. In: The Probability and Meaning Conference (PaM 2020), pp. 62–68. Association for Computational Linguistics (2020). https://aclanthology.org/2020.pam-1.8
24. Lewis, M., Steedman, M.: Combined distributional and logical semantics. Trans. Assoc. Comput. Linguist. **1**, 179–192 (2013)
25. Liu, X., He, P., Chen, W., Gao, J.: Multi-task deep neural networks for natural language understanding. In: The 57th Annual Meeting of the Association for Computational Linguistics, pp. 4487–4496. Association for Computational Linguistics (2019). https://doi.org/10.18653/v1/P19-1441, https://aclanthology.org/P19-1441
26. Luo, Z., Pollack, R.: Lego proof development system: user's manual (1992)
27. Martin-Löf, P.: Intuitionistic Type Theory, vol. 17. Bibliopolis, Naples, Italy (1984)
28. Palmgren, E.: Bishop's set theory (2005). http://www.cse.chalmers.se/research/group/logic/TypesSS05/Extra/palmgren.pdf
29. Palmgren, E.: From type theory to setoids and back (2019). https://doi.org/10.48550/arXiv.1909.01414
30. Richardson, K., Hu, H., Moss, L., Sabharwal, A.: Probing natural language inference models through semantic fragments. In: The AAAI Conference on Artificial Intelligence, vol. 34, pp. 8713–8721 (2020). https://doi.org/10.1609/aaai.v34i05.6397, https://ojs.aaai.org/index.php/AAAI/article/view/6397

31. Rocktäschel, T., Grefenstette, E., Hermann, K.M., Kocisky, T., Blunsom, P.: Reasoning about entailment with neural attention (2016). http://arxiv.org/abs/1509.06664
32. Rocktäschel, T., Riedel, S.: End-to-end differentiable proving. In: The 31st International Conference on Neural Information Processing Systems. Curran Associates, Inc. (2017)
33. Selsam, D., Lamm, M., Bünz, B., Liang, P., de Moura, L., Dill, D.L.: Learning a SAT solver from single-bit supervision. In: ICLR (Poster) (2019)
34. Takahashi, R., Tian, R., Inui, K.: Interpretable and compositional relation learning by joint training with an autoencoder. In: Proceedings of the 56th Annual Meeting of the Association for Computational Linguistics, vol. 1 (Long Papers), pp. 2148–2159. Association for Computational Linguistics (2018). https://doi.org/10.18653/v1/P18-1200, https://aclanthology.org/P18-1200
35. The Agda Team: Agda user manual: release 2.6.2.1 (2021). https://agda.readthedocs.io/en/v2.6.2.1/
36. The Coq Development Team: The Coq reference manual: release 8.14.1 (2021). https://coq.github.io/doc/v8.14/refman/
37. Šourek, G., Aschenbrenner, V., Železný, F., Schockaert, S., Kuželka, O.: Lifted relational neural networks: efficient learning of latent relational structures. J. Artif. Int. Res. **62**(1), 69–100 (2018)
38. Wang, M., Tang, Y., Wang, J., Deng, J.: Premise selection for theorem proving by deep graph embedding. In: Proceedings of the 31st International Conference on Neural Information Processing Systems, pp. 2783–2793. Curran Associates Inc. (2017)
39. Williams, A., Nangia, N., Bowman, S.: A broad-coverage challenge corpus for sentence understanding through inference. In: The 2018 Conference of the North American Chapter of the Association for Computational Linguistics: Human Language Technologies, vol. 1 (Long Papers), pp. 1112–1122. Association for Computational Linguistics (2018). https://aclanthology.org/N18-1101, https://doi.org/10.18653/v1/N18-1101
40. Yanaka, H., Mineshima, K., Bekki, D., Inui, K.: Do neural models learn systematicity of monotonicity inference in natural language? In: the 58th Annual Meeting of the Association for Computational Linguistics, pp. 6105–6117. Association for Computational Linguistics (2020). https://doi.org/10.18653/v1/2020.acl-main.543, https://aclanthology.org/2020.acl-main.543

Meaning-Driven Selectional Restrictions in *Remember* Versus *Imagine Whether*

Kristina Liefke

Abstract Fiction verbs (e.g., *imagine*, *dream*) typically reject *whether*-clauses. My paper traces the semantic source of this rejection: I argue that the difference between *remember whether* and **imagine whether* cannot be explained through the familiar properties (e.g., semantic type, neg-raising, non-veridicality) that have been used to explain the rejection of polar interrogative complements for other clause-embedding predicates (e.g., *believe*). I propose that the difference between *remember* and *imagine whether* is instead explained through (i) the parasitic dependence of the reported attitude content on an underlying experience, (ii) the counterfactuality (resp. non-counterfactuality) of the matrix attitude, and (iii) an actuality assumption in the lexical semantics of *whether*. I observe that *whether*-clauses are typically licensed only when the underlying experience is veridical and the attitude content true in the actual world. This is the case for most uses of *remember*, but only for few exceptional uses of *imagine*. I give a compositional semantics for *remember* and *imagine* that captures their licensing conditions.

Keywords Clausal selection · Selectional restrictions · Polar interrogatives · Fiction verbs · Inquisitive semantics · Factivity alternation · Cross-attitudinal parasitism

1 Introduction

Much recent work at the syntax/semantics-interface has discussed the selection properties of clause-embedding predicates. This holds especially for the inability of many of these predicates to take interrogative complements [73]. This property is often called 'anti-rogativity'. Anti-rogative predicates include neg-raising predicates like *believe* [15, 49, 66] (see (1a)), emotive factives like *regret* [20] (see (1b)), non-

K. Liefke (✉)
Department of Philosophy II, Ruhr-Universität Bochum, Universitätsstraße 150, 44780 Bochum, Germany
e-mail: kristina.liefke@ruhr-uni-bochum.de

R. Loukanova et al. (eds.), *Logic and Algorithms in Computational Linguistics 2021 (LACompLing2021)*, Studies in Computational Intelligence 1081,
https://doi.org/10.1007/978-3-031-21780-7_12

veridical preferential predicates like *hope* [69] (see (1c)), and truth-evaluating predicates like *be true* [20, 66] (see (1d)).

(1) John { a. believes {that, *why/*whether/*with whom}
b. regrets {that, *why/*whether/*with whom}
c. hopes {that, *why/*whether/*with whom} }
d. It is true {that, *why/*whether/*with whom} } a woman danced.

Surprisingly, work on clause-embedding predicates has largely neglected the (partial) anti-rogativity of representational counterfactual attitude verbs.[1] The latter are verbs like *imagine*, *hallucinate*, and *dream* that represent mental pictures or images (see [8, 74]). In contrast to non-counterfactual representational attitude verbs (e.g., *remember*, *notice*, *observe*; see (2b))—and like the predicates in (1)—, these verbs typically reject *whether*-clauses (see (2a), judgement due to D'Ambrosio and Stoljar [17]):

(2) a. John imagined {i. that, ii. *whether} a woman danced.
b. John remembers {i. that, ii. whether} a woman danced.

My paper provides an account of the difference in acceptability between *imagine whether* ($*$) and *remember whether* ($\checkmark$) that explains this difference through the lexical semantics of the (explicit and implicit) constituents of these constructions and through the way in which the semantic values of these constituents interact during semantic composition. My account falls squarely within recent attempts to derive the selectional restrictions of clause-embedding predicates from "independently motivated features of their lexical semantics" [65, p. 413] (see also [49, 66, 68, 69]).

To explain why *imagine*—unlike *remember*—typically rejects *whether*-clauses, my account assumes (i) the parasitic dependence of the reported attitude content on an underlying experience (see [7, 45, 46]), (ii) the counterfactuality (resp. non-counterfactuality) of the matrix attitude, and (iii) an actuality requirement in the lexical semantics of *whether* (see [29, 38, 40]). I observe that *whether*-clauses are typically only licensed when the underlying experience is veridical and when the reported attitude content is true in the actual world. This is the case for most uses of *remember*, but only for few exceptional uses of *imagine*. I give a compositional semantics for *remember* and *imagine* that captures their interaction with *whether*. To make this possible, I generalize the two-dimensional semantics from Blumberg [6, 7] to experience-dependence (see [45, 46]) and to inquisitive attitudes (see [13, 65, 68]).

The paper is structured as follows: To motivate the need for a 'new' account of **imagine whether*, I first show that none of the existing accounts (esp. familiar type-based accounts [31, 38, 39], neg-raising accounts [49, 66, 76], and veridicality- or factivity-based accounts [20, 32, 35]) can explain why *imagine* rejects *whether*-complements (see Sect. 2). I then introduce the driving notion behind my account:

[1] In [25], these verbs are called *fiction verbs*.

experiential parasitism (in Sect. 3) and use it to capture the varying validity of veridicality inferences in declarative *remember*-reports (Sect. 4). With this explanation in hand, I provide the lexical entries for interrogative uses of *remember* and *imagine* that account for the different acceptability of different uses of *remember whether* and for the deviance of (most occurrences of) *imagine whether* (in Sect. 5). The paper closes with a summary of my results and with pointers to future work.

Before I move to existing accounts of '*whether*-rejection', I want to express a caveat about the naturalness of reports like *John remembers whether a woman danced* (i.e., (2b-ii)). Arguably, even in the complement of veridical (or factive) uses of *remember*, *whether*-clauses are slightly deviant. I attribute this deviance to pragmatic factors—in particular, to the fact that the complementizers *that* and *whether* form a two-value Horn scale [34], where *that* is the stronger member of the scale (see (3)):

(3) that > whether

Given Grice's maxim of quantity ('Make your contribution as informative as is required'), a speaker's use of (2b-ii) then triggers a pragmatic inference to the negation of (2b-i). But this may clash with the report's linguistic or real-world context (viz. if this context establishes (2b-i)).

2 Non-explanations

An obvious route to explaining the difference between *remember whether* and **imagine whether* lies in existing accounts of '*whether*-rejection'. These include the restriction of *whether*-clauses to the complements of rogative and responsive predicates (see [28, 39]; discussed in Sect. 2.1), the restriction of *whether*-clauses to non-neg-raising predicates (see [49, 66, 76]; discussed in Sect. 2.2), and the restriction of *whether*-clauses to the complements of factive (see [5, 26, 33]) or veridical predicates (see [20]; discussed in Sect. 2.3).

2.1 Anti-rogativity

Importantly, the unacceptability of *imagine whether* cannot be attributed to the general inability of *imagine* to accept interrogative complements. This strategy is based on the general classification of clause-embedding predicates into rogative [= only interrogative-embedding], anti-rogative [= only declarative-embedding], and responsive [= declarative- and interrogative-embedding] predicates [28, 39, 65]. Rogative predicates include the verbs *wonder*, *be curious*, and *inquire* (see (4a)); anti-rogative predicates include the verbs *believe*, *hope*, and *fear* (see (4b)). Responsive predicates include the verbs *know*, *realize*, and *report* (see (4c)):

(4) a. John wonders {i.*that, ii. why/whether/with whom} Mary danced.

b. John believes {i. that, ii. *why/*whether/*with whom} Mary danced.
c. John knows {i. that, ii. why/whether/with whom} Mary danced.

Using the above classification, one could try to explain the unacceptability of (2a-ii) by identifying *imagine* as an anti-rogative predicate. However, in contrast to more 'standard' anti-rogatives, *imagine* only rejects polar interrogatives (see (5d)).[2] In line with the selection behavior of responsive predicates, it felicitously combines with constituent questions (in (5a-c): with *why*-, *how*-, and *who*-interrogatives):

(5) John is imagining { a. why / b. how [= in what manner] / c. with whom / d. *whether } a woman danced.

Since the wh-words in (a–c) fail the diagnostics for a 'free relative'-interpretation (i.e., they allow (!) accenting and coordination [70]; see (6a)–(6b)), the acceptability of (5a–c) cannot be explained by the ability of *imagine* to license direct objects (e.g., *a man* in (7a)): if the expression *with whom a woman danced* in (5c) were a free relative, it could straightforwardly combine with the intensional transitive [= direct object taking-]use of *imagine* (along the lines of (7b)), and would not need to be treated as an interrogative clause-taking expression.

(6) a. John remembers WHY the woman danced.
b. John is imagining *why*, *how*, and *with whom* a woman danced.

(7) a. John is imagining a man.
b. John is imagining (the man) with whom Mary danced.

The anti-rogativity of *imagine* is further challenged by the observation that *imagine whether* is, in fact, acceptable in certain contexts. The latter include contexts that aim to 'track reality' (see (8a), due to Peterson [60, p. 59])[3] and contexts that embed *imagine* under a (negated) ability modal or under the control predicate *try* (see (8b), German version due to [62, p. 1, ex. (2a)]):

(8) a. I am **imagining whether** the new sofa will fit into my living room.
b. Anna {was trying to, could not} **imagine whether** the lid would fit the kettle.

In contrast to most uses of *imagine*, the verb *remember* generally licenses both polar interrogative and wh-complements (see (9)):

---

[2] *imagine* thus shares the selection behavior of emotive factives like *be surprised* (see (⋆)):

(⋆) John was surprised {why, how, with whom, *whether} a woman danced.

[3] Since a detailed discussion of the selectional variability of *imagine whether* would exceed the scope of the present discussion (which focuses on the difference between *remember* and *imagine*), it is deferred to another paper, viz. [43].

(9) John remembers { a. why / b. how [= in what manner] / c. with whom / d. whether } a woman danced.

2.2 Neg-Raising

It has long been assumed that neg(ative)-raising predicates (e.g., *believe*, *expect*, *be likely*) reject interrogative complements (see [49, 66, 69, 76]). White [73], who does not endorse this assumption, calls it *Generalization* $\mathbb{NR}$ (for '$\mathbb{N}$eg-$\mathbb{R}$aising'). Neg-raising predicates are predicates V that license the inference from 'NP not-V S' to 'NP V not-S' [73, p. 3]. This form of inference is exemplified in (10):

(10) a. Bill does **not** believe that Mary danced.
⇒ b. Bill believes that Mary did **not** dance.

Recent work on clausal selection (esp. [15, 49, 66]) explains the anti-rogativity of neg-raising predicates by assuming that these predicates semantically involve an excluded middle (EM)-presupposition (see, e.g., [2, 23]). For the occurrences of *believe* in (10a) (copied in (11a)), this presupposition is given in (11b):

(11) a. Bill does **not** believe that Mary danced. (i.e., (10a))
b. *Presupposition:* 'Bill believes that Mary danced or
Bill believes that Mary did **not** dance'
⇒ c. Bill believes that Mary did **not** dance.

While the EM-presupposition yields the desired neg-raising effect for reports with declarative complements (see, e.g., (11)), its contribution is already asserted by reports in which a neg-raising predicate combines with a polar interrogative complement. In these reports, the asserted content of the complement is trivial relative to the presupposition of the relevant occurrence of *believe* [66, p. 106]:

(12) a. *Bill **believes whether** Mary danced.
[$\equiv$ Bill believes: Mary danced or Mary did **not** dance]
b. *Presupposition:* 'Bill believes that Mary danced or
Bill believes that Mary did **not** dance'
⇒ c. $\top$

Since this triviality is systematic [= is independent of the sentence's lexical material], it is perceived as a grammatical unacceptability [22, 66].

The above suggests that, if *imagine* were a neg-raising verb (s.t. its semantics would involve an EM-presupposition), its combination with a polar interrogative complement would likewise result in a logical triviality. Some work on neg-raising (e.g., [16]) assumes that *imagine* in fact "license[s] the neg-raising inference, at least for some speakers, in the present simple" [55, p. 93] (see (13), due to Özyıldız [55]):

(13) a. I **don't** imagine that dragons will invade Wisconsin.
$\Rightarrow$ b. I imagine that dragons **won't** invade Wisconsin.

However, as Özyıldız acknowledges, *imagine* often loses its neg-raising property when it occurs in progressive aspect (see (14)):

(14) a. I'm **not** imagining that dragons will invade Wisconsin.
$\not\Rightarrow$ b. I'm imagining that dragons will **not** invade Wisconsin.

Özyıldız attributes this change in neg-raising behavior to the fact that progressive occurrences of *imagine* saliently have an eventive reading. On this reading, *imagine* denotes the agent's relation to a counterfactual event or (visual) scene (see the paraphrase of (14a) in (15a)), rather than to a possible proposition or a possible fact (see the paraphrase of (14a) in (15b)).

(15) a. I'm not {visualizing, forming a mental image of} **a (future) event in which** dragons are invading Wisconsin.
$\not\equiv$ b. I'm not conjecturing **the possibility** that dragons will invade Wisconsin.

Since the occurrence of *imagine* in (2a-i) [*John imagined that a woman danced*] saliently has an eventive reading (see the paraphrase of the present-tense counterpart of (2a-i) in (16b)), it does not allow neg-raising (see (17)):

(16) a. John is imagining that a woman is dancing.
$\equiv$ b. John is {visualizing, forming a mental image of} **a (counterfactual) event/visual scene in which** a woman is dancing.

(17) a. John is **not** imagining that a woman is dancing.
$\not\Rightarrow$ b. John is imagining that a woman is **not** dancing.

The salience of the eventive reading of (2a)/(16a) also explains why this report sounds more natural when *imagine*—as well as the embedded predicate, *dance*—has progressive aspect (as in (16a)).

2.3 Veridicality

To account for the fact that *imagine* rejects polar interrogative complements (see (5d)), Egré [20] has proposed to restrict *whether*-clauses to the complements of veridical verbs. The latter are verbs like *know* and *prove* that "entail the truth of [their] complement when used in the positive declarative form, namely if [they] satisfy the schema $Vp \Rightarrow p$, where p is a *that*-clause" [20, p. 13]. For the verb *know*, this schema is exemplified in (18):

(18) a. Kim knows that a dog barked. $\Rightarrow$ b. 'A dog barked (in @)'

(19) a. Ron remembers that a dog barked. $\Rightarrow$ b. 'A dog barked (in @)'

White [73] calls the biconditional connection between veridicality and *whether*-licensing *Generalization* $\mathbb{V}$ (for '$\mathbb{V}$eridicality': "a predicate is responsive iff it is veridical"). In [72], he calls entailments like (18) and (19) *veridicality inferences*. This term is also used by Jeong (in [35]) and Mayr (in [49]).

Since Egré's definition of veridicality identifies the occurrence of *remember* from (19a) as veridical, it predicts that *remember* also licenses *whether*-clauses. This prediction is in line with the intuitive acceptability of (2b-ii). However, the veridicality of *remember* varies with the particular linguistic and real-world context.[4] Thus, while *remember* triggers the expected entailment in the (veridical) perception context from (20), it does not trigger this entailment in the dream context from (21).[5] Below, the contextually provided underlying experience (in (20)/(21): visual perception resp. dreaming) is highlighted in grey:

(20) *Context:* During last week's picnic in the park, John saw a woman dance.

a. Now, he remembers that the woman was dancing.

$\Rightarrow$ b. 'A woman was dancing in @'

(21) *Context:* After the picnic, John dozed off and dreamt of a hippo singing.

a. Now, he remembers that the hippo was singing.

$\not\Rightarrow$ b. 'A hippo was singing in @'

The non-veridicality of the occurrence of *remember* in (21a) is supported by the observation that this sentence allows denying the truth of its complement (see (22)). This differs from the veridical occurrence of *remember* in (20a), whose combination with the negation of the sentence's complement yields a contradiction (see (23)):

(22) John remembers that a hippo was singing (in his dream), but in fact, no hippo sang.

---

[4] The contextual variance of veridicality properties is not specific to *remember*. Rather, it holds for most recollection and attention verbs, as is evidenced in (†) and (‡):

(†) *Context:* Last week, John {imagined, hallucinated, dreamt of} a hippo singing.

a. Now, he {recalls, recollects, reminisces, has forgotten} that the hippo was singing.

$\not\Rightarrow$ b. 'A hippo was singing in @'

(‡) *Context:* John {is imagining, hallucinating, dreaming of} a hippo singing.

a. He {notices, observes} that the hippo is holding a microphone.

$\not\Rightarrow$ b. 'A hippo is holding a microphone in @'.

[5] This phenomenon is close in spirit to the phenomenon of factivity alternation (see [9, 41, 50, 54]). The behavior of (20)–(21) differs from factivity alternation in being conditioned neither by the type of the complement [9, 41, 50] nor by prosody [35, 53].

(23) #John remembers that a woman was dancing (in the park), but in fact, no woman danced.

The non-veridicality of (21a) suggests that, when it is interpreted against the context from (21), *remember* also does not—or not readily—accept *whether*-complements. This is indeed the case, as is shown in (25) (contrasted with the standard/expected case in (24)):

(24) John remembers whether a woman danced. (✓ given context (20))

(25) John remembers whether a hippo sang. (# given context (21))

Note, however, that *Generalization* $\mathbb{V}$ would still fail to predict the acceptability of the sofa-sentence from (8a). This is so since the occurrence of *imagine* in the *that*-clause counterpart of this sentence (in (26a)) is not obviously veridical (see (26)):

(26) a. I am **imagining that** the new sofa will fit into my living room.
$\not\Rightarrow$ b. 'The new sofa will fit into my living room (in @)'

3 Experiential Parasitism

My discussion of the difference in veridicality between (20a) and (21a) has already suggested that the experience whose content is remembered (in (20)/(21): visual perception resp. dreaming) is relevant to the veridicality of the respective use of *remember*—and attendantly, to the question whether this occurrence licenses *whether*-complements. I trace the difference between (24) [acceptable *remember whether*] and (25)/(2a-ii) [deviant {*remember*, *imagine*} *whether*] to the fact that the semantic complements of the relevant uses of *remember* and *imagine* take their content from the content of an underlying experience. In my work on experiential attitudes (see [45, 46]), I have called this dependence *experiential parasitism*.

Note: Since it relies on experience-dependence, my account of '*whether*-rejection' is restricted to experiential (!) uses of *remember* and *imagine* [= to uses that have an eventive reading]: While this account can be generalized to experiential uses of other fiction verbs (e.g., *dream*, *hallucinate*) and to verbs of recollection and attention (e.g., *notice*, *observe*; see fn. 4), it cannot explain the deviance of suppositional uses of *imagine whether* (see (27b)):

(27) a. Ida imagines that there are infinitely many twin primes.
b. *Ida imagines whether there are infinitely many twin primes.

However, since suppositional imagining—like semantic remembering—is commonly taken to be a distinct attitude from experiential imagination [1, 58] (s.t. it is associated with a different lexical entry [64]), this does not pose a problem for my account.

3.1 Experientially Parasitic Attitudes

Experientially parasitic attitudes are mental states like episodic remembering and 'referential' imagining whose content depends on the content of another experience. To make this dependency explicit, I follow Maier [47, 48] in calling the dependent attitude (in (20a): remembering) the *parasite* (or *parasite attitude*). I call the underlying experience (in (20a): visual perception) the *host* (or *host experience*). For perspicuity, I hereafter mark the predicate for the parasite attitude with a grey frame. The predicate for the host experience is highlighted in grey, as in (20) and (21).

In the contexts from (20) and (21), (20a) and (21a) can be paraphrased by reports that make explicit reference to the host experience (see (28) resp. (29)):

(28) a. John remembers that *the* woman whom he had seen at the park was dancing in the park.

b. John remembers a particular fact about a certain visual scene (from the park), viz. that a woman was dancing in this scene.

(29) a. John remembers that the hippo from his dream was singing in his dream.

b. John remembers a particular fact about a certain oneiric scene, viz. that a hippo was singing in this scene.

Attitudinal parasitism is a well-documented phenomenon. However, following Karttunen [37] and Heim [30], its investigation has focused on doxastic parasitism (i.e., an attitude's parasitic dependence on the agent's *belief*). This kind of parasitism is exemplified in (30):

(30) a. Mary hopes that the man whom she believes to be the murderer is arrested. [48, p. 142]

b. Mary is imagining what it would be like if the man whom she believes to be the murderer was arrested. [48, p. 142]

The literature on parasitic attitudes even contains some few cases of experiential parasitism. These include Ninan's example from (31) [52, p. 18] and Blumberg's example from (32) [7, ex. (102)]:

(31) Ralph is imagining that the man whom he sees sneaking around on the waterfront is flying a kite in an alpine meadow.

(32) John is imagining that the woman who threatened him in his dream last night is swimming in the sea.

In (28)–(32), the parasitic behavior of the reported attitude (there: remembering resp. imagining) is made explicit by the presence of predicates for the host experience (there: *believe*, *see*, resp. *dream*). However, a parasitic analysis can also be triggered in the absence of such predicates. To see this, consider the imagination report in (33a) (modelled on Blumberg's [6] 'burgled Bill'-example):

(33) *Context:* Last night, Paul dreamt of a tattooed woman (no particular one whom he has come across in real life)

a. Now, he is imagining that she has clear, untattooed skin.

$\not\equiv$ i. *de re:* There exists a specific tattooed woman (in @) of whom Paul is imagining that she has clear, untattooed skin. (✗)

$\not\equiv$ ii. *de dicto:* Paul is imagining an inconsistent fact, viz. that some woman simultaneously does and does not have tattoos. (✗)

$\equiv$ iii. *de credito:* Paul is imagining that the tattooed woman from his dream has clear, untattooed skin. (✓)

The parasitic interpretation of *imagine* in (33a) is triggered by the observation that, given the context from (33) and the attendant non-existence of said woman in the real world, (33a) is false on its *de re*-reading (which gives the pronoun *she* [= *a tattooed woman*] a specific interpretation; see (33a-i)) and that (33a) is contradictory on its *de dicto*-reading (see (33a-ii)). The parasitic interpretation is then prompted by the observation that (33a) has plausible truth-conditions on a reading that evaluates *she* at some other world or (possible) situation that is different both from the actual world/situation and from Paul's imagination alternatives (see (33a-iii)). The name for this reading, i.e., *de credito*, is due to [75].

To capture parasitic dependencies like the above, Blumberg [6] has proposed to use Percus' [59] *Index Variables-approach* with distinct variables for the alternatives that are introduced by the parasite attitude (in (33): Paul's imagining), s_2, and for the alternatives that are introduced by the host experience (in (33): Paul's dreaming), s_1. The different readings of the imagination report in (33a) are then given by the LFs in (34). The relevant LF—on which (34a) is true—is given in (34c). This LF assumes that some specific individual who is a woman in s_1 has clear skin in s_2.[6]

(34) a. [a woman-in-**@**][λt. Paul imagines -in-@ [λs_1 [λs_2. t has-clear-skin-in-s_2]]]

b. Paul imagines -in-@ [λs_1 [λs_2. a woman-in-$\boldsymbol{s_2}$ has-clear-skin-in-s_2]]

c. Paul imagines -in-@ $\underbrace{[\lambda s_1\ [\lambda s_2.\ \text{a woman-in-}\boldsymbol{s_1}\text{ has-clear-skin-in-}s_2\ ]\ ]}_{\text{(the designator of) a paired proposition}}$

[6] To make this interpretation possible, I use Kripke-style possible worlds (with a fixed-domain modal logic), in which the same individual can inhabit more than one world (*Haecceitism*; see [36]).

The semantic complement of *imagine* in (34a–c) is a *paired proposition* (i.e., a function from host experience situations to sets of parasite attitude situations, type $\langle s, \langle s, t\rangle\rangle$). While this paired proposition is a constant function in (34a) and (34b) (this holds since the *de re*- and *de dicto*-reading can already be captured by using only @ and s_2), it is a non-constant function in (34c). This function sends host experience alternatives, s_1, to a proposition [= the set of situations denoted by '$\lambda s_2. \ldots$'] that depends on s_1.[7]

3.2 Experientially Parasitic Remembering

Using Blumberg's extended *Index Variables-approach*, the most plausible readings of (20a)/(28) and (21a)/(29) are given in (35), respectively in (36):

(35) [something-in-@] [λt. John remembers -in-@
[λs_1 [λs_2. t is a woman-in-s_1 and dances-in-s_2]]]

(36) John remembers -in-@ [λs_1 [λs_2. a hippo-in-s_1 sings-in-s_2]]

Note that the analysis in (35) raises the indefinite article *a* (in *a woman*) [analyzed as 'something'] outside the scope of *remember*. This move is justified by the observation that perception verbs (paradigmatically: veridical uses of *see*) require a specific interpretation of their embedded subject (see the validity of the inference in (37); based on [3, 18]) and that the same individual (of which 'woman' and 'dances' are attributed) exists in both John's perception- and in his memory-alternatives.

(37) a. John sees a woman dance.
$\Rightarrow$ b. There is someone/some thing (in @), viz. a particular woman, whom/that John sees dance.

---

[7] An anonymous reviewer has suggested to generalize the arity of paired propositions from 2 to $n \in \mathbb{N}$ situation arguments. This generalization would facilitate the interpretation of 'doubly parasitic' reports like (%), whose content depends on two (or more) experiences:

(%) *Context:* Last week, Paul dreamt of a tattooed woman. Yesterday, he hallucinated a goat-headed man.

a. Now, he is imagining that she is playing chess with him [$\neq$ Paul].

$\equiv$ b. Paul is imagining that the tattooed woman from last week's dream is playing chess with the goat-headed man from yesterday's hallucination.

This observation is, in fact, in line with Blumberg's own proposal from [7, Ch. 5.2]. However, to keep my semantics as simple as possible—and since experiential memory reports (*contra* imagination reports) are typically parasitic on a single experience (see [10, 64, 67])—, I restrict my presentation to paired propositions. The possibility of, and limits on, extending paired propositions to sets of ordered n-tuples are discussed in some detail in [44].

Since, in his memory, John conceptualizes this individual as 'woman',[8] (35) analyzes the restrictor, *woman*, of the embedded subject as a referentially opaque expression (see [61]) that resists substitution by an extensional equivalent. (35) captures the referential opacity of *woman* by leaving this predicate inside the scope of *remember*. The opaque interpretation of *woman* in (20a) is supported by the observation that, in a scenario in which John did not recognize the woman from the park as a member of the local fencing club, the content of his remembering could only be correctly described by (38a), but not by the result, (38c), of replacing *woman* in (38a) by *member of the fencing club*:

(38) a. John remembers that a/the woman was dancing (in the park). **(T)**
b. The woman (who was dancing in the park) is a member of the fencing club. **(T)**

$\not\Rightarrow$ c. John remembers that a member of the fencing club was dancing. **(F)**

Note that, in (35) and (36), the embedded predicate (i.e., *dances* resp. *sings*) is evaluated at the parasite [= memory-]alternative, s_2, while the restrictor of the embedded subject (i.e., *woman* resp. *hippo*) is evaluated at the host alternative, s_1. The evaluation of the embedded subject at s_1 is needed to 'anchor' the actual referent of *a/something* in (35) to the host experience. The evaluation of the embedded predicate at s_2 is required by Percus' *Generalization X*. The latter demands that the situation variable that a predicate selects for must be co-indexed with the nearest lambda above it (see [59, p. 201]).

4 Veridicality Inferences in *Remember*-Reports

4.1 Semantics for Declarative Remember

Admittedly, the interpretation of the restrictor and the predicate in (35) at different attitudinal alternatives goes against the intuition from (28) (which interprets both *woman* and *dance* w.r.t. John's visual scene from the park). My semantics for declarative uses of *remember* (in (39)) solves this problem by filling both argument slots of the paired proposition R with the same situation variable (see '$R(s_1, s_1)$' resp. '$R(s_2, s_2)$'). To stay as close as possible to familiar attitude semantics (which treats *remember* as a relation between an attitudinal agent, z, and a proposition/set of situations, $\lambda s_2. \ldots$; see [51]), the semantics in (39) 'converts' the host-parameter of R (in (34): s_1) into an 'experientiality presupposition' on the parasite attitudinal alternatives s_2 (underlined in (39)). This presupposition demands that, in the world, $w_@$, of which @ is a spatio-temporal and/or informational part, z has (had) an experience with content $p_{\langle s,t\rangle}$. I abbreviate this requirement as '$exp_{w_@}(z, p)$':

[8] The predicate *woman* is thus epistemically positive in the sense of Dretske [19] (see [3, 4]).

(39) $[\![\text{remember}_{\text{DECL}}]\!]^{@}$
$= \lambda R^{\langle s,\langle s,t\rangle\rangle} \lambda z^{e}.\, \mathit{remember}_{@}\big(z, \lambda s_2\colon \underline{\mathit{exp}_{w_@}(z, \lambda s_1.\, R(s_1, s_1))}.\, R(s_2, s_2)\big)$

The presuppositional status of the experientiality requirement, $\mathit{exp}_{w_@}(z, \lambda s_1.\, R(s_1, s_1))$, is supported by the observation that this requirement is preserved under negation (see (40))[9]:

(40) a. John does **not** remember that a woman danced.
$\Rightarrow$ b. A woman danced.

My semantics for declarative *remember* enables the compositional interpretation of (36) and (35) as (41) respectively as (42). In line with my observation from Sect. 3.2, these interpretations provide a semantics for the *de dicto*-reading of (2a-i) [*John imagines that a woman danced*] and the *de re*-reading of (2b-i) [*John remembers that a woman danced*], respectively. Obviously, given the 'right' LFs (with the embedded subject or its quantifier raised out of the scope of *remember*), (39) can also be used to obtain a *de re*-reading of (2a-i) (assuming that there is a specific such hippo in @) and a *de dicto*-reading of (2b-i).

(41) $[\![\text{John}\ \boxed{\text{remembers}_{\text{DECL}}}\text{-in-@}\ [\lambda s_1\ \boxed{[\lambda s_2.}\ \text{a hippo-in-}s_1\ \text{sings-in-}s_2\boxed{]}\ ]]\!]$
$\equiv [\![\text{remember}_{\text{DECL}}]\!]^{@}\big([\![\text{John}]\!], \lambda s_1 \lambda s_2\, (\exists x)[\mathit{hippo}_{s_1}(x) \wedge \mathit{sing}_{s_2}(x)]\big)$
$= \mathit{remember}_{@}\big(\mathit{john}, \lambda s_2\colon \underline{\mathit{exp}_{w_@}(\mathit{john}, \lambda s_1 \exists y.\, \mathit{hippo}_{s_1}(y) \wedge \mathit{sing}_{s_1}(y))}.$
$\exists x.\, \mathit{hippo}_{s_2}(x) \wedge \mathit{sing}_{s_2}(x)\big)$

(42) $[\![[\text{sth.-in-@}][\lambda t.\, \text{John}\ \boxed{\text{remembers}_{\text{DECL}}}\text{-in-@}$
$[\lambda s_1\ \boxed{[\lambda s_2.}\ t\ \text{is a woman-in-}s_1\ \text{and dances-in-}s_2\boxed{]}\ ]\ ]]\!]$
$\equiv [\![\text{sth.}]\!]^{@}\big(\lambda x.\, [\![\text{remember}_{\text{DECL}}]\!]^{@}$
$\big([\![\text{John}]\!], \lambda s_1 \lambda s_2.\, \mathit{woman}_{s_1}(x) \wedge \mathit{dance}_{s_2}(x)\big)\big)$
$= (\exists x)\big[\mathit{remember}_{@}\big(\mathit{john}, \lambda s_2\colon \underline{\mathit{exp}_{w_@}(\mathit{john},}$
$\underline{\lambda s_1.\, \mathit{woman}_{s_1}(x) \wedge \mathit{dance}_{s_1}(x))}.\, \mathit{woman}_{s_2}(x) \wedge \mathit{dance}_{s_2}(x)\big)\big]$

The interpretations in (41) and (42) only have classical propositional (type-$\langle s, t\rangle$) arguments. As is expressed in their intuitive paraphrases in (28) and (29), the semantic content of the complement ('a woman dances' resp. 'a hippo sings') is true at both the memory- and the experience-alternatives.

[9] *Note:* The status of the experientiality requirement changes (from a presupposition to an assertion) when the declarative complement is replaced by a gerundive small clause (in (§); see [46]):

(§) a. John does **not** remember a woman dancing.
$\not\Rightarrow$ b. A woman was dancing.

This is a consequence of the epistemic positiveness of *that*-clause reports (compared to the epistemic neutrality of gerundive small clauses; see [3, 19]), rather than of the eventive interpretation of the complement (see Sect. 2.2). As has been argued in [64, 71], even *that*-clause complements can be used to report eventive remembering.

4.2 Capturing Factivity Variation

Notably, by itself, the interpretation of (20a) in (42) does not (yet) capture the veridicality inference in (20). To validate this inference, we need one of the following:

- a context like (20) that identifies the particular mode of the underlying experience (together with assumptions about the veridicality of this mode)
- an explicit linguistic specification of the particular mode of the experience (along the lines of (28a) and (28b); together with assumptions about the veridicality of this mode)
- an assumption about the default veridicality of the underlying experience.

Each of these assumptions further specifies the experience predicate, *exp*, in (39) with respect to its veridicality. I assume that this specification proceeds through Jeong's [35] veridicality operator $\mathcal{V}$ (in (43)). This operator adds an assumption about the relation between the propositional content, p, of the experience and the relevant part, $w_@$, of the actual world at which the report is evaluated (see the highlighted conjunct, '$p_{w_@}$', in (43)).

(43) $\mathcal{V}\big(exp_{w_@}(z, p)\big) := exp_{w_@}(z, p) \wedge p_{w_@}$

In (42), the specification of veridicality is triggered by the context from (20)—especially by the verb *see* and a default assumption about the veridicality of visual perception. This specification adds a 'veridicality conjunct', $p_{w_@}$, in the presupposition of the agent's (in (42): John's) attitudinal alternatives, s_2 (see (44a)). This conjunct validates the veridicality inference on the global interpretation of the experientiality presupposition (in (44b)):

(44) a. (42) + the 'veridicality-inducing' context from (20) $\equiv$ (42) + $\mathcal{V}$

$$= (\exists x)\big[remember_@\big(john, \lambda s_2\colon \underline{\mathcal{V}\ (exp_{w_@}(john, \lambda s_1 .\, woman_{s_1}(x) \wedge dance_{s_1}(x)))}.\, woman_{s_2}(x) \wedge dance_{s_2}(x)\big)\big]$$

$$\equiv (\exists x)\big[remember_@\big(john, \lambda s_2\colon \underline{exp_{w_@}(john, \lambda s_1 .\, woman_{s_1}(x) \wedge dance_{s_1}(x)) \wedge (woman_{w_@}(x) \wedge dance_{w_@}(x))}\,.\, woman_{s_2}(x) \wedge dance_{s_2}(x)\big)\big]$$

b. $(\exists x)\big[\big((woman_{w_@}(x) \wedge dance_{w_@}(x)) \wedge exp_{w_@}(john, \lambda s_1 .\, woman_{s_1}(x) \wedge dance_{s_1}(x))\big) \wedge remember_@(john, \lambda s_2 .\, woman_{s_2}(x) \wedge dance_{s_2}(x))\big]$

$$\Rightarrow (\exists x .\, woman_{w_@}(x) \wedge dance_{w_@}(x)) = [\![\text{A woman dance(d)}]\!]^@$$

Let us now turn to the 'hippo memory'-report in (21a): Since the context from (21) does not show any of the veridicality-inducing factors from the beginning of this subsection (i.e., context, linguistic specification, or default veridicality), (21) does not validate a veridicality inference. To capture the non-validity of (21a) $\not\Rightarrow$ (21b) in the context from (21), one can proceed in either of two ways, viz. (A) by assuming that the context from (21) does not further specify the (positive or negative) veridicality properties of the presupposed experience or (B) by assuming that the context from (21) (in particular, the verb *dreamt*) introduces an anti-veridicality operator, $\mathcal{A}$ (in (45)).

(45) $\mathcal{A}\big(exp_{w@}(z, p)\big) := exp_{w@}(z, p) \wedge \neg p_{w@}$

Option (A) is the simplest route to explaining the non-validity of the inference in (21). However, for reasons that will become clear below, I instead use option (B). This option is independently motivated by the fact that dreaming—like most instances of imagining—is commonly classified as an anti-veridical (as opposed to a merely non-veridical) attitude (see, e.g., [21, 25, 27]). The non-validity of (21) is captured in (46a):

(46) a. (41) + the 'anti-veridicality inducing' context from (21) $\equiv$ (41) + $\mathcal{A}$

$= remember_{@}\big(john, \lambda s_2\text{: } \underline{\mathcal{A}\ (exp_{w@}(john, \lambda s_1 \exists y.\, hippo_{s_1}(y) \wedge sing_{s_1}(y)))}.\ \exists x.\, hippo_{s_2}(x) \wedge sing_{s_2}(x)\big)$

$\equiv remember_{@}\big(john, \lambda s_2\text{: } \underline{exp_{w@}(john, \lambda s_1.\, \exists y.\, hippo_{s_1}(y) \wedge sing_{s_1}(y)) \wedge \neg(\exists z.\, hippo_{w@}(z) \wedge sing_{w@}(z))}\ .\ \exists x.\, hippo_{s_2}(x) \wedge sing_{s_2}(x)\big)$

b. $\big(\neg(\exists z.\, hippo_{w@}(z) \wedge sing_{w@}(z)) \wedge exp_{w@}(john, \lambda s_1 \exists y.\, hippo_{s_1}(y) \wedge sing_{s_1}(y))\big) \wedge remember_{@}(john, \lambda s_2 \exists x.\, hippo_{s_2}(x) \wedge sing_{s_2}(x))$

$\Rightarrow \neg(\exists z.\, hippo_{w@}(x) \wedge sing_{w@}(x))$

$= [\![\text{It is not the case that a hippo sings/sang}]\!]^{@}$

Since (46a) entails the negation of the embedded *that*-clause from (21a) (see (46b)), option (B) is strictly stronger than option (A).

Note that, while $\mathcal{V}$ and $\mathcal{A}$ are required to capture the discussed veridicality- resp. anti-veridicality inferences, one need not introduce these operators in all cases. This is especially so when none of the requirements (in bullet points) from the beginning of this subsection are in place. Waiving the introduction of $\mathcal{V}$ (resp. of $\mathcal{A}$) further facilitates the modelling of 'non-commital' uses of *remember* like (47b).[10]

(47) a. A: Was it raining last week?

b. i. B_1: I remember that it was (but I could be wrong).

ii. B_2: I'm not sure, but John remembers that it was.

[10] I owe this example to Reviewer 6.

4.3 Semantics for Interrogative Remember

My previous discussion has focused on declarative occurrences of *remember* that combine with a *that*-clause complement. To account for the acceptability of *remember whether*-constructions (see (24) vs. (25)), I generalize the semantics from (39) to a semantics for interrogative uses of *remember*. This is easily accomplished: to allow that *remember* accepts interrogative complements, I first generalize the clausal argument of $remember_{\text{DECL}}$ (in (39): a paired proposition, R) to a paired question, Q (type $\langle\langle s, \langle s, t\rangle\rangle, t\rangle$). The latter is a characteristic function of a set of paired propositions (see [42]). My generalized semantics for *remember* in (48) provides the 'right' [= correctly typed] arguments for this function.

(48) $[\![\text{remember}_{\text{INQ}}]\!]^{@}$
$= \lambda Q^{\langle\langle s,\langle s,t\rangle\rangle,t\rangle} \lambda z^{e}.\, remember'_{@}(z, \lambda p : \underline{exp_{w_{@}}(z, p)}.\, Q(\lambda\langle s_1, s_2\rangle.\, p_{s_2}))$

The different-type complements of declarative and interrogative *remember* require that *remember* is treated as a different-type relation in (48). In particular, while (39) analyzes *remember* as a relation to a proposition [= a set of situations, $\lambda s_2. \ldots$], the entry in (48) analyzes *remember* as a relation to a question [= a set of **propositions**, $\lambda p^{\langle s,t\rangle}. \ldots$]. To capture this difference, (48) uses a different non-logical constant, i.e., *remember'* (with a prime,'; type $\langle\boldsymbol{\langle}\langle s, t\rangle, \boldsymbol{t}\boldsymbol{\rangle}, \langle e, \langle s, t\rangle\rangle\rangle$), than (39) (whose constant, *remember*, is of type $\langle\langle s, t\rangle, \langle e, \langle s, t\rangle\rangle\rangle$).

Assuming that paired propositions can be lifted to paired questions, (48)—like (39)—can be used for the interpretation of declarative *remember*-reports. This interpretation uses the type-shifter E- PARA in (49) (based on [42, p. 334]). The latter sends sets of ordered pairs of situations, R [= coded paired propositions], to their powersets $\mathcal{P}(R)$, i.e., to downward-closed sets of paired propositions. (This move follows [13, 14], who propose this idea for the conversion of classical propositions into questions). The downward closure of these sets effects that, if these sets include the 'paired proposition'-interpretation of a declarative R, they also include every paired proposition that is informationally stronger than R. My use of E- PARA (rather than of a suitably-typed version of Partee's shifter IDENT; see [68]) is motivated by my wish to capture inferences from *that*-clause complements to *whether*-clause complements with a more general lexical content (e.g., (50); see Sect. 5.1).

(49) $\text{E- PARA} := \lambda S^{\langle s,\langle s,t\rangle\rangle} \lambda R^{\langle s,\langle s,t\rangle\rangle} (\forall\langle s_1, s_2\rangle)[R(s_1, s_2) \rightarrow S(s_1, s_2)]$

(50) a. John remembers that a woman sang and danced.
⇒ b. John remembers whether a woman danced.

When applied to the 'paired propositional'-analysis of the declarative clause *that a woman danced* (see (35)), this shifter yields the paired question in (51):

(51) $\text{E- PARA}([\![\, [\lambda s_1\ \boxed{[\lambda s_2.}\ t \text{ is a woman-in-} s_1 \text{ and dances-in-} s_2 \boxed{]}\ ]\,]\!])$
$= \lambda R\, (\forall\langle s_1, s_2\rangle)[R(s_1, s_2) \rightarrow (woman_{s_1}(x) \wedge dance_{s_2}(x))]$

(52) uses the above to account for the veridicality inference in (20):

(52) a. $[\![\text{something}]\!]^{@}\big(\lambda t.\,[\![\text{remembers}_{\text{INQ}}]\!]^{@}$
$\big(\text{E-PARA}\big([\![\,[\lambda s_1\ \boxed{[\lambda s_2.}\ t \text{ is a woman-in-}s_1 \text{ dances-in-}s_2\boxed{]}\ ]\,]\!]\big)\big)\big) + \mathcal{V}$

$= (\exists x)\big[\mathit{remember}'_{@}\big(\mathit{john}, \lambda p\colon\ \mathcal{V}\ (\mathit{exp}_{w@}(\mathit{john}, p)).$
$(\forall \langle s_1, s_2\rangle .\ p_{s_2} \rightarrow (\mathit{woman}_{s_1}(x) \wedge \mathit{dance}_{s_2}(x)))\big)\big]$

$\equiv (\exists x)\big[\mathit{remember}'_{@}\big(\mathit{john}, \lambda p\colon \mathit{exp}_{w@}(\mathit{john}, p) \wedge\ p_{w@}\,.$
$(\forall \langle s_1, s_2\rangle .\ p_{s_2} \rightarrow (\mathit{woman}_{s_1}(x) \wedge \mathit{dance}_{s_2}(x)))\big)\big]$

b. $(\exists x)(\exists q)\big[\big((\ q_{w@} \wedge \mathit{exp}_{w@}(\mathit{john}, q)) \wedge$
$(\forall \langle s_1, s_2\rangle .\, q_{s_2} \rightarrow (\mathit{woman}_{s_1}(x) \wedge \mathit{dance}_{s_2}(x)))\big) \wedge$
$\big[\mathit{remember}'_{@}\big(\mathit{john}, \lambda p \forall \langle s_1, s_2\rangle .\ p_{s_2} \rightarrow (\mathit{woman}_{s_1}(x) \wedge \mathit{dance}_{s_2}(x))\big)\big]\big]$

$\Rightarrow (\exists x.\, \mathit{woman}_{w@}(x) \wedge \mathit{dance}_{w@}(x)) \ = \ [\![\text{A woman danced}]\!]^{@}$

5 Remembering and Imagining *Whether*

With the semantics for interrogative occurrences of *remember* in place, we are now ready to specify the semantics for the polar interrogative complementizer *whether*. Based on my previous discussion—and on my wish to derive the acceptability of *remember whether* and the deviance of *imagine whether* from "independently motivated features of their lexical semantics" [65, p. 413]—, I expect that this semantics compositionally combines with the semantics of interrogative *remember* as desired, but generates a systematic triviality or contradiction when combining with the semantics of *imagine*.

5.1 Remembering Whether

In providing a semantics for the polar complementizer, I start from the classical entry for *whether* (due to [29, 38, 40]). This entry interprets *whether* through the disjunction of a proposition and its negation. In particular, on this semantics, a phrase of the form *whether q* denotes the set of propositions $\{p : p \text{ is true } \& \ (p = q \vee p = \neg q)\}$ (see (53)). I hereafter call the first conjunct in this term, $p_{w@}$ [read: 'p is true'], the 'actuality assumption':

(53) $[\![\text{whether}]\!]_{\text{CLASSIC}} = \lambda q^{\langle s,t\rangle} \lambda p^{\langle s,t\rangle} [\underbrace{p_{w@}}_{\text{actuality assumption}} \wedge (p = q \vee p = \neg q)]$

To allow that the entry for *whether* accepts paired propositions, I generalize this entry (incl. the actuality assumption) to (54). In line with recent attempts to capture the entailment behavior of interrogative complements (see [13, 14], discussed in Sect. 4.3), this generalization replaces the identity relation between the members, R, of the range set and the 'paired proposition'-denotation, S, of the embedded TP by a semantic inclusion relation.

(54) $[\![\text{whether}]\!]_{\text{PAIRED}}$
$$= \lambda S^{\langle s,\langle s,t\rangle\rangle} \lambda R\big[\underbrace{R(w_@, w_@)}_{\textbf{generalized}\text{ actuality assumption}} \wedge \big((\forall\langle s_1, s_2\rangle.\, R(s_1, s_2) \to S(s_1, s_2)) \vee (\forall\langle s_1, s_2\rangle.\, R(s_1, s_2) \to \neg S(s_1, s_2))\big)\big]$$

For a paired proposition (e.g., the denotation, (55a), of the TP in (50a))—enriched by the assumption of its truth-at-$w_@$—and its polar variant (i.e., the denotation, (55b), of the TP in (50b)), this inclusion relation enables the entailment from 'true *that*-(55a)' to '*whether*-(55b)' (see (56)):

(55) a. $[\lambda s_1\ [\lambda s_2.\ t$ is a woman-in-s_1 sings-in-s_2 and dances-in-s_2]]
b. $[\lambda s_1\ [\lambda s_2.\ t$ is a woman-in-s_1 dances-in-s_2]]

(56) a. E- PARA($[\![$ $[\lambda s_1\ [\lambda s_2.\ t$ woman-in-s_1 sings-in-s_2 and dances-in-s_2]] $]\!]$) + TRUE
$$= \lambda R\big[R(w_@, w_@) \wedge \big(\forall\langle s_1, s_2\rangle.\, R(s_1, s_2) \to (woman_{s_1}(x) \wedge sing_{s_2}(x) \wedge dance_{s_2}(x))\big)\big]$$

$\subseteq$ b. $[\![\text{whether}]\!]_{\text{PAIRED}}$($[\![$ $[\lambda s_1\ [\lambda s_2.\ t$ woman-in-s_1 dances-in-s_2]] $]\!]$)
$$= \lambda R\big[R(w_@, w_@) \wedge \big((\forall\langle s_1, s_2\rangle.\, R(s_1, s_2) \to (woman_{s_1}(x) \wedge dance_{s_2}(x))) \vee (\forall\langle s_1, s_2\rangle.\, R(s_1, s_2) \to \neg(woman_{s_1}(x) \wedge dance_{s_2}(x)))\big)\big]$$

In (56a), the additional assumption $R(w_@, w_@)$ [obtained by the enrichment with 'TRUE'] is required by the (generalized) actuality assumption in my semantics for *whether*. As a consequence of this assumption, declarative *remember*-reports can only entail their polar interrogative counterparts if the TP in the declarative complement is true at $w_@$. In (44), this assumption is added through the contextually supplied veridicality operator $\mathcal{V}$.

The interpretation of *whether a woman danced* from (56b) felicitously combines with my proposed entry for interrogative uses of *remember*, as expected:

(57) $[\![(24)]\!]^@ \equiv [\![$John **remembers whether** a woman danced$]\!]^@$
$\equiv [\![\text{remember}_{\text{INQ}}]\!]^@([\![\text{John}]\!],$ $[\![\text{whether}]\!]_{\text{PAIRED}}$($[\![$ $[\lambda s_1\ [\lambda s_2.\ t$ woman-in-s_1 dances-in-s_2]] $]\!]$))
$$= (\exists x)\big[remember'_@\big(john, \lambda p\colon \overline{exp_{w_@}}(john, p).\ \big[p_{w_@} \wedge \big((\forall\langle s_1, s_2\rangle.\, p_{s_2} \to (woman_{s_1}(x) \wedge dance_{s_2}(x))) \vee (\forall\langle s_1, s_2\rangle.\, p_{s_2} \to \neg(woman_{s_1}(x) \wedge dance_{s_2}(x)))\big)\big]\big)\big]$$

Note the actuality assumption, $p_{w_@}$, on the set of propositions $\lambda p.\,[\ldots]$ that serves as the semantic complement of *remember* in (57). This assumption (which is already assumed in early Montague-style semantics for *whether*; see (53) is consonant with the veridicality of the relevant use of *remember* (given the context in (20), which anchors John's remembering to John's experience of an actual, real-world event). In the explicitly veridical experience-based interpretation of (24), i.e., (58a), the veridicality of *remember* gives rise to a decidability assumption, intuitively: 'a particular woman either did or did not dance (in the particular spatio-temporal part of $w_@$ that shares the coordinates of John's visual scene from the park)'. In (58b), this assumption is expressed by the conjunction of $p_{w_@}$ and $(\forall \langle s_1, s_2\rangle . p_{s_2} \rightarrow (woman_{s_1}(x) \wedge dance_{s_2}(x))) \vee (\forall \langle s_1, s_2\rangle . p_{s_2} \rightarrow \neg(woman_{s_1}(x) \wedge dance_{s_2}(x)))$. However, this conjunction is already included in (57). As a result, the veridicality assumption leaves the meaning of (24) unchanged.

(58) a. $[\![\text{John } \textbf{remembers whether} \text{ a woman danced}]\!]^@ + \text{Context (20)}$

$$= (\exists x)\big[remember'_@\big(john, \lambda p\colon \underline{\mathcal{V}\,(exp_{w_@}(john, p))}.\; \big[p_{w_@} \wedge ((\forall\langle s_1, s_2\rangle . p_{s_2} \rightarrow (woman_{s_1}(x) \wedge dance_{s_2}(x))) \vee (\forall\langle s_1, s_2\rangle . p_{s_2} \rightarrow \neg(woman_{s_1}(x) \wedge dance_{s_2}(x)))\big]\big)\big]$$

$$\equiv (\exists x)\big[remember'_@\big(john, \lambda p\colon \underline{(exp_{w_@}(john, p) \wedge p_{w_@})}.\; \big[p_{w_@} \wedge \big((\forall\langle s_1, s_2\rangle . p_{s_2} \rightarrow (woman_{s_1}(x) \wedge dance_{s_2}(x))) \vee (\forall\langle s_1, s_2\rangle . p_{s_2} \rightarrow \neg(woman_{s_1}(x) \wedge dance_{s_2}(x)))\big)\big]\big)\big]$$

$\Rightarrow$ b. $(\exists x)\big[(woman_{w_@}(x) \wedge dance_{w_@}(x)) \vee \neg(woman_{w_@}(x) \wedge dance_{w_@}(x))\big]$

$$= [\![\text{A woman did or did not dance}]\!]^@$$

Things are different for the interpretation of *John remembers whether a hippo sang* (i.e., (25)). In this interpretation, the anti-veridicality conjunct, $\neg p_{w_@}$, that the contextually supplied operator $\mathcal{A}$ introduces into the presupposition of *remember′* clashes with the veridicality assumption, $p_{w_@}$, of the lexical entry for *whether*. As a result, the interpretation of (21a) not only fails to give rise to a veridicality inference—it comes out contradictory (and is, hence, false at all evaluation indices). It is this contradictoriness that explains the markedness of (25).

(59) $[\![\text{John } \textbf{remembers whether} \text{ a hippo sang}]\!]^@ + \text{Context (21)}$

$$= remember'_@\big(john, \lambda p\colon \underline{\mathcal{A}\,(exp_{w_@}(john, p))}.\; \big[p_{w_@} \wedge \big((\forall\langle s_1, s_2\rangle . p_{s_2} \rightarrow (\exists x . hippo_{s_1}(x) \wedge sing_{s_2}(x))) \vee (\forall\langle s_1, s_2\rangle . p_{s_2} \rightarrow \neg(\exists y . hippo_{s_1}(y) \wedge sing_{s_2}(y)))\big)\big]\big)$$

$$\equiv remember'_@\big(john, \lambda p\colon \underline{(exp_{w_@}(john, p) \wedge \neg p_{w_@})}.\; \big[p_{w_@} \wedge \big((\forall\langle s_1, s_2\rangle . p_{s_2} \rightarrow (\exists x . hippo_{s_1}(x) \wedge sing_{s_2}(x))) \vee (\forall\langle s_1, s_2\rangle . p_{s_2} \rightarrow \neg(\exists y . hippo_{s_1}(y) \wedge sing_{s_2}(y)))\big)\big]\big)$$

$$\equiv\ \lightning$$

Note that, in contrast to the triviality of *believe whether* (see Sect. 2.2) and the contradictoriness of *imagine whether*, the contradictoriness of (59) is not systematic [= a case of L-analyticity] in the sense of Gajewski [22] (more on which in Sect. 5.2). This is so since this contradiction is brought about by external factors (here: by the anti-veridicality of the host experience which, in turn, introduces the operator $\mathcal{A}$). The anti-veridicality of this experience can even be lexically realized (see (60a)). The difference between a contradictory interpretation (e.g., (60a)) and an admissible, non-contradictory interpretation (e.g., (60b)) is then a question of lexical content alone.

(60) a. #John **remembers whether** the hippo from his dream was singing.

b. John **remembers whether** the hippo which he had been watching during his last visit to San Diego Zoo was singing.

The difference in systematicity between the contradiction in (60a) and the contradiction in (2a-ii) [*John imagined whether a woman danced*] can be used to explain why anti-veridical cases of *remember whether*-reports are often taken to be less 'bad' than *imagine whether*-reports. I will return to this question in the next subsection.

5.2 Imagining Whether

My previous discussion suggests that the veridicality of the host experience is crucial for the acceptability of *remember whether*. To preserve maximal similarity to uses of *remember* that are based on an anti-veridical experience (e.g., (59)), I hardcode '*whether*-rejection' directly into the experientiality presupposition of *imagine* (viz. through the anti-veridicality conjunct, $\neg p_{w_@}$). The relevant entry is given in (61):

(61) $$[\![\text{imagine}]\!]^{@} = \lambda Q^{\langle\langle s,\langle s,t\rangle\rangle,t\rangle}\lambda z.\, imagine'_{@}\big(z, \lambda p \colon \underline{\exists q.\, ((\forall s.\, p_s \rightarrow q_s) \wedge exp_{w_@}(z,q)) \wedge \neg p_{w_@}}\,.\, Q(\lambda\langle s_1, s_2\rangle.\, p_{s_2})\big)$$

Since *imagine* also embeds constituent questions (as I have illustrated in (5)), I model the entry in (61) on the entry for interrogative *remember* from (48). To capture the fact that imagination typically does not preserve all content of the host experience,[11]

[11] Of course, this only holds if imagining is parasitic on an underlying [= host] experience. The latter need not be the case, as is shown by (∗):

(∗) a. Inger is imagining that a fairy is flying above.

≡ b. Inger is imagining that a fairy in her imagined visual scene is flying above (in this scene).

Since, in (∗), Inger's experience (viz. her non-veridical visualization of a fairy and her flying) [= the parasite] depends on her imagination [= the host], the order of the lambdas over Inger's imagination- and her experience-alternatives is inversed w.r.t. (36). In (∗∗), the interpretation of *is flying above* at s_1 is required by Percus' *Generalization X*:

(∗∗) Inger imagines -in-@ $[\lambda s_2$ $[\lambda s_1.$ a fairy-in-s_1 flies-above-in-s_1]]

The above differences notwithstanding, the entry for *imagine* from (61) still provides an intuitive semantics for (∗∗) (in (∗ ∗ ∗)). This is so since this entry does not specify the particular dependency relation between imagination and the experience (as I have observed for *remember* in Sect. 4.1).

I replace 'p' in the experientiality presupposition by the variable for a more general proposition q (s.t. $\forall s.\, p_s \rightarrow q_s$). This replacement enables the evaluation of *the hippo* in *John is imagining whether the hippo is pantomining* (see (62b)) at John's oneiric alternatives, without forcing the evaluation of *is pantomiming* at these alternatives. For example, it is possible for John to imagine the hippo from his dream pantomiming, even if—in his dream—the hippo was not pantomiming.[12]

The entry in (61) explains the markedness of (62b) as desired (see the contradiction between the anti-veridicality conjunct in the experientiality presupposition of *imagine* and the veridicality conjunct that is introduced by the semantics of *whether*):

(62) *Context:* At last week's picnic in the park, John dozed off and dreamt of a hippo singing.

a. Now, he is imagining that the hippo is pantomiming.

b. $[\![$*John is **imagining whether** a/the hippo is pantomiming$]\!]^{@}$

$$\begin{aligned}
&\equiv [\![\text{imagine}]\!]^{@}([\![\text{John}]\!], \\
&\qquad [\![\text{whether}]\!]_{\text{PAIRED}}([\![\,[\lambda s_1\ \boxed{[\lambda s_2.}\ \text{a hippo-in-}s_1 \text{ pantomimes-in-}s_2 \boxed{]}\ ]\,]\!])) \\
&= \mathit{imagine}'_{@}\big(\mathit{john}, \lambda p\colon \underline{\exists q.\,((\forall s.\, p_s \rightarrow q_s) \wedge \mathit{exp}_{w@}(\mathit{john}, q)) \wedge} \\
&\qquad \underline{\neg p_{w@}} \cdot [\,p_{w@} \wedge ((\forall\langle s_1, s_2\rangle.\, p_{s_2} \rightarrow (\exists x.\, \mathit{hippo}_{s_1}(x) \wedge \mathit{mime}_{s_2}(x))) \vee \\
&\equiv\ \text{↯} \qquad\qquad (\forall\langle s_1, s_2\rangle.\, p_{s_2} \rightarrow \neg(\exists y.\, \mathit{hippo}_{s_1}(x) \wedge \mathit{mime}_{s_2}(x))))]\big)
\end{aligned}$$

Note that, in contrast to the contradiction in 'non-veridical experience'-based *remember whether*-reports (see (59), (60a)), the contradiction in (62b) is not a lexical or contextual coincidence. Rather, it is a systematic contradiction that arises for all *imagine whether*-reports—independently of the lexical material in the complement and the veridicality (or anti-veridicality) of the underlying experience. This observation supports the assumption that the meaning of (62b) is L(ogically)-analytic in the sense of Gajewski [22, 24]. Gajewski [22] has argued that L-analyticity manifests itself at the level of grammar (see also [11, 12]). This explains why we perceive L-analytic sentences as ungrammatical, rather than just as logically deviant. The intuitive unacceptability of the report in (62b) corroborates this reasoning.

(∗ ∗ ∗) $[\![\text{imagine}]\!]^{@}([\![\text{Inger}]\!], \text{E-PARA}([\![\,\boxed{[\lambda s_2}\ [\lambda s_1.\ \text{a fairy-in-}s_1 \text{ flies-above-in-}s_1\ ]\ \boxed{]}\,]\!]))$

$$\begin{aligned}
&\equiv [\![\text{imagine}]\!]^{@}([\![\text{Inger}]\!], \lambda R(\forall\langle s_1, s_2\rangle)[R(s_1, s_2) \rightarrow (\exists x.\mathit{fairy}_{s_1}(x) \wedge \mathit{fly}_{s_1}(x))]) \\
&= \mathit{imagine}'_{@}\big(\mathit{inger}, \lambda p\colon \underline{\exists q.\,((\forall s.\, p_s \rightarrow q_s) \wedge \mathit{exp}_{w@}(z, q)) \wedge \neg p_{w@}} \,. \\
&\qquad\qquad (\forall\langle s_1, s_2\rangle)[p_{s_2} \rightarrow (\exists x.\mathit{fairy}_{s_1}(x) \wedge \mathit{fly}_{s_1}(x))]\big).
\end{aligned}$$

[12] A helpful reviewer has pointed out that the first condition of the experientiality presupposition in (61), i.e., $(\exists q)[(\forall s.\, p_s \rightarrow q_s) \wedge \mathit{exp}_{w@}(z, q)]$, is vacuously satisfied by $\lambda s.\top$ (s.t. $\forall s.\, p_s \rightarrow \top$). My semantics blocks such vacuous satisfaction by assuming that agents can only stand in the experience relation *exp* to non-trivial contents. (For ease of exposition, I assume that the latter are propositions that are false at ≥ 1 possible situation).

6 Outlook

I have argued that the deviance of *imagine whether* is, at core, a semantic phenomenon that arises from the particular interaction of the semantics of *imagine* and *whether*. I have contrasted the deviance of *imagine whether* with the acceptability of most uses of *remember whether*, and have identified a difference in the kind of contradiction that arises in marked cases of *remember whether* and *imagine whether*.

My proposed semantics explains the unacceptability of *imagine whether* through a contradiction between the anti-veridicality presupposition of *imagine* and the actuality assumption in the semantics of *whether*. While the systematicity of this contradiction—and the attendant L-analyticity of the meaning of *imagine whether*-reports—captures the ungrammaticality of many uses of *imagine whether*, it fails to explain the availability of acceptable (!) uses of this construction. Such uses include uses in contexts that aim to 'track reality' (see (8a), due to [60]; copied in (63)) and in contexts that embed *imagine* under a (negated) ability modal or under the control predicate *try* (see (64a) and (64b), copied from (8b)):

(63) I am **imagining whether** the new sofa will fit into my living room.

(64) a. I {am trying to, cannot} **imagine whether** the new sofa will fit into my living room.

b. Anna {could not, was trying to} **imagine whether** the lid would fit the kettle.

The selectional variability of some clause-embedding predicates (s.t. these predicates reject polar interrogative complements on some of their uses, but allow for such complements on other uses) has recently been observed by White [73]. In response, Özyıldız, Qing, Roelofsen, Romero, & Uegaki [57] have suggested that—for doxastic attitude verbs like *believe* and non-veridical preferential predicates like *hope*—this variability may be explained through familiar tools like highlighting (see [57]), a lack of the Excluded Middle-presupposition (see [56, 66]), or multiple predicate senses (see [63]). Future work will need to assess whether the acceptability of (63) and (64) can be explained in a similar fashion. A first step in this direction is made in [43].

Acknowledgements I thank the anonymous reviewers from LACompLing21 for valuable comments on an earlier version of this paper. Special thanks go to Reviewer 6, whose comments and suggestions have significantly improved the final version of this paper. The paper has profited from discussions with Maria Aloni, Kyle Blumberg, Chungmin Lee, Deniz Özyıldız, Frank Sode, Wataru Uegaki, Markus Werning, Simon Wimmer, and Aaron Steven White. Earlier versions of this paper have been presented at the MECORE kickoff workshop (Konstanz/virtual, October 2021), the Bochum LLI Colloquium (Bochum, November 2021), and at LACompLing2021 (Montpellier/virtual, December 2021). The research for this paper is supported by the German Research Foundation [DFG] as part of the research unit *FOR 2812: Constructing Scenarios of the Past* (DFG grant 397530566) and by the German Federal Ministry of Education and Research [BMBF] (through Kristina Liefke's WISNA professorship).

References

1. Arcangeli, M.: The conceptual nature of imaginative content. Synthese **199**(1), 3189–3205 (2021)
2. Bartsch, R.: 'Negative transportation' gibt es nicht. Linguist. Berichte **27**(7), 1–7 (1973)
3. Barwise, J.: Scenes and other situations. J. Philos. **78**(7), 369–397 (1981)
4. Barwise, J., Perry, J.: Situations and Attitudes. MIT Press, Cambridge, MA (1983)
5. Berman, S.: On the semantics and logical form of wh-clauses. Ph.D. thesis, University of Massachusetts, Amherst, Mass (1991)
6. Blumberg, K.: Counterfactual attitudes and the relational analysis. Mind **127**(506), 521–546 (2018)
7. Blumberg, K.: Desire, imagination, and the many-layered mind. Ph.D. thesis, New York University (2019)
8. Bolinger, D.: Postposed main phrases: an English rule for the Romance subjunctive. Can. J. Linguist. **14**(1), 3–30 (1968)
9. Bondarenko, T.: Factivity from pre-existence. Glossa **109**, 1–35 (2020)
10. Cheng, S., Werning, M., Suddendorf, T.: Dissociating memory traces and scenario construction in mental time travel. Neurosci. & Biobehav. Rev. **60**, 82–89 (2016)
11. Chierchia, G.: Broaden your views: implicatures of domain widening and the 'logicality' of language. Linguist. Inq. **37**(4), 535–590 (2006)
12. Chierchia, G.: Logic in Grammar: Polarity, Free Choice, and Intervention. Oxford University Press, Oxford (2013)
13. Ciardelli, I., Groenendijk, J., Roelofsen, F.: Inquisitive Semantics. Oxford University Press, Oxford (2018)
14. Ciardelli, I., Roelofsen, F., Theiler, N.: Composing alternatives. Linguist. Philos. **40**(1), 1–36 (2017)
15. Cohen, M.: A note on belief, question embedding and neg-raising. In: Baltag, A., Seligman, J., Yamada, T. (eds.) Proceedings of LORI 2017, pp. 648–652. Springer, Berlin (2017)
16. Collins, C., Postal, P.M.: Classical NEG Raising: An Essay on the Syntax of Negation. Linguistic Inquiry Monographs. The MIT Press (2014)
17. D'Ambrosio, J., Stoljar, D.: Vendler's puzzle about imagination. Synthese **199**, 12923–12944 (2021)
18. van der Does, J.: A generalized quantifier logic for naked infinitives. Linguist. Philos. **14**, 241–294 (1991)
19. Dretske, F.I.: Seeing and Knowing. Routledge and Kegan Paul, London (1969)
20. Egré, P.: Question-embedding and factivity. Ms (2008). https://jeannicod.ccsd.cnrs.fr/ijn_00226386/
21. Farkas, D.: On the semantics of subjunctive complements. In: Romance Languages and Modern Linguistic Theory, pp. 69–104 (1992)
22. Gajewski, J.: L-analyticity and natural language. Manuscript, MIT (2002)
23. Gajewski, J.: Neg-raising and polarity. Linguist. Philos. **30**(3), 289–328 (2007)
24. Gajewski, J.: L-triviality and grammar. Handout of a talk given at the UConn Logic Colloquium (2009)
25. Giannakidou, A., Mari, A.: Truth and Veridicality in Grammar and Thought: Mood, Modality, and Propositional Attitudes. University of Chicago Press, Chicago (2021)
26. Ginzburg, J.: Resolving questions, II. Linguist. Philos. **18**(6), 567–609 (1995)
27. Giorgi, A., Pianesi, F.: Tense and Aspect: From Semantics to Morphosyntax. Oxford University Press, Oxford (1996)
28. Grimshaw, J.: Complement selection and the lexicon. Linguist. Inq. **10**(2), 279–326 (1979)
29. Hamblin, C.: Questions in Montague grammar. Found. Lang. **10**(1), 41–53 (1973)
30. Heim, I.: Presupposition projection and the semantics of attitude verbs. J. Semant. **9**(3), 183–221 (1992)
31. Heim, I.: Interrogative semantics and Karttunen's semantics for *know*. In: IATL 9, pp. 128–144. Academon, Jerusalem (1994)

32. Hintikka, J.: Semantics for propositional attitudes. In: Models for Modalities, pp. 87–111. Springer, Dordrecht (1969)
33. Hintikka, J.: Different constructions in terms of the basic epistemological verbs. In: The Intentions of Intensionality, pp. 1–25. Kluwer, Dordrecht (1975)
34. Horn, L.R.: On the semantic properties of logical operators in English. Ph.D. thesis, University of California, Los Angeles (1972)
35. Jeong, S.: The effect of prosody on veridicality inferences in Korean. In: Sakamoto, M., et al. (eds.) New Frontiers in Artificial Intelligence. JSAI-isAI 2020, pp. 133–147. Springer (2020)
36. Kaplan, D.: How to Russell a Frege-Church. J. Philos. **72**(19), 716–729 (1975)
37. Karttunen, L.: Presuppositions of compound sentences. Linguist. Inq. **4**(2), 169–193 (1973)
38. Karttunen, L.: Syntax and semantics of questions. Linguist. Philos. **1**(1), 3–44 (1977)
39. Lahiri, U.: Questions and Answers in Embedded Contexts. Oxford University Press, Oxford (2002)
40. Larson, R.K.: On the syntax of disjunction scope. Nat. Lang. Linguist. Theory **3**(2), 217–264 (1985)
41. Lee, C.: Factivity alternation of attitude *know* in Korean, Mongolian, Uyghur, Manchu, Azeri, etc. and content clausal nominals. J. Cogn. Sci. **20**(4), 449–503 (2019)
42. Liefke, K.: Modelling selectional super-flexibility. In: Proceedings of SALT 31 (2021)
43. Liefke, K.: The selectional variability of *imagine whether*. Semant. Linguist. Theory (SALT) **32**, 639–660 (2023)
44. Liefke, K.: Second-order experiential attitudes are not 'parasitic attitudes'. Slides from a talk at Sinn und Bedeutung 27 (2022)
45. Liefke, K., Werning, M.: Experiential imagination and the inside/outside-distinction. In: Okazaki, N., et al. (eds.) New Frontiers in Artificial Intelligence. JSAI-isAI 2020. Lecture Notes in Computer Science, vol. 12758, pp. 96–112. Springer, Cham (2021)
46. Liefke, K., Werning, M.: Factivity variation in experiential *remember*-reports. In: Butler, A., et al. (eds.) New Frontiers in Artificial Intelligence, JSAI-isAI 2021. Lecture Notes in Artificial Intelligence. Springer, Cham (2022)
47. Maier, E.: Parasitic attitudes. Linguist. Philos. **38**(3), 205–236 (2015)
48. Maier, E.: Referential dependencies between conflicting attitudes. J. Philos. Log. **46**(2), 141–167 (2017)
49. Mayr, C.: Triviality and interrogative embedding: context sensitivity, factivity, and neg-raising. Nat. Lang. Semant. **27**(3), 227–278 (2019)
50. Moulton, K.: Natural selection and the syntax of clausal complementation. Ph.D. thesis, University of Massachusetts, Amherst (2009)
51. Niiniluoto, I.: Perception, memory, and imagination as propositional attitudes. Log. Investig. **26**, 36–47 (2020)
52. Ninan, D.: Counterfactual attitudes and multi-centered worlds. Semant. Pragmat. **5**(5), 1–57 (2012)
53. Özyıldız, D.: Factivity and prosody in Turkish attitude reports. UMass generals paper. Master's thesis, University of Massachusetts, Amherst (2017)
54. Özyıldız, D.: Attitude reports with and without true belief. Semant. Linguist. Theory (SALT) **27**, 397–417 (2017)
55. Özyıldız, D.: The event structure of attitudes. Ph.D. thesis, University of Massachusetts Amherst, Amherst, Mass (2021)
56. Özyıldız, D.: Embedded questions and aspect. Ms. Universität Konstanz (2022)
57. Özyıldız, D., Qing, C., Roelofsen, F., Romero, M., Uegaki, W.: Cross-linguistic patterns in the selectional restrictions of preferential predicates. Slides from a talk at GLOW 45 (2022)
58. Peacocke, C.: Implicit conceptions, understanding and rationality. Philos. Issues **9**, 43–88 (1998)
59. Percus, O.: Constraints on some other variables in syntax. Nat. Lang. Semant. **8**, 173–229 (2000)
60. Peterson, E.M.: On supposing, imagining, and resisting. Ph.D. thesis, University of Kentucky (2017)

61. Quine, W.V.: Quantifiers and propositional attitudes. J. Philos. **53**(5), 177–187 (1956)
62. Sode, F.: Differences in the grammar of German *vorstellen* and *träumen*. Ms (2022)
63. Spector, B., Egré, P.: A uniform semantics for embedded interrogatives: an answer, not necessarily the answer. Synthese **192**(6), 1729–1784 (2015)
64. Stephenson, T.: Vivid attitudes: centered situations in the semantics of *remember* and *imagine*. Semant. and Linguist. Theory (SALT) **20**, 147–160 (2010)
65. Theiler, N., Roelofsen, F., Aloni, M.: A uniform semantics for declarative and interrogative complements. J. Semant. **35**(3), 409–466 (2018)
66. Theiler, N., Roelofsen, F., Aloni, M.: Picky predicates: why *believe* doesn't like interrogative complements, and other puzzles. Nat. Lang. Semant. **27**(2), 95–134 (2019)
67. Tulving, E.: Episodic and semantic memory. In: Tulving, E., Donaldson, W. (eds.) Organization of Memory, pp. 381–402. Academic Press, New York (1972)
68. Uegaki, W.: Content nouns and the semantics of question-embedding. J. Semant. **33**(4), 623–660 (2016)
69. Uegaki, W., Sudo, Y.: The **hope-wh* puzzle. Nat. Lang. Semant. **27**(4), 323–356 (2019)
70. Umbach, C., Hinterwimmer, S., Gust, H.: German *wie*-complements: manners, methods and events in progress. Nat. Lang. Linguist. Theory **40**, 307–343 (2022)
71. Werning, M., Cheng, S.: Taxonomy and unity of memory. In: Bernecker, S., Michaelian, K. (eds.) The Routledge Handbook of Philosophy of Memory, pp. 7–20. Routledge, New York (2017)
72. White, A.S.: Lexically triggered veridicality inferences. In: Östman, J.O., Verschueren, J. (eds.) Handbook of Pragmatics, pp. 115–148. John Benjamins Publishing Co. (2019)
73. White, A.S.: On believing and hoping whether. Semant. Pragmat. **14**(6), early access (2021)
74. White, A.S., Hacquard, V., Lidz, J.: Semantic information and the syntax of propositional attitude verbs. Cogn. Sci. **42**(2), 416–456 (2018)
75. Yanovich, I.: The problem of counterfactual *de re* attitudes. Semant. Linguist. Theory (SALT) **21**, 56–75 (2011)
76. Zuber, R.: Semantic restrictions on certain complementizers. In: Hattori, S., Inoue, K. (eds.) Proceedings of the 13th International Congress of Linguists, pp. 434–436. Proceedings Publication Committee, Tokyo (1982)

A Unified Cluster of Valence Resources

Lars Hellan

Abstract A 'Valence Catalogue' is a design for succinctly representing information about the valence frames of a language. It can be coordinated with in-depth annotated corpora and computational parsers. Such a cluster is realized by the verb Valence Catalogue *NorVal* of Norwegian Hellan (Natural Language Processing in Artificial Intelligence—NLPinAI 2021, Springer, 2022) together with a computational parser for Norwegian, *NorSource*, and the multi-lingual grammar system T*ypeGram*. Another such cluster is realized by a verb Valence Catalogue of the West African language *Ga* (*GaVal*) together with the language specific parser *GaGram* and with coverage also by TypeGram. The resources are all based on the general formalism of Typed Feature Structures, whereby grammars and valence catalogues are formally tied together, and specifications in the resources are succinctly comparable also cross-linguistically. In describing the multilingual aspects of these architectures, we thereby demonstrate that systems of this formal nature are fully attainable no matter where on the ladder of 'high-low-resourced' a language finds itself. In a comparative perspective, we also demonstrate discrepancies between the valence systems of the two languages, made visible through the succinctness and coverage of Valence Catalogues, and perspectives they open for typological research and for cross-linguistically informed resource generation.

Keywords Norwegian · Ga · Valence frame types · Verb construction types · Formal grammar · String-based formalisms · Typed feature structures

L. Hellan (✉)
Nidareid 5, 7017 Trondheim, Norway
e-mail: lars.hellan@ntnu.no

Norwegian University of Science and Technology (NTNU), Trondheim, Norway

R. Loukanova et al. (eds.), *Logic and Algorithms in Computational Linguistics 2021 (LACompLing2021)*, Studies in Computational Intelligence 1081,
https://doi.org/10.1007/978-3-031-21780-7_13

1 Introduction

By the *Verb Valence Profile* of a language, we understand the assembled types of valence frames used by the verbs of the language. By a *Valence Lexicon* of a language, we understand a lexicon where the frame types of the Valence Profile are specified relative to the verbs of the language. By a *Valence Catalogue*, we understand a valence lexicon with a specifically succinct formalism to facilitate the finding and statement of regularities across the lexicon.

A Valence Catalogue for Norwegian has been presented in [30], called *NorVal*, hosting valence specifications of about 7,400 verb lexemes. It is coordinated with a large-scale 'deep' computational grammar, called *NorSource* (cf. [36]). A further, although smaller Valence Catalogue is one for the Kwa language Ga, called *GaVal*. It uses the same formalisms as NorVal, and is associated with a grammar of the same type as NorSource, called *GaGram* (cf. [28]). A multilingual grammar called *TypeGram*, informed by Germanic languages, Kwa languages, Bantu languages, Ethio-semitic languages and Indic languages, is also based on these formalisms; the general system of TypeGram for specifying valence frames and construction types underlies the formalism of both Valence Catalogues (cf. [26]).

A basic design feature of the Catalogues reflects the circumstance that a verb can have more than one valence frame. This means that a valence resource needs two kinds of entries, viz. entries for *verb lexemes*, and entries for *a verb relative to a given frame*; the latter we refer to as a '*lexically instantiated valence frame*', for short '*lexval*'. In *NorVal* there are currently 17,300 lexvals, and in GaVal 1,980 lexvals. In the computational grammar applications, lexvals are the most relevant objects, while in lexicons, representations at lexeme level are of course also crucial. A lexeme together with its potentially many valence frames, i.e., lexvals, we call a *valpod* ('pod' as one among many words for assembly). A Valence Catalogue is organized both according to its lexemes, thus, as a list of valpods, and according to its lexvals, as a list of lexvals. The Frame Type encoding for lexvals is the same whether they are part of specifications in the lexvals list or in the enumeration inside a valpod, so that cross-referring between lexvals and valpods is straightforward. The lexvals list is accompanied by example sentences, in GaVal properly annotated with grammatical information, English glossing and translation, while in Norval the example sentences have so far no annotation. Resources for connecting lexvals to corpora have been partly implemented, cf. [38].

A lexval is specified by a standard written form of the verb in question and a formula representing the valence frame in question, the *Frame Type*. In these terms, a *Valence Profile* of a language is a list of those Frame Types which occur in lexval specifications for the language. Frame Types are specified in grammatical terms definable across languages, so that Valence Profiles can in principle be compared across languages. The Valence Profile of NorVal counts 320 Frame Types, while the Valence Profile of GaVal, for mono-verbal verbal expressions (and with the same types of parameters as are distinguished for Norwegian), counts 280. We illustrate in 3.2 how a comparison between the Frame Types of NorVal and GaVal is made in these terms.

The general format of Valence Catalogues is demonstrated in the initial parts of Sect. 2, illustrated from NorVal. The other Norwegian resources are outlined in the remainder of Sect. 2. Section 3 presents the Ga resources. In Sect. 4 we address some of the 'added' values that two such parallel resource clusters represent, seeing the clusters as a unified resource cluster in turn.

2 The Norwegian Valence Catalogue *NorVal* and the Computational Grammars *NorSource* and *TypeGram*

In Sects. 2.1 and 2.2 we present the basic concepts, designs and formalisms of the Valence Catalogue *NorVal*. In Sect. 2.3 we illustrate the general formalism in which the Catalogue is constructed. As mentioned, this is done within the system *TypeGram*, which also acts a computational grammar, and In Sect. 2.4 we present this grammar and the larger computational grammar *NorSource*. A summary of the resources and the features unifying them is given in Sect. 2.5.

The development of the resources started in the mid 80ies through the Norwegian formal lexicon project *TROLL*, to which NorVal may now be seen as a successor.[1] In the early 2000 the development of the lexicon was partly continued in the computational grammar project *NorSource*, continued until now and with interactions with TypeGram. Developments in NorVal have in turn been used to enrich NorSource.

2.1 Basic Constructs of a Valence Catalogue: Lexvals and Valpods

Given that a verb can have more than one valence frame, a valence resource needs two kinds of entries, viz. entries for *verb lexemes*, and entries for *a verb with a given frame*; the latter we refer to as a '*lexically instantiated valence frame*', for short '*lexval*'. In the Norwegian resource there are currently 17,300 lexvals, and in the Ga resource 1,980 lexvals.

The format of a *lexval* entry is illustrated by one of the frame environments for the Norwegian verb lemma *huske* 'remember' exemplified in (1); the lexval entry is stated as in (2), instantiating a general coding pattern indicated in (3):

[1] Prime initiators to the project were Margaret Magnus, Lars G. Johnsen and the present author. Cf. [42].

(1)

Han	husket	på	at	det	var	søndag
He	remember-PAST	'on'	that	it	was	Sunday
Pron	V	Prep	Compl	Pron	V	N

'He remembered that it was Sunday'

(2) `huske-på__intrObl-oblDECL`

(3) Lemma – selected Item (if any) __ Frame Type

(2) reads as an entry with the lexeme *huske*, the selected item *på* ('on'), and the frame type 'intransitive with oblique', coded as `intrObl`, where the oblique argument consists of the preposition *på* and a declarative clause, the latter encoded as `oblDECL`.

The verb *huske* has many possible frames, and a compact way of summarizing the lexvals representing them is shown in (4), illustrating the format of a multivalent *verb lexeme* entry. Each lexval is represented with 'V' as a placeholder for the lemma; the structure we call a *valpod*. It is a set represented with ordering conventions (e.g., intransitives before transitives), and such that the whole specification is represented as *one string* (see the following subsections for explanations of labels):

(4) *huske*:`{V__intr & V-på__intrObl-oblDECL &`
`V-på__intrObl-oblEqSuInf & V-på__intrObl-oblINTERR &`
`V-på__intrObl-oblN & V__tr & V__tr-obDECL &`
`V__tr-obEqSuInf & V__tr-obINTERR}`

The number of *multi-membered* valpods in NorVal is about 3840, and the number of *uni-membered valpods* is about 3570.

2.2 Frame Types and Their Notation

2.2.1 Frame Types

Each lexval includes a unique *Frame Type,* where the notion 'Frame Type' (with capitals 'F' and 'T') is defined in terms of:

- Traditional valence notions like *intransitive, transitive, ditransitive, copular*, and more;
- Grammatical functions such as 'subject', 'object', etc.;
- Argument parameters such as 'direct' vs. 'oblique';
- For an argument to be noun-headed vs. being an *embedded clause* (*declarative, interrogative*, or *infinitival*, in canonical or 'extraposed'[2] position);

[2] The phenomenon of 'extraposition' we see as lexically governed.

- For an argument to relate syntactically and semantically to the same predicate or not;
- Occurrence of particles;
- Basic structures of what we define in 2.3 as 'Logical Form' as far as argument structures go.

The notation for Frame Types uses the system *Construction Labeling* ('CL') (cf. [21, 26, 34]). The system sees verb-headed constructions and verb valence frames as very similar constructs, although here we will refer to them only as valence frames. Valence frames are represented through strings of symbols built up in the following way from-left-to-right: first a construction's head category, then the valence of the head encoded holistically through what we call a Global Label, then syntactic properties of the argument constituents, then semantic properties of these arguments, and then the construction's semantic properties holistically, through aspect, aktionsart and situation type. In NorVal mainly grammatically relevant aspects are represented, so that in the schematic view in (5), 'Argument Label1- Argument Label2,...' represent grammatical information, although in many cases with 'Logical Form' information included (cf. [30]); in GaVal, also the semantic parameters aspect, aktionsart and situation type are marked:

(5) POS_{head} – Global Label – Argument Label1- Argument Label2-...

2.2.2 Global Labels

A *Global Label* (with capital 'G' and 'L') categorizes the *frame as a whole*. It consists minimally of a symbol for overall valence, such as `tr` for 'transitive', in many cases with additional symbols indicating further structure. Thus, the frame representation `intrObl-oblDECL` in (2) has `intrObl` as Global Label. Table 1 exemplifies some Global Labels.

The left column states the label itself, the next column indicates the GFs that are relevant ('declared') in a frame coded by that label.

The column 'Number of logical arguments' indicates arguments that would be represented in the 'logical form' of a linguistic construction coded by the label.

A mark in the column 'Subject Expletive' means that the subject in a construction coded by the label has an expletive subject.

The column 'Target of Predicative' indicates for constructions containing a secondary predicate ('`Scpr`') whether it is predicated of the subject or the object, and, marked by '`Arg`' vs. '`Nrg`', whether this predication target also serves as a semantic argument of the matrix verb or not, marked by '`suArg`' or '`obArg`' if the subject or object predication target *is* semantically an argument of the verb, or by '`suNrg`' or '`obNrg`' if the subject or object predication target is *not* semantically an argument of the verb (if the predicate is an infinitive, corresponding to what is traditionally called 'raising' constructions).

Relevant for constructions containing a so-called 'extraposed' clause, such as *to swim* in *It is pleasant to swim*, the rightmost column indicates whether this clause is

Table 1 Global Labels and the parameters they represent

Global label	GFs declared	Number of logical arguments	Subject expletive	Target of Predicative scSuNrg= target is subject being not semantic argument of verb'	Correlate of 'extraposed' clause
intr	su	1			
tr	su, ob	2			
ditr	su, iob, ob	3			
impers	su	0	X		
intrObl	su, obl	2			
trObl	su, ob, obl	3			
impersObl	su, obl	1	X		
intrScpr	su, sc	1		scSuNrg	
intrScpr	su, sc	2		scSuArg	
trScpr	su, ob, sc	2		scObNrg	
trScpr	su, ob, sc	3		scObNrg	
intrPresnt	su, pres	1	X		
trPresnt	su, ob, pres	2	X		
intrExpn	su, expn	1	X		subject
trExpnSu	su, ob, expn	2	X		subject
trExpnOb	su, ob, expn	2	X		object
copAdj	su, sc	1			
copIdN	su, id	2			

correlated with the subject (as in the example) or the object (as, as also possible in English, in *They made it pleasant to swim*).

The exact usage of the Global Labels is defined and discussed in [30], being well rooted in general and typological linguistics. For instance, the use of 'transitive' is purely formal, meaning that the construction consists of only a subject and an object, no matter what semantic content is expressed.[3]

[3] The notion can nevertheless be linked to a semantic notion of transitivity, if seen as taking as a 'prototypical' basis a relation where force emanates from one participant targeted at another

Table 2 GF interpretation of argument label prefixes

Notion of GF	Prefix in argument label
Subject	su
Direct object	ob
Indirect object	iob
Complement	comp
Oblique	obl
Oblique2	obl2
Presented[4]	pres
Secondary predicate	sc
Extraposed	expn
Extralinked[5]	exlnk
Identifier[6]	id
Adverb	adv
Particle	prtcl

2.2.3 Argument Labels

An *Argument Label* describes a *constituent* of the construction or frame. It is composed of a *prefix* indicating the grammatical function (GF) of the constituent, and a *body*, indicating salient properties of the constituent. For instance, obDECL is an Argument Label with ob as GF-indicating prefix and DECL as body indicating that the constituent is a declarative clause. Table 2 indicates prefixes corresponding to the grammatical functions mentioned in the left column.

To illustrate, in the Frame Types in (4), repeated as (4'),

(4') *huske:* {V__intr & V-på__intrObl-oblDECL & V-på__intrObl-oblEqSuInf & V-på__intrObl-oblINTERR & V-på__intrObl-oblN & V__tr & V__tr-obDECL & V__tr-obEqSuInf & V__tr-obINTERR}

the Argument Label oblINTERR stands for 'the governee in the oblique is an interrogative clause', oblEqSuInf stands for 'the governee in the oblique is an infinitive clause equi-controlled by subject', and oblN stands for 'the governee in the oblique is a noun-headed phrase'. The Global Label tr stands for 'transitive',

---

participant, and counting the *formal* configuration generally used in the language for expressing such a relation as the *grammatical* transitivity notion valid for that language (cf. [11]). The label tr stands for the formal notion involving a subject and an object.

[4] For the'presented' NP in a presentational construction, normally indefinite, as in *det sitter en katt her*'there sits a cat here'.

[5] An oblique clause with an expletive subject as correlate, as in *det later til at det blir vår*'it looks_P that it becomes spring'; here the verb *later* governs the preposition *til,* which in turn governs the 'at'-clause.

[6] The post-verbal item of an identity-copula.

Table 3 Clausal arguments of type 'declarative' (of 15,700 lexvals)

Argument label	Instances
suDECL	87
obDECL	460
oblDECL	485
expnDECL	89
oblExlnkDECL	5
DECL	**1142**

Table 4 Clausal arguments of type 'interrogative' (of 15,700 lexvals)

Argument label	Instances
suINTERR	22
obINTERR	235
compINTERR	77
expnINTERR	48
oblExlnkINTERR	1
oblINTERR	432
INTERR	**849**

and its accompanying Argument Labels obDECL, obEqSuInf and obINTERR stand for the object being clausal with the respective values 'declarative', 'infinitive equi-controlled by subject', and 'interrogative'.

When not occurring with qualifying Argument Labels, the Global Labels intr, tr and ditr are understood as having nominal dependents. When ...Obl occurs in a Global Label, in contrast, there is always an accompanying Argument Label obl... specifying the dependent of the preposition, also when the dependent is nominal.

The system presented so far allows for a fine-grained classification of argument types. A complete list of Norwegian Frame Types, with exemplifying sentences with English translations, is found in Appendix 1 of [30], and in a continuously updated version at https://typecraft.org/tc2wiki/NorVal_resources.

2.2.4 Examples

From [30] we repeat some of the results obtained in the version available in July 2021.[7] For instance, Table 3 states that among the totality of the then 15,700 lexvals in the resource, 1140 lexvals contain an argument specified with DECL as a defining label; all searches relate to the Argument Label indicated in the left column (Tables 4, 5, 6 and 7).

[7] Expansions from that version are partly due to cooperation with Margaret Magnus.

Table 5 Clausal arguments of type 'controlled infinitive' (of 15,700 lexvals)

Argument label	Instances
suEqObInf ('subject is an infinitive controlled by object')	21
obEqSuInf ('object is an infinitive controlled by subject')	135
obEqIobInf ('object is an infinitive controlled by indirect object')	51
oblEqSuInf ('oblique is an infinitive controlled by subject')	291
oblEqObInf ('oblique is an infinitive controlled by object')	476
expnEqObInf ('extraposed is an infinitive controlled by object')	31
EqInf	**1066**

Table 6 Clausal arguments of type 'absolute infinitive' (of 15,700 lexvals)

Argument label	Instances
suAbsinf	17
obAbsinf	35
oblAbsinf	160
expnAbsinf	28
Absinf	**267**

Table 7 Clausal arguments of type 'bare infinitives' (of 15,700 lexvals)

Argument label	Instances
obEqIobBareinf	2
scBareinf	18
obEqBareinf	2
Bareinf	**22**

An immdediate observation concerning the clausal arguments is that they to a large extent obtain as governees of prepositions, i.e., as obliques. This is summarized in the following table (Table 8).

Of the by July 2021 15,700 lexvals, the overall number of lexvals declared for including an oblique argument is around 4600 (relevant global labels including trObl, intrObl, ditrObl, impersObl, ...PrtclObl), distributed

Table 8 Oblique clausal arguments summarized (of 15,700 lexvals)

Argument label	Instances
oblDECL	485
oblINTERR	432
oblEqSuInf	291
oblEqObInf	476
oblAbsinf	160
Oblique clausal arguments	**1844**

over 113 Frame Types, thus nearly one third of all lexvals, and involving more than one third of all Frame Types, underlining the role of oblique arguments in general.

These and similar results in terms of descriptive statistics are efficiently obtained in the formats described, with lexvals and valpods represented as simple text strings.

2.3 Frame Types Represented in AVM Format, Combining Grammatical Functions and Logical Form

2.3.1 AVMs and Typed Feature Structures (TFS)

Frame Types can be alternatively represented in terms of attribute value matrices (AVMs). They combine a representation of the grammatical functions (GFs) realized in the Frame Type with a representation of 'Logical Form' (a concept to be defined shortly) and further semantic parameters. An example including GF and 'Logical Form' is shown in (6):

(6) $$tr\begin{bmatrix} \text{GF} \begin{bmatrix} \text{SUBJ } signe\begin{bmatrix}\text{INDX } \boxed{1}\end{bmatrix} \\ \text{OBJ } signe\begin{bmatrix}\text{INDX } \boxed{2}\end{bmatrix} \end{bmatrix} \\ \text{ACTNT} \begin{bmatrix} \text{ACT1 } \boxed{1} \\ \text{ACT2 } \boxed{2} \end{bmatrix} \end{bmatrix}$$

The attribute 'GF' introduces *grammatical functions* and the attribute 'ACTNT' introduces attributes used in representing the 'actants' of 'logical form'. The boxed numbers '[1]' and '[2]' encode identical values, serving as co-reference or 'reentrancy' symbols, here expressing that ACT1 ('actant 1') is identical with the referent of the subject, and that ACT2 ('actant 2') is identical with the referent of the object. The AVM follows the design of *Typed Feature Structures* (TFS), according to which attributes are declared by specific types, and in turn take types as values, following the general conventions [A] and [B][8]:

[A] A given type introduces the same attribute(s) no matter in which environment the type is used.
[B] A given attribute is declared by one type only (but occurs with all of its subtypes).

In (6), *tr* and *signe* are the only type symbols explicitly entered. [26] gives an introduction to the type system assumed in the present AVM system (referred to as TypeGram). In the AVMs, attributes corresponding to grammatical functions are the same GFs as indicated in Table 2, but now represented as AVM *attributes* rather than CL-string Argument Label *prefixes*.

[8] See Copestake [8].

Table 9 'Logical form' attributes

Attribute label	Description of attribute label
ACT0	'Event index' for a sentence and 'thing index' for an NP
ACT1	Index of subject of verb used in active voice
ACT2	Index of direct object of verb used in active voice
ACT3	Index of indirect object of verb used in active voice
ACTobl	Index of the governee of a PP functioning as oblique
XACT	Index of subject or 'external argument' of a predicate[9]
ACTloc	Index of valence-bound locative specification
ACTdir	Index of valence-bound directional specification

2.3.2 Logical Form and the Attribute 'ACTNT'

Attributes used inside the attribute 'ACTNT' are in turn as listed in Table 9.

The attributes ACT1, ACT2 etc. as used here can be seen partly as coarse *role labels*, and partly as *enumeration markers*. As enumeration markers, they list the participants present in the situation expressed (including implicit ones), starting with ACT1, using ACT2 only if there is an ACT1, and using ACT3 only if there is an ACT2. As coarse role markers, when there is an ACT1 and an ACT2, ACT1 is the role associated with emanation of force, and ACT2 is the 'impacted' part relative to the force; an ACT3 would then express a slightly less directly involved participant, such as the recipient or benefactive in a ditransitive sentence.[10] When there is only one actant, it will be marked as ACT1, regardless of its role; here one is thus back to the enumeration modus.[11] 'Logical' aspects of the representation include the circumstance that also implicit participants receive an ACT, and that, although the default linking is for *subject* to represent *ACT1*, in a passive sentence a subject can correspond to an ACT2 or ACT3 participant. (In some ways, the representations at

[9] The 'external argument' of a predicate is the argument which can be semantically bound from outside; this is the 'logical subject' of an adjective or a preposition, and also of an infinitival verb in active voice, while if an infinitival verb in passive voice, its external argument is whichever argument corresponds to its subject position.

[10] The Paninian *k-system* is the earliest in this tradition (cf. [47]). PropBank (http://verbs.colorado.edu/~mpalmer/projects/verbnet.html) use of ARG0, ARG1, … is similar to the k-roles in that both represent *fixed roles*, contrary to the ACTNT system. The present system is used in some grammars belonging to the 'Matrix' family mentioned below.

[11] ACTNT specifications can be enriched with role and other semantic information, as explored in [2, 33]. Representations at the ACTNT level often closely mirror syntactic structure, but a notable exception is an analysis of comparative constructions built on a Montague-style analysis first proposed in [22].

this level correspond to what would count as *Deep Structures* in early 'consolidated' Transformational Grammar.)

Traditionally, 'Logical Form' represents aspects of sentences which are standardly seen as important to logical reasoning. The relevant patterns are traditionally *lexically robust*, in that words are usually treated as they are, without semantic decomposition,[12] and *syntactically economical*, in that sentence sets whose members have the same truth value will standardly be represented by just one of these members, such as the active form in a set of sentences involving an active and a passive version. Standard logic has much of its focus on quantification and variable-binding operators, with factors like those mentioned constituting more of a practical background. Most linguistic studies involving 'logical form' indeed also relate to quantifier scope, but the notion of *Deep Structure* in early transformational grammar may count as representing sentence internal factors, as do significant branches of linguistic semantics, with analytic representations formed on the patterns of predicate-argument structures in Predicate Logic, or modeled as Attribute-Value structures, either way commonly referred to under a label like *Semantic Argument Structure* (with [25] counting as an early work in the tradition).

Aspects of semantics represented at this level, observing lexical robustness and syntactic economy, are features like 'being semantically an argument of', 'being clausally embedded within', causality, and implicitness of arguments, all following patterns that could in principle be overtly expressed in a language or coming close to overt expression, and all morpho-syntactically retrievable in a grammatical construction for which the representations are proposed in a given case. Thus, a representation of causality could only be invoked when grammatically retrievable, morphologically or syntactically.

2.3.3 Co-habitation of GF and Logical Form Representation Exemplified

The system of GF attributes comes close to what is used in the f-structures of *Lexical Functional Grammar* (LFG, cf. [6]), and also to dependency relations in *Universal Dependency Grammar* (UDG, cf. [43]) and valence specifications in *Head-Driven Phrase Structure Grammar* (HPSG, cf. [44]). However, the explicit co-habitation of GF representations and Logical Form representations is specific to the present TFS architecture. Thus, LFG and HPSG expose the counterparts to these dimensions in different representations and components, while UDG does not have the Logical Form counterpart.

---

[12] Thus, not of the kind involved if, for instance, 'kill' is considered rephrased as 'cause to die'.

The items in the list of 320 Frame Types applicable to Norwegian[13] comprise each a CL string representation together with an AVM representation and an example sentence. This is exemplified in (7)–(9) below (with ordering between the examples corresponding to the ordering in the list, the ordering being largely alphabetic according to the CL Frame Type labels).

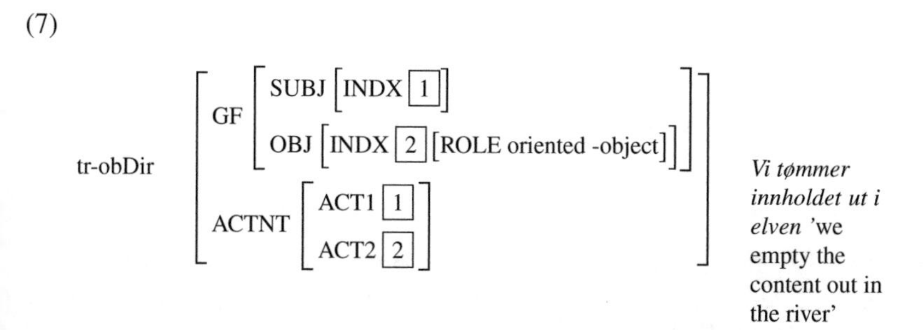

The OBJ ('*innholdet*' in the example sentence), whose referent is the value of ACT2, is here specified for the role of a 'directed object'. Whether the formal representation of the direction (*ut I elven*) should count as an argument or as an adjunct is a point of discussion in the general literature; the present design chooses the latter, thus not including the directional expression as a specific GF or ACTNT, but rather as a feature, as exemplified in (7).[14]

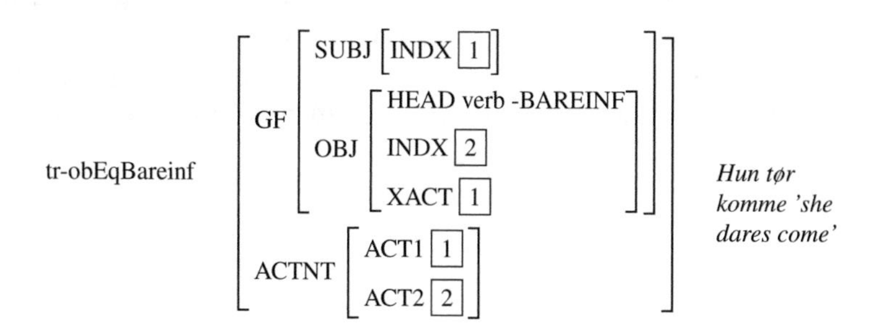

[13] cf. https://typecraft.org/tc2wiki/NorVal_resources.

[14] For discussion of the feature 'ROLE' as used in (7), see Sect. 3.3. Locational and directional roles are represented in NorVal and NorSource, as developed in [2, 33].

(9)

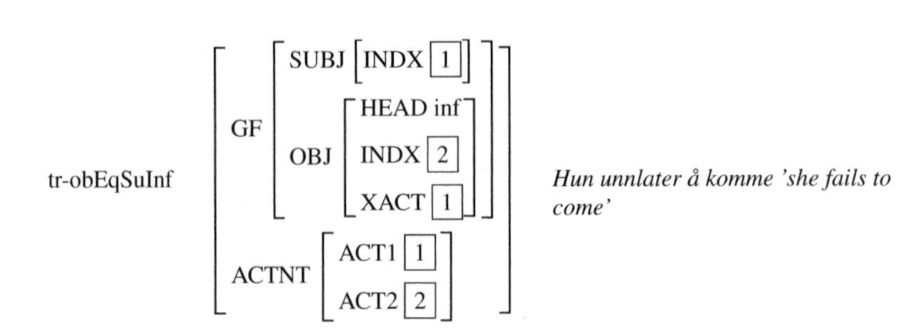

In (8) and (9), the embedded clause is represented by the GF OBJ and the ACTNT ACT2, as indicated by the coindexing symbol '[2]' for ACT2 and the referent INDX of OBJ. The attribute 'XACT' inside OBJ identifies that argument of the embedded clause which is 'equi'-controlled by SUBJ. The only difference between the structures in (8) and (9) is in the distinction between *å*-marked infinitive ('Inf' in CL-type label, '*inf*' in value of 'HEAD' in *(9)*) and 'bare' infinitive.

2.3.4 Strings of CL-Labels and Unification of AVMs

A CL-string and an AVM, as said, can represent the same Frame Type. Just as a Frame Type is in many cases defined by a string of minimal CL labels, the corresponding AVM representation of that Frame Type is in such a case built up from simpler AVMs corresponding to the minimal CL labels, joined through unification. To illustrate, the AVM in (9) corresponds to the CL-string (10a). Its constituent CL-labels have defined AVM counterparts as indicated in (10b, c). The unification of these AVMs is the AVM in (9).

(10)

a. tr-obEqSuInf

b. tr - [GF [SUBJ [INDX [1]], OBJ [INDX [2]]], ACTNT [ACT1 [1], ACT2 [2]]]

c. obEqSuInf- $\begin{bmatrix} \text{GF} \begin{bmatrix} \text{SUBJ} \begin{bmatrix}\text{INDX} \boxed{1}\end{bmatrix} \\ \text{OBJ} \begin{bmatrix} \text{HEAD inf} \\ \text{XACT} \boxed{1} \end{bmatrix} \end{bmatrix} \end{bmatrix}$

Relative to the TFS system, the AVMs belong to *types*, here labeled by the same strings as serve as CL labels, so that the AVM in (b) has *tr* as its type and the AVM in (c) has *obEqSuInf* as its type. The unification of the AVMs has the string corresponding to the full CL label as its type, thus, the AVM in (9) has *tr-obEqSuInf* as its type (for more on these points, see Sect. 3.2).

Using a term defined in [21], we refer to units like those in (b) and (c) as *Minimal Construction Information Units*, abbreviated *MCU*s. They are to be seen not as segmental units (morphs, words, phrases or the like) but as properties of clausal constructions, thus, information units. They can be represented either in the form of a 'minimal' CL-label, or in the form of an AVM. The inventory of Frame Types is constructed out of a general inventory of MCUs, such that for any CL-string representing a Frame Type, the AVMs representing the component MCUs unify, as just illustrated; the formal system, as said, is defined in TypeGram.[15]

2.4 Grammars

To sustain a grammar of a language, the MCUs and Frame Types need to be anchored in lexicons and defined for the morpho-syntax of the language, which can be done on a smaller, demo-like, basis, or on a larger basis. This applies whether one builds such a grammar for the sole purpose of creating an abstract model of the language, or one wants to have computationally implemented system.

For Norwegian, the grammars defined in the TFS formalism are a smaller grammar included in the TypeGram package, to be referred to as TG_NOR, and a large scale computational grammar based on HP, called *NorSource*. In this subsection we describe and compare the two.

The parse outputs for TG_NOR include AVMs, which for the 300 illustrating sentences in the list are exactly the AVMs used for illustrations in the list, as exemplified in the AVMs in (7)–(9). NorSource (cf. [36]) also produces AVMs for Logical Form, but it also produces structures called 'Minimal Recursion Semantics' (MRS) (cf. [9]). (11) illustrates the MRS of a parse of (9), *Hun unnlater å komme*'she fails to come'.[16]

[15] The essential elements of TypeGram can be downloaded from https://typecraft.org/tc2wiki/TypeGram.

[16] From parse at http://regdili.hf.ntnu.no:8081/linguisticAce/parse, 10.05.2022. NorSource belongs to the family of HPSG-grammars referred to as 'Matrix'-grammars; cf. [5]. As semantic output they generally produce MRSs.

(11)

```
ltop=h0, index=e1
h3:hun_pron_rel([arg0:x2])
h4:_pronoun_q_rel([arg0:x2, rstr:h5, body:h6])
h7:_unnlate_v-tr_rel([arg0:e1, arg1:x2, arg2:h8])
h8:_inf-mark_rel([arg0:u9, arg1:h10])
h10:_komme_v-intr_rel([arg0:e11, arg1:x2])
< qeq(h5,h3) >
e1, sort=verb-act-specification, sf=prop, e.tense=pres, e.mood=indicative,
e.aspect=semsort
x2, wh=-, png.ng.num=sing, png.ng.gen=f, png.pers=thirdpers
```

'arg1' corresponds to 'ACT1' in the AVM in (9), etc. A formal difference is that the attribute paths for clausal embedding exemplified in the AVMs in (8) and (9) are matched in the MRS by independent-standing predications, tied together through a system of extra labels such as 'h8' and 'h10', making the formalism more amenable to computational parsing. This aspect aside, the ACTNT representations in the TypeGram format are consonant with the type of semantic representation as produced in HPSG parsers. Moreover, the NorVal lexicon and the NorSource verb lexicons are synchronized, using the same type labeling code and with the NorSource lexical entries coinciding with the set of lexvals used in NorVal; that is, for each lexical entry in NorSource, its type corresponds to a Frame Type in a lexval in NorVal with the same lexeme.

As far as valence type specifications are concerned, the respective overall feature structures of TG_NOR and NorSource can be illustrated as in (12), for the valence type 'transitive'.

(12)

a. TG_NOR AVM for transitive verb; the top type of the TFS is *signe*:

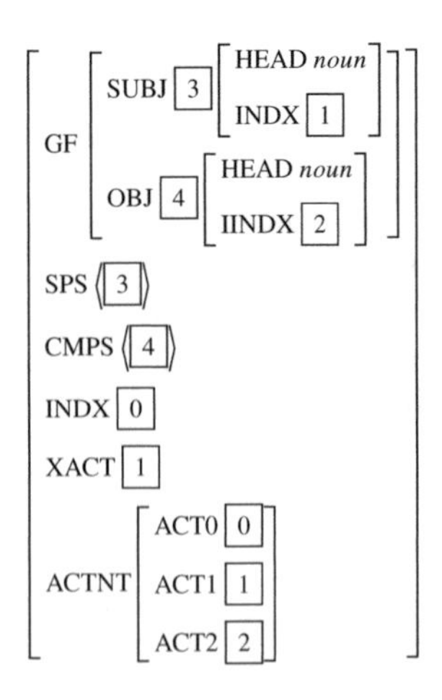

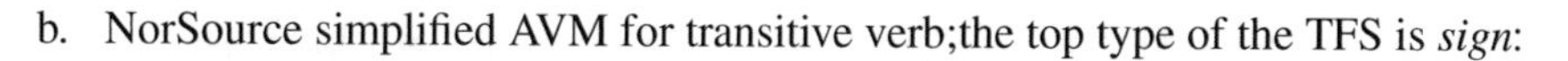
b. NorSource simplified AVM for transitive verb;the top type of the TFS is *sign*:

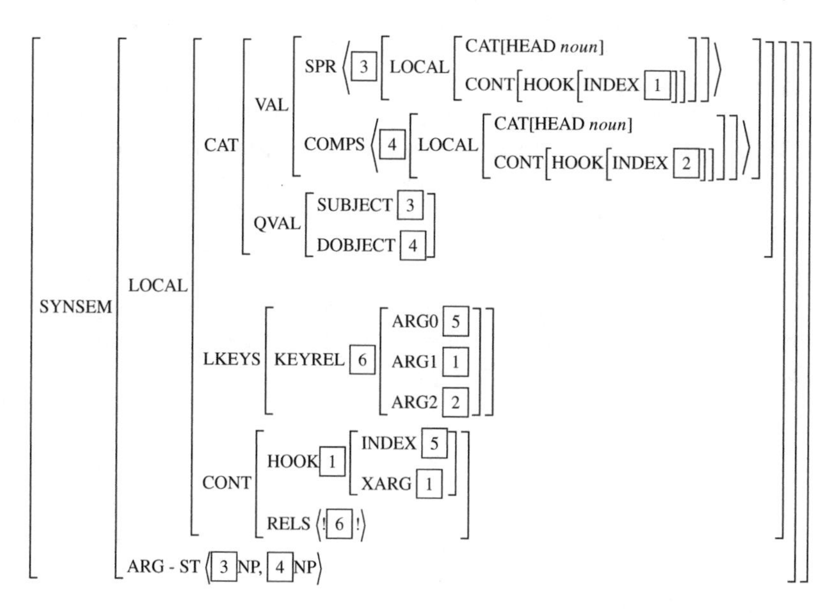

In (12a), two further items occur than in the AVM in (9), namely ‘SPS’ and ‘CMPS’; these are lists of items with which the verb needs to combine; corresponding attributes in (12b) are ‘SPR’ and ‘COMPS’. In the representation of a construction where these requirements are fulfilled, as in (9), these lists are empty, and can thus be omitted in a construction display. In (12b), the attribute ‘QVAL’ corresponds to ‘GF’ in (10a). ‘RELS’ in (12b) is a ‘difference list’ assembling the elementary predications (EPs) in a clause, of which the item ‘[6]’ is the predication tied to the verb; (11) exemplifies what a full list of such elementary predications may look like, being the items prefixed with ‘h’ (‘h3’, ‘h4’, ‘h7’, etc.). TypeGram does not have this facility in-built.[17]

The feature structure hierarchy under the type *signe*, of which (12a) is a subtype, defines types corresponding to an overall inventory of possible Frame Types (i.e., members of possible Valence Profiles) as defined in the TypeGram formalism. For each language, *its* inventory of Frame Types (i.e., *its* Valence Profile) is a subset of that overall inventory. The feature structure hierarchy under the type *sign,* of which (12b) is a subtype, in turn defines types corresponding to the same overall inventory of possible Frame Types, but now encoded according to the feature geometry illustrated in (12b). If one establishes a Valence Profile of a language L, that means that one can automatically deduce the feature structures of either a TypeGram design grammar or a

[17] A further item shown in (12b) but not in (12a) is ‘ARG-ST’, which holds information about the items processed from the valence lists in the final output. In this information-preserving capacity relative to item cancellation in valence lists, ARG-ST is thus a correspondent to GF/QVAL, and is for that reason not needed in the presence of ‘GF’ in (12a). (The same holds relative to QVAL in NorSource, but ‘ARG-ST’ is here retained as inherited from the Matrix original system.).

Matrix-design grammar, the former giving minimally complex structural displays, the latter a well developed processing grammar with MRS, given the morpho-syntactic rules and lexical items specific to L.

As the 'main standard' of HPSG one may count [44], which also underlies the feature architecture adopted in a number of grammar implementations, most notably ERG [23, 24], and in the 'Matrix' grammars.[18] A difference between 'standard' versions of HPSG and NorSource resides in the latter including a dimension of valence itemizing GF notions represented as SUBJECT, DOBJECT, IOBJECT, OBL, etc., introduced within the attribute 'QVAL', as seen in (12b). This point of parallelism between NorSource and TG_NOR is what enables the general interplay between the resources just mentioned, in particular it allows for a mapping between the CL formalism and the NorSource formalism analogous to the mapping illustrated in (10) between the CL formalism and the TypeGram formalism.[19,20]

[18] Cf. [5], and http://moin.delph-in.net/MatrixTop.

[19] The status of NorSource and TG_NOR as parallel systems is highlighted by the circumstance that their feature structures can be merged in a joint TFS. Relative to the displays in (12), for instance, the structure (12a) could be placed inside the attribute LOCAL in (12b) (with coindexations suited, for convenience). This is illustrated in the TFS grammar for Ga, *GaGram*, in the file 'matrix.tdl', at lines 377-385 and lines 622-629; which can be downloaded at https://typecraft.org/tc2wiki/Ga_computational_HPSG_grammar#; the overall 'merged' design now described is implemented in this grammar.

[20] It may be noted that the LKB formalism sustaining the HPSG-based grammars disallows the specification of a list within the specification of a list. There are cases in Norwegian (and presumably other languages) where one however descriptively wants to state dependencies corresponding to such a situation; thus, in Norwegian it is frequently the case that in the COMPS list of a verb selecting a PP, the verb's selection concerns not only the head preposition but also whether the item governed by the preposition is a noun, an infinitive, a declarative subordinate clause or an interrogative subordinate clause (the latter as in *Jeg lurer på om hun kommer* (lit.'I wonder about whether she comes')). Such specifications represent selection by the verb, but must be done inside the COMPS list of the preposition, thus a case of specification in a COMPS list inside of a COMPS list. The availability of GF specifications allow for an alternative feature path to the PP as introduced by the GF *OBL* in the verb's GF, where the head of the preposition's governee is reached by specifications within the GF *OBJ* in the path initiated by OBL. (This specification operates in tandem with the standard list cancellation relative to the verb. Note that since the construct 'ARG-ST' is list based, it cannot serve this function of GFs, even if it otherwise has the capacity of holding 'pre-cancelation' information in a way similar to GFs.).

Table 10 Norwegian valence resources summarized

Name	Description	Type of formal object	Size	Published	Online
NorVal	Valence catalogue for Norwegian	Files with simple text	7400 verb lexemes, 17,300 lexvals, 320 Frame Types	[30]	
NorSource	HPSG large computational grammar of Norwegian	TFS computational grammar	90,000 lexical entries	[36]	[22,23]
TypeGram	Universal grammar of grammatical relations, based on HPSG formalisms	TFS computational grammar	5000 lines type def		[24]
TG_NO	*AVMs with GF andLogical Form of 300 Norwegian Valence Frames*		330 AVM types		[25]
Norwegian Valence corpus	22,000 sentences annotated for valence and morphology	Based on NorSource parses, via XML export to TypeCraft as an IGT project		Hellan et al. (2021)	[26]
MultiVal	Valence frames for 55,000 lexvals from Ga, Norwegian, Spanish, Bulgarian			[35]	[27]

2.5 *Summary of a Unified Cluster of Valence Resources for Norwegian*

The valence resources for Norwegian now described are summarized below; we here also include items not mentioned in the above exposition but with presentations as indicated (Table 10)[21]:

They constitute a unified cluster in that the CL specifications tie the resources descriptively together, and in the ways in which a formal entity in one resource can play a role as a formal entity in another, such as an independent formal lexicon serving as a lexical module in a computational system. A further aspect of unification is that all structures are defined in a unitary TFS system.

3 Valence Catalogue for Ga (GaVal)

3.1 *Overall Design of GaVal*

The Valence Catalogue for Ga, referred to as *GaVal*,[28] is a further development of resources for Ga developed by the late Prof. Mary Esther Kropp. Based on the ToolBox file prepared for the Ga-English dictionary [15], she in [16] added valence specification to all the verb entries in this file. The valence specifications here use the Frame Type format developed in the CL system. [17] is a parallel introduction to the system based on the Gur language Gurene. The Ga system, together with the Norwegian system, is used as illustration of the general exposition of the CL system published in [34]. Later expositions of the Ga valence system include [18, 20, 21].

Around 2005 a computational grammar of Ga was developed using the LKB platform mentioned above, called *GaGram* (cf. [19]). In subsequent steps, the Toolbox lexicon was converted to the lexicon format of the LKB grammars, with the CL notation being used as type labels in the verb lexicon of this grammar (cf. [7, 28,

[21] An essential area of valence representation not yet formalized in NorVal is the valence patterns of other word classes, and derivational relationships between, e.g., verbs and nouns. A sketch of how NorVal can be extended to include the representation of such phenomena is given in [31].

[22] http://regdili.hf.ntnu.no:8081/linguisticAce/parse.

[23] https://github.com/Regdili-NTNU/NorSource/tree/master.

[24] https://typecraft.org/tc2wiki/TypeGram.

[25] https://typecraft.org/tc2wiki/NorVal_resources.

[26] https://typecraft.org/tc2wiki/Norwegian_Valency_Corpus.

[27] http://regdili.hf.ntnu.no:8081/multilanguage_valence_demo/multivalence.

[28] Cf. https://typecraft.org/tc2wiki/Ga_Valence_Profile.

39]). This resource counts 480 verb lemmas, organized as 1980 verb entries, corresponding to the present notion of *lexvals*. The conversion to the present 'catalogue' format was in turn done by the present author.

As witnessed by the word 'construction' in the label 'Construction Labeling' ('CL'), the intended objects of classification were originally conceived as *construction* types as much as valence types. The expressions classified were largely fairly short sentences, many of them used in the Ga-English lexicon as illustrations of uses of the verbs represented, and of grammatical phenomena. In the latter respect it was thus constructions that were classified, but in a more restricted sense of 'constructions' than in the then dominant framework of 'Construction Grammar'. The notion 'valence' has less tradition in the description of West African languages than for European languages. Reasons for using 'valence' in the present context were nevertheless factors like the following:

- The sentences were short, and mainly such that no part could be grammatically omitted;
- Most of the sentences had a single verb, which could count as head;
- Some sentences had many verbs, but the English translation would contain only one verb.

The last point refers to Multiverb constructions, generally known as being frequent in the Kwa languages, and in Ga including the following types (cf. [3, 12–14, 18, 19]):

- 'Extended Verb Complex', i.e., one or more 'pre-verbs' preceding the main verb (EVC);
- 'Serial Verb Construction' (SVC);
- 'Verbid' phrase, i.e., a clause-final VP with locative or a similar function.

Generally subscribing to the principle of 'immanent' language description, candidates for counting as valence- or construction heads would in such cases be the Ga verbs involved (not an English verb), but with due consideration as to whether one of the verbs would take the other as an attribute, maybe as part of a 'Multiword Expression' (MWE), in which case the notion of *valence* would be relevant, or the two (or more) verbs would have more equal status, in which case the notion of *construction* would be the more relevant. Either way, the multiverb constructions would not fall within the inventory of European type valence frames.

For mono-verbal sentences, there would also be cases where English-type valence frame types would seem inadequate. These are cases where, say, the subject and an object involve grammatical binding or control relations, or also semantic relations, that in English-type constructions would require rather different grammatical configurations than in the Ga constructions. Such cases will warrant specifications also for the Ga constructions, and potentially counting as valence frame types specific to Ga. Examples of the latter kind are given in (13) below (in describing salient factors, we use 'possessor' and 'specifier' interchangeably, relevant since these sentences all involve a pronominal possessor; in the description of each sentence, we start with features of the syntactic configuration, then semantic features)[29]:

(13)

a.	Ee-la	e-daa-ŋ
	3S.PROG-sing	3S.POSS-mouth-LOC"
	V	N

Gloss-literally: 'he sings his mouth'

Free translation: "He's murmuring incoherently to himself."

Salient factors: Identity between subject and possessor of object, object denotes a body-part of the possessor of the object, the object expresses the location of the activity.

(Example is from entry-ID in GaVal: 956.)

b.	E-ŋmra	e-toi-ŋ
	3S.AOR-scrape	3S.POSS-ear-LOC
	V	N

Gloss-literally: 'she scraped his ear'.

Free translation: "She slapped him."

Salient factors: Object denotes a body-part of the possessor of the object, the object expresses the location of the activity.

(Example is from entry-ID in GaVal: 1166.)

c.	E-tsuinaa	mii-funta	lɛ
	3S.POSS-desire	PROG-nauseate	3S
	N	V	PN

Gloss-literally: 'her desire nauseates her'.

Free translation: "She feels sick, nauseous."

Salient factors: Subject's possessor is identical to object, subject is effector, object is experiencer.

(Example is from entry-ID in GaVal: 401.)

[29] Examples of multiverb constructions appear massively in the literature, for Ga see especially [34], and so we won't give illustrations here.

d.	Mi-yitso	mii-gba	mi
	1S.POSS-head	PROG-split	1S
	N	V	PN

Gloss-literally: 'my head splits me'.

Free translation: "My head is aching."

Salient factors: Subject is a body-part of subject's possessor, subject's possessor is identical to object, subject is locus of the event, object is experiencer.
(Example is from entry-ID in GaVal: 415.)

e.	E-tsui	naa	wa
	3S.POSS-heart	edge	AOR.hard
	N	N	V

Gloss-literally: 'his heart's edge hardened'.

Free translation: "He is brave".

Salient factors: Subject has a relational head, subject's specifier is a possessor phrase, subject's specifier is a body-part of subject's specifier's specifier, subject is locus of the event.
(Example is from [34], Ga Appendix.)

In [34], the 'salient factors' in these examples are all assigned CL labels, retained as MCUs in the Valence Catalogue. Some of the CL labels are illustrated in the CL representation of (13d), given in (14):

(14)

Mi-yitso	mii-gba	mi
1S.POSS-head	PROG-split	1S
N	V	PN

Gloss-literally: 'my head splits me'; Free translation: "My head is aching."
CL representation:
`v-tr-suPossp_suBPsuSpec_suSpecIDob-suLocus_obExp`
`-EXPERIENCE`
CL labels for 'salient factors':
Subject is a body-part of subject's possessor (`suBPsuSpec`),
subject's possessor is identical to object (`suSpecIDob`),
subject is locus of the event (`suLocus`),
object is experiencer (`obExp`),
overall situation type is experience (`EXPERIENCE`).

Table 11 Sample argument labels in GaVal

Argument label	Explanation
suPostP	The subject is a 'postpositional' nominal phrase, i.e., an NP with a spatial-relational noun as head and an NP specifier
suPossp	The subject has a possessor (NP) phrase as specifier
suSpecPossp	The subject's specifier has a possessor (NP) phrase as specifier
suBPsuSpec	The referent of the subject is a bodypart of the referent of the subject's specifier
suSpecBPsuSpecSpec	The referent of the subject's specifier is a bodypart of the referent of the subject's specifier's specifier
obUnif	Object is an 'inherent complement', i.e., unifies with the verb to determine the verbal meaning
obNomvL	Object is a nominalization of a verbal expression, in which the verb occurs last, i.e., following its arguments
suIDcompSu	The subject is identical to the complement's subject
suSpecIDobSpec	The specifier of the subject is identical to the specifier of the object
suIDobSpec	The subject is identical to the specifier of the object

Some of the overall most frequently occurring Argument Labels are listed in Table 11, including some of those mentioned in (14).

The overall number of Frame Types for mono-verbal constructions represented in GaVal is 280, when leaving out specifications of semantic roles. The MCUs of the frames are described in detail in [34]. The comparable number of Frame Types represented in NorVal is 320. The overlap between these sets of Frame Types is rather small, as is the overlap between the sets of MCUs constituting the respective sets of Frame Types; some figures are presented in Sect. 3.2.

The resources thus provide specifications on which one can conduct a comparative linguistic investigation between the frames and MCUs, not just relative to the two languages but also relative to families of which they are part.

3.2 *Lexvals and Valpods in GaVal*

The general definition of lexvals and valpods is as in NorVal, but with some differences in the detailed design. The structure of lexvals in GaVal is illustrated in (15), and the structure of valpods in (16), for the lexeme shared by the lexvals in (15).

(15)
```
  jo_640: = v-tr-suAg_obTh & LEXEME <"jo″> & PHON
<"dʒò″> & ENGL-GLOSS <"shake″> & EXAMPLE "A-jo-ɔ
Bra″ & GLOSS "3-dance-HAB Bla″ & FREE-TRANSL "they
perform the "Bra Joo″ ceremony." &
  jo_641: = v-intr-suPostp-suLoc-MOTION & LEXEME
<"jo″> & PHON <"dʒò″> & ENGL-GLOSS <"shake″>
```

```
& EXAMPLE "E-he jo-ɔ" & GLOSS "3S.POSS-self
dance-HAB" & FREE-TRANSL "he shakes (in place)." &
  jo_642: = v-tr-obUnif-suAg_obTh-MONOTONIC_DEV
-MOTION & LEXEME <"jo"> & PHON <"dʒò"> & ENGL-GLOSS
<"run"> & EXAMPLE "E-jo foi" & GLOSS "3S.AOR-run
speed" & FREE-TRANSL "he ran, bolted." &
  jo_643: = v-ditr-obPostp_ob2Unif-suAg_obLoc_ob2Th
-MONOTONIC_DEV-MOTION & LEXEME <"jo"> & PHON
<"dʒò"> & ENGL-GLOSS <"run"> & EXAMPLE "E-jo mi-naa
foi" & GLOSS "3S.AOR-run 1S.POSS-edge speed" &
FREE-TRANSL "she avoided me." &
  jo_644: = v-tr-obPostp-suAg_obLoc-MONOTONIC_DEV
-MOTION & LEXEME <"jo"> & PHON <"dʒò"> & ENGL-GLOSS
<"run"> & EXAMPLE "E-jo sɛɛ" & GLOSS "3S.AOR-run
PRO back" & FREE-TRANSL "he committed adultery" &
  jo_645: = v-trVid-obUnif_vidObPostp-suAg_obTh
_vidObLoc-MONOTONIC_DEV-MOTION & LEXEME <"jo"> &
PHON <"dʒò"> & ENGL-GLOSS <"run"> & EXAMPLE "E-jo
foi yɛ kpabuu lɛ mli" & GLOSS "3S.AOR-run running
AOR.be.at prison DEF" & FREE-TRANSL "he committed
adultery" &
```

In each lexval, the symbol ': = ' interconnects *LexvalEntry-ID* and *Frame-Type*. The symbol is taken from the computational grammar lexicons, meaning 'is a subtype of'. While in a NorVal lexval the whole expression '*lemma__ Head-POS—Global Label—Argument Label1- Argument Label2—...*' serves as a lexval identifier by itself, the identifier role in GaVal is carried by just 'orth_number' as in 'jo_645' above. This is a reflex of provenance and the difference has no substantial significance.[30]

As for the provenance of the other elements of a Ga lexval, the last four elements have exactly the content of the original ToolBox file.

Given that tone is a lexical feature in Ga, what identifies a word as a lexeme is not just the string 'LEXEME <"jo" > <"jo">, but also the tone entered in the subsequent string PHON <"dʒò">.

The valpod corresponding to the lexvals in (15) is shown in (16). As lexeme identifier is used 'jo, PHON <"dʒò">', converted from 'LEXEME <"jo"> & PHON <"dʒò">' in the lexvals. Of the last four elements in the lexval, the last three are dropped in the valpod, but ENGL-GLOSS is retained for each member. Note that this item, which might be taken to represent 'meaning', is not part of the lexeme definition, since among those in (16), the first two have the gloss 'shake', the others 'run'.

```
(16) jo, PHON <"dʒò"> : {V__640_tr-suAg_obTh &
     ENGL-GLOSS <"shake">, V__641_intr-suPostp-suLoc
```

[30] The NorVal lexicon notation also at some point was created by conversion from the Norwegian grammar, and here the ': = ' sign was converted away yielding to the present format of the NorVal entry names.

```
-MOTION & ENGL-GLOSS <"shake">, V_642_tr-obUnif
-suAg_obTh-MONOTONIC_DEV-MOTION & ENGL-GLOSS
<"run">, V_643_ ditr-obPostp_ob2Unif-suAg_obLoc
_ob2Th-MONOTONIC_DEV-MOTION & ENGL-GLOSS <"run">,
V_644_tr-obPostp-suAg_obLoc-MONOTONIC_DEV-MOTION &
ENGL-GLOSS <"run">, V_645_trVid-obUnif_vidObPostp
-suAg_obTh_vidObLoc-MONOTONIC_DEV-MOTION &
ENGL-GLOSS <"run">}.
```

In the question of 'meanings' or 'senses', NorVal leaves that point open in its present version. NorVal also leaves phonology open, and provides examples only in the shape of Norwegian strings, not in the IGT format and without glossing. In these respects, the Ga resource is thus significantly richer than NorVal.

In the Catalogue, labels of the 'NP-internal' kind illustrated in 3.1 are quite many, and they combine with each other in many ways; in addition come semantic specifications as exemplified. Even ignoring the latter, and ignoring CL-formulas for the multiverb expression types SVCs and EVCs (which number 180), the number of Frame Types for Ga verbs is 280. In comparison, as mentioned, NorVal has 320 Frame Types, all without semantics and with only monoverbal expressions.

Comparing the GaVal and NorVal inventories as they are given, it turns out that they have only *eight* Frame Types in common—that is, only 2–3% of the respective total inventories. Of these, only 'intransitive', 'transitive' and 'ditransitive' are used with any regularity in both. As an isolated figure, this might sound dramatic—can languages, in their Valence Profiles, be that different?

Looking more closely at the matter, the labels '`tr`', '`intr`' and '`ditr`' used as non-augmented ('bare') Global Labels in the lexvals of the two languages, *with no specific object modifications*, distribute as shown in Table 12 (out of totals of 1980 lexvals in GaVal and 17,300 lexvals in NorVal, on current counts):

If one eyes through the list of lexvals in Ga, a good 25% of them will thus have a Frame Type that one also finds in NorVal (modulo semantic specification, which, as mentioned, NorVal doesn't have), and eying through the list of lexvals in NorVal, 45% of them have a FrameType which one would also find in GaVal. So seen, the Frame Type inventories are not that dramatically different.

If one looks at the FrameType specifications *including* complex CL strings (i.e., '`tr-obDECL`' and the like), the corresponding figures are as shown in Table 13.

Table 12 Number of Frame Types with 'tr', 'intr' and 'ditr' as Global Labels and no further constituent specifications

Global Label (non-augmented)	Occurrences in Lexvals in GaVal	% of all lexvals in GaVal	Occurrences in Lexvals in NorVal	% of all lexvals in NorVal
`tr`	359	18	5404	31
`intr`	146	7	2163	13
`ditr`	38	2	162	1

Table 13 Number of frame types with 'tr', 'intr' and 'ditr' as global labels, serving as frame labels alone or with further constituent specifications

Global label (non-augmented)	Occurrences in lexvals in GaVal	Occurrences in lexvals in NorVal
`tr`	1034	7162
`intr`	240	2464
`ditr`	132	380

The number of Frame Types with `tr` as global label here more than *doubles* in GaVal relative to Table 12, while only half of the original amount of Frame Types with `tr` as global label is added in NorVal, relative to Table 12. What follows `tr` in the Frame Type labels in Ga are for the most part the kinds of 'NP-internal' specifications already illustrated, while in NorVal, the CL-labels following `tr` often indicate light reflexives, and many indicate clausal objects or subjects; cf. Tables 3, 4, 5, 6 and 7.

It may be noted that many of the Ga construction types annotated with complex Frame Type labels with `tr` as global label and Argument Labels like `obPossp` or `obPostP` would have as counterparts in Norwegian (or English) *oblique* constructions, of which there are many, cf. Tables 3, 4, 5, 6, 7 and 8. With systematic correspondences also in these respects, there would be still less of a dramatic discrepancy between the Frame Type inventories of the two languages.

The discrepancy between the inventories of the Valence Profiles mentioned—with intersection of only 2–3% of the sets—is nevertheless noteworthy. And the figures observed demonstrate that comparisons based on just Frame Type inventories in the abstract may be far remote from the picture one gets when having an overall view of the full valence lexicons.

3.3 Interacting CL Labels, AVMs and Semantic Specification

Like in the Norwegian resource, the MCUs in GaVal are expressed as CL labels as well as AVMs. We illustrate for Ga with the AVM composition for (14), repeated in (17a) with the examples together with the assigned MCU specifications linked to the 'salient features'. (17b, c, d) represent three of these MCUs, and (17e) represents the AVM of the whole CL string resulting from unification of these MCUs together with appropriate AVMs for the semantic features; comments on the semantic features follow in turn:

(17)

a.

`v-tr-suPossp_suBPsuSpec_suSpecIDob-suLocus_obExp-EXPERIENCE`

Mi-yitso	mii-gba	mi
1S.POSS-head	PROG-split	1S
N	V	PN

Gloss-literally: 'my head splits me'; free translation: "My head is aching."; salient factors: Subject is a body-part of subject's possessor (`suBPsuSpec`), subject's possessor is identical to object (suSpecIDob), subject is locus of the event (suLocus), object is experiencer (obExp), overall situation type is experience (EXPERIENCE).

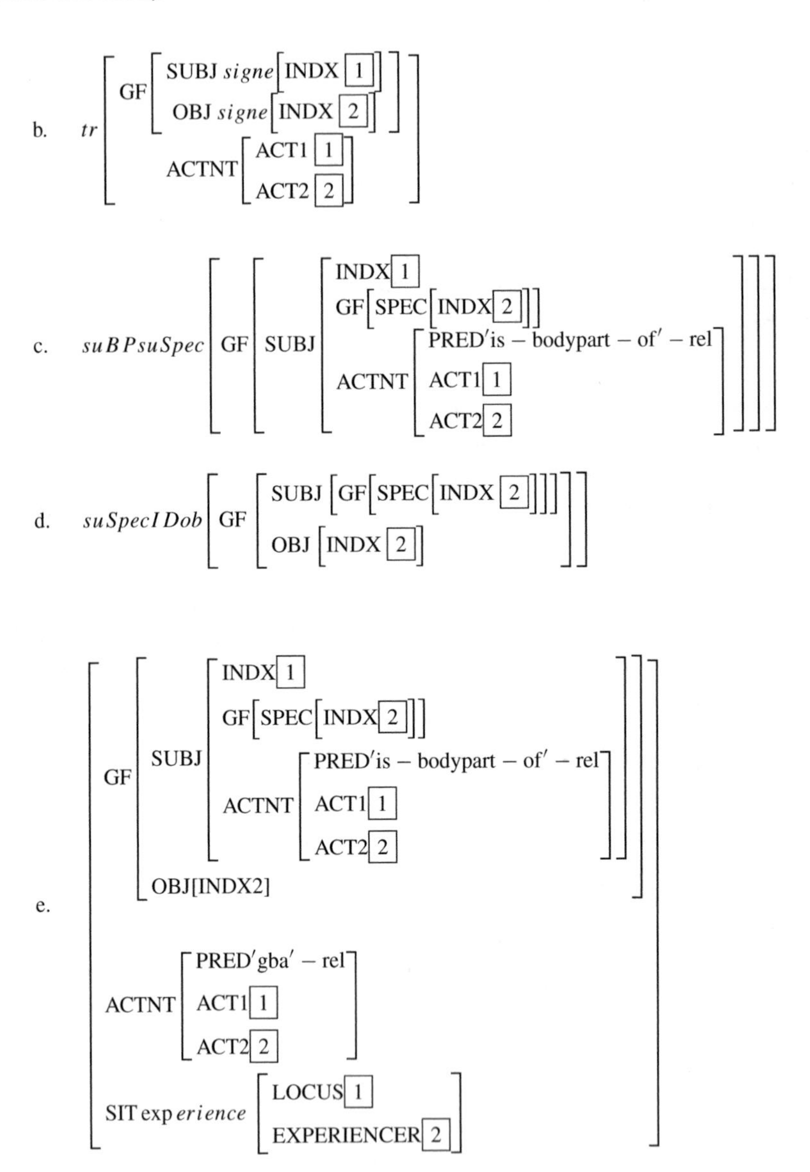

The MCU unification procedure works in just the same way as indicated in (10) above. The only new aspect is the lower attribute in (17e), namely SIT, and the way in which it reflects the CL string part 'suLocus_obExp-EXPERIENCE' in the

Frame in (17a). We now outline how this aspect of the system works (cf. [26], and also [27, 29]).

The CL-string 'suLocus_obExp-EXPERIENCE', and its counterpart in the AVM in (17e), represents 'semantic roles' in the case of 'suLocus_obExp 'and *situation types* in the case of 'EXPERIENCE'. We ssume that verbs have in common that they express *states of affairs* or *situations*, notions that may be subsumed under Wittgenstein's notion *Sachverhalt* (cf. [48]), and we use the notion *Situation Type* for this kind of content. In a TFS system, situation types, like types generally, can be expressed in hierarchies, following the principles [A] and [B] stated above, repeated:

[A] A given type introduces the same attribute(s) no matter in which environment the type is used.
[B] A given attribute is declared by one type only (but occurs with all of its subtypes).

Figure 1 shows such a hierarchy, in principle with a type *sit* on top representing 'Sachverhalt', here showing the lower types *locomotion* and *effort* as the uppermost types, introducing the respective attributes MOVER and ACTOR. These are inherited down to the type *launching*, which introduces three new attributes LAUNCER, LAUNCH-MECH(anism) and LAUNCHED, with a definition stating that the value of LAUNCHER is the same as that of the inherited attribute MOVER. The type *eject* is stated as a subtype of *launching*, with the specification that the inherited attributes MOVER and LAUNCHED have the same value.

Both of the verbs *throw* and *sling* express acts that may be called 'ejections'. Salient differences are that throwing may involve a better aiming than slinging, and that the curved arm movement typically associated with 'throw', by which the kinetic energy behind the ejection is built up, typically unfolds in a vertical plane, while with 'sling' that plane is more horizontal. As type labels reflecting the latter difference we

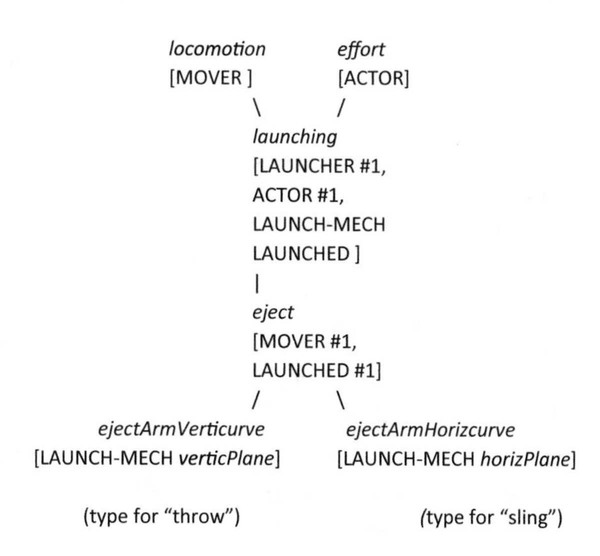

Fig. 1 Part of a situation type hierarchy

use the subtypes *ejectArmVerticurve* and *ejectArmHorizcurve*, respectively, distinguished by the attribute LAUNCH-MECH defined accordingly, as indicated as the lowest nodes in Fig. 1.

As minimal as this example of a situation type hierarchy is, it illustrates how the common feature of *throw* and *sling* residing in a moving item being ejected can be attached to one supertype *eject*, while the further common feature of being agentive events is traced to another, still higher node, *effort*, this node of course with far more descendants.

To integrate such information in an AVM, we add an attribute SIT in the architecture so far, on the same level as GF and ACTNT. To make it an AVM reflecting the verb *throw*, or *fɔ* in Ga (cf. (19)), we draw out the information accumulated in the left lowest node in Fig. 1, and link the role-bearers here to the appropriate subattributes of GF and ACTNT:

(18)

$$tr-EJECTIONarmVerticurve\begin{bmatrix} \text{GF} & \begin{bmatrix} \text{SUBJ}\ signe\begin{bmatrix}\text{INDX}\ \boxed{1}\end{bmatrix} \\ \text{OBJ}\ signe\begin{bmatrix}\text{INDX}\ \boxed{2}\end{bmatrix}\end{bmatrix} \\ \text{ACTNT} & \begin{bmatrix} \text{ACT1}\ \boxed{1} \\ \text{ACT2}\ \boxed{2}\end{bmatrix} \\ \text{SIT} & \begin{bmatrix} eject \begin{bmatrix} \text{ACTOR}\ \boxed{1} \\ \text{LAUNCH -MECH}\ verticPlane \\ \text{LAUNCHED}\ \boxed{2} \\ \text{MOVER}\ \boxed{2}\end{bmatrix}\end{bmatrix}\end{bmatrix}$$

The type *tr-EJECTIONarmVerticurve* in (18) belongs to the *signe*-hierarchy (cf. (12a) in 2.4) in the TFS analysis.

The interconvertibility between CL-strings and TFS-AVMs extends to the situation types, so that the *signe* type *tr-EJECTION* will reflect the CL string type `tr-EJECTION`, and *tr-EJECTIONarmVerticurve* will reflect the CL string type `tr-EJECTIONarmVericurve`.

In GaVal, not only situation types but also semantic *role* indicators are used. These can also be construed off the situation hierarchy as illustrated in the following, for the CL formula in the lexval (19):

(19) `fɔ_361: = v-tr-suAg_obMover-EJECTION` & ENGL-GLOSS <"throw"> & EXAMPLE "Efɔ bɔɔlu lɛ" & GLOSS fɔ_361 & FREE-TRANSL "she threw the ball."

In this case the situation type label is `EJECTION`, the corresponding *sit* type thus being the one corresponding to the second lowest node in Fig. 1.

In representing how a string of CL-labels can be said to technically unify to constitute the TFS object corresponding to the CL string, we will be a step more careful than in the demonstrations in (10) above, and make one point explicit: it is the *TFS signe objects* that unify, not the CL-strings (cf. [26] for a general outline of the system). But there are 1-to-1-correspondences all the way. Thus, the individual CL strings representing the respective parts of the full CL string are linked to the *signe* types indicated at the outermost bracket at each of the AVMs listed in (20), and it is *their* unifications which yield the type to which the full CL string corresponds, namely (21):

(20)

CL-string		TFS *signe* object.
v	---	[HEAD verb]
tr	---	*tr* [GF [SUBJ [INDX [1]], OBJ [INDX [2]]], ACTNT [ACT1 [1], ACT2 [2]]]
suAg	---	*suAg* [GF [SUBJ [INDX [1]]], SIT [ACTOR [1]]]
obMover	---	*obMover* [GF [OBJ [INDX [2]]], SIT [MOVER [2]]]
EJECTION	---	*EJECTION*[SIT *eject*]

(21)

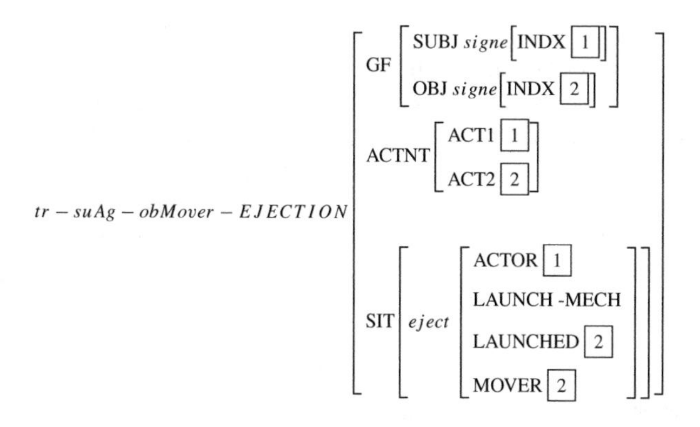

As shown, the role-representing *signe*-type *suAg* is here construed as a join of specifications involving the GF and the SIT spaces, using the attribute ACTOR in the capacity that it has in Fig. 1; correspondingly for the attribute MOVER in the SIT specification and its corresponding role-representing *signe*-type *obMover*. Through such joins of specifications we get the *signe*-type names to which the CL role labels in turn correspond.

GaVal includes more than 100 situation type labels, the following being the ones most frequently used, rendered to give an impression of the various levels of generality represented by the labels[31]; also aspectual notions are here included:

(22)

```
ABSENT, ACQUISITION, CARETAKING, CAUSATIVE, CAUSED, CLOSING,
COGNITION, COMMUNICATION, COMPARISON, COMPLETED-MONODEVMNT,
CONTACT, CREATION, CUTTING, EJECTION, EMOTION, EXPERIENCING,
MAINTAINPOSITION, MOTIONDIRECTED, PHENOMENON, PLACEMENT,
POSTURE, PROPERTY, DYNAMIC-PROPERTY, PSYCHSTATE, REMOVAL,
SENSATION, TRANSFER, USINGVEHICLE
```

The idea of such a dimension of semantic description has been entertained in many semantic frameworks. A challenge in making a comprehensive inventory relative to a language is that for many verbs, situation types are far less obvious than in cases like directed movements or throwing, and that, given that there are potentially extremely many types, one may easily lose the classificational orientation points that one started out with. A hierarchical system of situation types thus may be helpful, and still more helpful it is if the classificatory system is explicitly linked to a system of formal grammatical representation, as here. Even though the present system has not been applied at a larger scale, it may thus hold a possibility of becoming useful, both in linguistic description snd in formal analysis.

3.4 Summary: A Unified Cluster of Valence Resources for Ga

The valence resources for Ga are summarized in Table 14.

Like the Norwegian resources, the Ga resources constitute a unified cluster in that the CL specifications tie them descriptively together, and in the ways in which a formal entity in one resource can play a role as a formal entity in another, such as an independent formal lexicon serving as a lexical module in a computational system. A further aspect of unification is that all structures are defined in a unitary TFS system.

[31] In developing such a situation label system, the possibility of using the 'frame' inventory in FrameNet—cf. https://framenet.icsi.berkeley.edu/fndrupal—was discarded, judged as being too close to English lexicalization to serve for such multilingual purposes as including Ga.

Table 14 The valence resources for Ga

Name	Description	Type of formal object	Size	Published	Online
GaVal	Valence Catalogue for Ga	Files with simple text and Excel sheets	500 verb lexemes, 19,800 lexvals, 280 mono-verbalFrame Types and 185 multiverb Frame Types, semantics excluded		32
GaGram	HPSG computational grammar of Ga	TFS computational grammar	10,000 lexical entries	[28]	33
TypeGram		TFS computational grammar	5000 lines type def		
TG_GA	*AVMs with* GF and Logical Form of selected Ga Valence Frames (ex: (14d))		300		34
Ga Annotated corpus	construction types annotated for valence and morphology	TypeCraft IGT project	100 construction types		35
MultiVal	See Table for NorVal				

4 Final Remarks

Both of the clusters of resources described had their start of development in the 1980ies, and represent large amounts of linguistic analysis accumulated, partly as published in articles and books, partly as contributions to resources such as dictionaries and grammars, using digital tools as relevant. Partly parallel resources have been developed, although of different sizes, and with partly identical formalisms, in both cases with lexicons as the initial steps. The previously available resources differ strongly, Ga counting as a digitally less resourced language, Norwegian not (although also not as highly resourced), and Ga also research-wise having a much shorter tradition than Norwegian. A conclusion is that even with large differences in terms of previous resources, clusters like those described can in principle be developed for any language. While for instance, to our knowledge, no other Sub-Saharan

[32] https://typecraft.org/tc2wiki/Ga_Valence_Profile.

[33] https://typecraft.org/tc2wiki/Ga_computational_HPSG_grammar.

[34] To appear: https://typecraft.org/tc2wiki/Ga_Valence_Profile.

[35] https://typecraft.org/tc2wiki/Ga_annotated_corpus.

African language has this kind of assembly of resources, the methodologies realized in the case of Ga may thus be of interest more widely.

A motivation point for some investigations into valence systems is the idea of what we may call 'semantic footprints' in the inventory of valence frames in a language, namely that the meaning of a verb may contribute to its selection of Frame Types, and that for larger sets of verbs, common selections of sets of Frame Types may reflect semantic factors shared between the verbs. This is a motivational factor for studies into 'verb classes' and 'valency classes' such as [41], *VerbNet* (cf. [40][36]), and the *Leipzig Valency Classes Project* (cf. [42][37]). These issues are not explicitly addressed in the present valence resources; however, the construct of *valpods* can conceivably function as a unit in terms of which to investigate the aspect of whether for larger sets of verbs, common selections of sets of Frame Types reflect semantic factors shared between the verbs.

From a general prespective, Valence Profiles, i.e., the sets of Frame Types employed by a language, can be represented in grammatical terms partly shared across descriptions of languages, and thus constitute an aspect of valence description that can be explored cross-linguistically. Frameworks like *Lexical Lunctional Grammar* (LFG, cf. [6]), *Head-Driven Phrase Structure Grammar* (cf. [44]) and *Universal Dependency Grammar* (UDG, cf. [43]) are formal frameworks relevant to such a perspective.[38] An interesting point of comparison between the valence systems of Norwegian and Ga can be found at this level, residing in the minimal intersection of their sets of Frame Types noted in Sect. 3.2. Are the differences between the verbal syntax and principles of verb lexicalization between the two languages really so dramatic? Volta Basin Languages on the whole having much in common with Ga, is this a somewhat underestimated domain of typological diversity?[39]

The lexical specification aspects of Valence Catalogues can less readily be brought into cross-linguistic comparison, since, of course, the *words* of which they consist are language specific. As our discussion in Sect. 3.2 shows, however, information readily accessible over large sets of entries in the Valence Catalogues can serve for arriving at far more balanced conclusions over valence system differences than adherence to Valence Profiles alone. A common perspective in contrastive language studies is, moreover, that of 'minimal pairs', comparing patterns in one language with partially similar patterns in another, such as, for valence, e.g., studies on formally and semantically related reflexive constructions across related languages (see, e.g., [37]). Such studies often include an appeal to semantic factors, but invoked only for the particular domain of investigation.

---

[36] http://verbs.colorado.edu/~mpalmer/projects/verbnet.html.

[37] Cf. the online interface http://valpal.info/.

[38] To a large extent the present comparative figures are based on HPSG investigations of both languages, such as [19] and [28]. It would be an interesting task to see how the present MCUs would come out in terms of UDG relations. Concerning Ga in particular, one may note that UDG so far has only relation covering multiverb constructions (viz.,'compound:svc'), as against the 180 SVC- and EVC construction types catalogued in GaVal.

[39] Relevant studies on other West African languages include [1, 10, 46].

A more broadbent involvement of meanings cross-linguistically is the use of a 'Swadesh list' of English verbs conducted in the Leipzig Valency Classes Project (cf. [42]), where 80 English verbs were used to serve as cross-linguistically represented meanings for an investigation of valence diversity between 30 languages. In an exploration of possible parallels between the'valence footprints' of given meanings across a set of languages, some cross-linguistic meaning denominators would be needed although, as noted above, the denominators would hardly be at word level; and a very large number of words per language would have to be addressed.

A project more concretely involving similarity of word meanings across languages can be conceived as follows, for a scenario where one language is specified for valence valpods, another not, lexical meaning equivalences between the languages are established, and their verb grammars are not too different. In such a scenario one could endeavor essentially copying the valpods hosted by the verbs in the first language to the meaning-equivalent verbs in the second language, with adjustments for the general grammar differences between the languages. The verbs in the second language would thus be assigned a system of valpods being a mirror-image of those in the first language. The task will then be, verb by verb in the target language, to see if the 'mirror-vision' holds up. To the extent that it does, one may have found an effective way of creating a valence resource for the language; and for the conceivably many verbs where it doesn't hold up, one has an excellent source for observing and describing differences in the lexical grammars between the languages.[40]

Back to the basis: the creation of a Catalogue is foremost a matter of extensive, painstaking philological and linguistic work, with very few shortcuts possible. 'Intra'- and 'inter' developer consistency is always a factor, still more when resources are brought together. Aside from careful explanations of analyses and terms, and 'legends' of employed and possible alternative interpretations of terms, the presentation of the resources should ideally include the possibility of sharing, for users to develop them further, possibly along different branches. An overall condition for such endeavours to be possible, however, is that one has a welldefined formal frame.

Acknowledgements Parts of this paper were presented at the LACompLing 2021 conference, and at University of Florida, Gainesville; I am grateful for invitations to both occasions. I am grateful to Dorothee Beermann and the reviewers for comments and advice.

References

1. Ameka, F.K.: Three place predicates in West African serializing languages. Stud. Afr. Linguist. **42**(1), 1–32 (2013)
2. Beermann, D., Hellan, L.: A treatment of directionals in two implemented HPSG grammars. In: Müller, S. (ed.) Proceedings of the HPSG04 Conference, Katholieke Universiteit Leuven. CSLI Publications http://csli-publications.stanford.edu/ (2004)

[40] The possibility of such a project involving Norwegian and German is mentioned in [45], and a project involving Ga and Akan along similar lines is considered in [4].

3. Beermann, D., Hellan, L.: West African serial verb constructions: the case of Akan and Ga. In: Agwuele, A., Bodomo, A. (eds.) The Routledge Handbook of African Linguistics, pp. 207–221. Routledge, London and New York (2019)
4. Beermann, D., Hellan, L.: Enhancing grammar and valence resources for Akan and Ga. In: West African languages. Linguistic theory and communication. Warzawa: Wydawnictwa Uniwersytetu Warszawskiego 2020, pp. 166–185 (2020) ISBN 978–83–235–4623–8
5. Bender, E.M., Drellishak, S., Fokkens, A., Poulson, L., Saleem, S.: Grammar customization. Res. Lang. Comput. **8**(1), 23–72 (2010)
6. Bresnan, J.: Lexical Functional Syntax. Blackwell, Oxford (2001)
7. Bruland, T.: Ga_verb_dictionary_for_digital_processing. (2011). https://typecraft.org/tc2wiki/Ga_Valence_Profile
8. Copestake, A.: *Implementing Typed Feature Structure Grammars*. CSLI Publications (2002)
9. Copestake, A., Flickinger, D., Sag, I., Pollard, C.: Minimal recursion semantics: an introduction. J. Res. Lang. Comput. 281–332 (2005)
10. Creissels, D.: Valency properties of Mandinka verbs. In [42], pp. 221–260 (2015)
11. Creissels, D.: Transitivity, valency, and voice. Ms. European Summer School in Linguistic Typology. Porquerolles (2016)
12. Dakubu, M.E.K.: The Ga preverb *kɛ* revisited. In: Dakubu M.E.K., Osam, K. (eds.) *Studies in the Languages of the Volta Basin* **2**, 113–134. Legon: Linguistics Dept. (2004)
13. Dakubu, M.E.K.: Ga clauses without syntactic subjects. J. Afr. Lang., Linguist. **25**(1), 1–40 (2004)
14. Dakubu, M.E.K.: Ga verb features. In: Ameka, F., Dakubu M.E.K. (eds.) Aspect and Modality in Kwa Languages. Amsterdam & Philadelphia: John Benjamins Publishing Co., pp. 91–134 (2008)
15. Dakubu, M.E.K.: Ga-English Dictionary with English-Ga Index. Black Mask Publishers, Accra (2009)
16. Dakubu, M.E.K.: Ga Toolbox project expanded with Construction Labeling valence information. (2010) https://typecraft.org/tc2wiki/Ga_Valence_Profile
17. Dakubu, M.E.K.: Gurene Verb Constructions. Unpubl ms, University of Ghana. (2010). https://typecraft.org/tc2wiki/Gurene_verb_constructions
18. Dakubu, M.E.K.: Ga Verbs and their constructions. Univ. of Ghana, Monograph ms (2013)
19. Dakubu, M.E.K., Hellan, L., Beermann, D.: Verb Sequencing Constraints in Ga: Serial Verb Constructions and the Extended Verb Complex. In: St. Müller (ed.) Proceedings of the 14th International Conference on Head-Driven Phrase Structure Grammar. Stanford: CSLI Publications (2007). http://csli-publications.stanford.edu/
20. Dakubu, M.E.K., Hellan, L.: Verb Classes and Valency classes in Ga. Paper read at Symposium on West African Languages (SyWAL) II, Vienna (2016)
21. Dakubu, M.E.K., Hellan, L.: A labeling system for valency: linguistic coverage and applications. In [37]. (2017)
22. Davis, C., Hellan, L.: An integrated analysis of comparatives. Unpubl, Notre Dame University and University of Trondheim (1975)
23. Flickinger, D.: On building a more efficient grammar by exploiting types. In: Oepen, S., Flickinger, F., Tsujii, J., Uszkoreit, H. (eds.) Collaborative Language Engineering, pp. 1–17. CSLI Publications, Stanford (2002)
24. Flickinger, D.: Accuracy vs. Robustness in Grammar Engineering. In: Bender, E.M., Arnold J.E. (eds.) Language from a Cognitive Perspective: Grammar, Usage, and Processing, pp. 31–50. CSLI Publications, Stanford. (2011)
25. Grimshaw, J.: Argument Structure. MIT Press, Cambridge, MA (1995)
26. Hellan, L.: Construction-based compositional grammar. J. Logic Lang. Inform. (2019). https://doi.org/10.1007/s10849-019-09284-5
27. Hellan, L.: Situations in Grammar. In: Essegbey, J., Kallulli, D., Bodomo, A. (eds.) The grammar of verbs and their arguments: a cross-linguistic perspective. Studies in African Linguistics. Berlin: R. Köppe (2019)
28. Hellan, L.: A computational grammar of Ga. LREC workshop RAIL (2020)

29. Hellan, L.: Interoperable Semantic Annotation. LREC workshop ISA-16, 6th Joint ACL-ISO Workshop on Interoperable Semantic Annotation (2020)
30. Hellan, L.: A valence catalogue for Norwegian. In: Loukanova, R. (ed.) Natural Language Processing in Artificial Intelligence—NLPinAI 2021. Springer (2022). https://doi.org/10.18710/8U3L2U
31. Hellan, L.: Unification and selection in Light Verb Constructions. A study of Norwegian, In: Light verb constructions as complex verbs. Features, typology and function (Series "Trends in Linguistics. Studies and Monographs"), Mouton de Gruyter. (2022)
32. Hellan, L., Johnsen, L.G., Pitz, A.: TROLL. Ms, University of Trondheim (1989). https://www.nb.no/sprakbanken/ressurskatalog/?_search=TROLL
33. Hellan, L., Beermann, D.: Classification of prepositional senses for deep grammar applications. In: Kordoni, V., Villavicencio, A. (eds.) Proceedings of the Second ACL-SIGSEM Workshop on The Linguistic Dimensions of Prepositions and their Use in Computational Linguistics Formalisms and Applications. University of Essex (2005)
34. Hellan, L., Dakubu, M.E.K.: Identifying verb constructions cross-linguistically. In: Studies in the Languages of the Volta Basin 6.3. Legon: Linguistics Department, University of Ghana (2010). https://typecraft.org/tc2wiki/Verbconstructions_cross-linguistically_-_Introduction
35. Hellan, L., Beermann, D., Bruland, T., Dakubu, M.E.K, Marimon, M.: MultiVal: Towards a multilingual valence lexicon. In: Calzolari et al. (eds.) LREC 2014. (2014) http://regdili.hf.ntnu.no:8081/multilanguage_valence_demo/multivalence
36. Hellan, L., Bruland, T.: A cluster of applications around a Deep Grammar. In: Vetulani et al. (eds.) Proceedings from The Language & Technology Conference, LTC2015, Poznan (2015). http://regdili.hf.ntnu.no:8081/linguisticAce/parse
37. Hellan, L., Malchukov, A.L., Cennamo, M. (eds.): Contrastive Studies in Valency. Amsterdam & Philadelphia: John Benjamins Publ. Co. (2017)
38. Hellan, L., Beermann, D., Bruland, T., Haugland, T., Aamot, E.: Creating a Norwegian valence corpus from a deep grammar. In: Human Language Technology. Challenges for Computer Science and Linguistics. 8th Language & Technology Conferene, LTC2017. Springer. (2020) ISBN 978–3–030–66526–5. https://typecraft.org/tc2wiki/Norwegian_Valency_Corpus
39. Hirzel, H.: "Porting lexicon files from Toolbox into LKB-grammars: A case study for a grammar of Ga." (2006). https://typecraft.org/tc2wiki/File:Toolbox-LKB-Link-slides_-_version_4.pdf
40. Korhonen, A., Briscoe, T.: Extended lexical-semantic classification of english verbs. In: Proceedings of the HLT/NAACL Workshop on Computational Lexical Semantics. Boston, MA (2004)
41. Levin, B.: English Verb Classes and Alternations. University of Chicago Press, Chicago (1991)
42. Malchukov, A.L., Comrie, B. (eds.): Valency classes in the world's languages. Mouton De Gruyter, Berlin (2015)
43. Marneffe, M.-C., Manning, C. D., Nivre, J., Zeman, D.: Universal Dependencies. Computational Linguistics. (2021). https://doi.org/10.1162/COLI_a_00402
44. Pollard, C., Sag, I.A.: Head-Driven Phrase Structure Grammar. Chicago University Press, Chicago (1994)
45. Quasthoff, U., Hellan, L., Körner, E., Eckart, T., Goldhahn, D., Beermann, D.: Typical Sentences as a Resource for Valence. LREC 2020 (2020). http://www.lrec-conf.org/proceedings/lrec2020/index.html
46. Schaefer, R.B., Egbokhare, F.O.: Emai valency classes and their alternations. In [42], pp. 261–298 (2015)
47. Staal, J.F.: A reader on the Sanskrit grammarians. MIT Press (1972)
48. Wittgenstein, L.: Tractatus Logico-Philosophicus. (First published in German 1921.) New York: Harcourt, Brace & Company/ London: Kegan Paul, Trench, Trubner & Co. (1922)

Printed in the United States
by Baker & Taylor Publisher Services